Bernhard Hofmann-Wellenhof
Helmut Moritz

Physical Geodesy

Second,
corrected edition

SpringerWienNewYork

Dr. Bernhard Hofmann-Wellenhof
Dr. Helmut Moritz
Institut für Navigation und Satellitengeodäsie
Technische Universität Graz, Graz, Austria

SpringerWienNewYork is a part of Springer Science + Business Media,
springeronline.com

Typesetting: Composition by authors
Cover illustration: Elmar Wasle, Graz

Printed on acid-free and chlorine-free bleached paper
SPIN 11737476

With 111 Figures

Library of Congress Control Number 2006931768

ISBN-10 3-211-33544-7 SpringerWienNewYork
ISBN-13 978-3-211-33544-4 SpringerWienNewYork
ISBN-10 3-211-23584-1 1st edn. SpringerWienNewYork

This book is dedicated to the memory of

Weikko Aleksanteri Heiskanen
(1895–1971)

Pioneer, initiator, and coauthor of "Physical Geodesy",
whose dreams have come true in a way unexpected to
all of us.

Foreword

Almost the period of one generation has passed since 1967, the year of the first release of Physical Geodesy by Weikko A. Heiskanen and Helmut Moritz. Soon this book became a bestseller. Today, when studying publications dealing with physical geodesy, not surprisingly the book is still frequently quoted. Have the clocks been stopped since then? Not at all, time has flown as fast as usual or maybe even faster – at least in someone's imagination. However, excellent quality is correlated with a long life expectation. This is the reason why "the book" still plays an important role in geodetic science and beyond.

In the last decades, nevertheless, geodesy has certainly continually developed further – on the one hand by new computational methods and ideas and on the other hand by modern measurement techniques. This is where the story of this book starts.

Several years ago, I tried to convince Helmut Moritz on the necessity of a new edition of Physical Geodesy. Even if I encountered some interest, I did not manage to completely succeed. "Steter Tropfen höhlt den Stein" (persistent drops hollow out the stone), I thought and started to repeat my request regularly. The reason for my somehow obstinacy originated from the past. In 1993, I got the chance to support Helmut Moritz in writing the book entitled Geometry, Relativity, and Geodesy. For me, this was a tremendously exciting time where we developed a great cooperation in any respect. Immediately after this experience, I manifested my desire of another chance for a cooperation. In these days, the idea of a new edition of Physical Geodesy matured.

Finally, the persistent drops succeeded. I cannot tell you the Why and the When; suddenly we had a contract with the Springer Publishing Company. To me it seemed as if the wheel of time had been turned back – thank you, Helmut!

Many persons deserve credit and thanks. Prof. Dr. Klaus-Peter Schwarz, retired from the Department of Geomatics Engineering of the University of Calgary, strongly influenced the balance between keeping, eliminating, updating, and adding topics.

Prof. Dr. Herbert Lichtenegger, retired from the Institute of Navigation and Satellite Geodesy of the Graz University of Technology, was a reviewer of the book. He has critically read and corrected the full volume. His many suggestions and improvements, critical remarks and proposals are gratefully acknowledged.

Prof. Dr. Norbert Kühtreiber from the Institute of Navigation and Satellite Geodesy of the Graz University of Technology has helped with constructive critique and valuable suggestions. Furthermore, he has strongly helped to shape Chap. 11 by providing numerical examples and his valuable experience on the practical aspects of geoid computation.

In several fruitful discussions, Prof. Dr. Roland Pail from the Institute of Navigation and Satellite Geodesy of the Graz University of Technology has provided his rich experience on the space gravity missions. Parts of his structured lecture notes are mirrored in the corresponding section. He also deserves thanks for a careful proofreading of this section.

The cover illustration was designed and produced by Dipl.-Ing. Elmar Wasle of TeleConsult Austria GmbH (www.teleconsult-austria.at). When presenting this illustration to the Springer Publishing Company, the response was extremely positive because of its eye-catcher quality.

The index of the book was produced using a computer program written by Dr. Walter Klostius from the Institute of Geoinformation of the Graz University of Technology. Also, his program helped in the detection of some spelling errors.

The book is compiled with the text system LaTeX2ε. One of the figures included is also developed with LaTeX2ε. The remaining figures are drawn by using CorelDRAW 11. Primarily Dr. Klaus Legat from the Institute of Navigation and Satellite Geodesy of the Graz University of Technology deserves the thanks for the figures. He was supported by Prof. Dr. Norbert Kühtreiber. The highly academic level of the producers assures a seal of quality. Many of these figures are redrawings of the originals in Heiskanen and Moritz (1967).

I am also grateful to the Springer Publishing Company for their advice and cooperation.

The inclusion by name of a commercial company or product does not constitute an endorsement by the authors. In principle, such inclusions were avoided whenever possible.

Finally, your ideas for a future edition of this book and your advice are appreciated and encouraged.

The selection of topics is certainly different from the original book written by Heiskanen and Moritz. However, basically we tried not only to keep the overall structure wherever possible but also to leave the text unchanged. The primary selection criteria of the topics were relevancy, tutorial content, and the interest and expertise of the authors. A detailed description of the contents is given in the Preface.

March 2005   B. Hofmann-Wellenhof

Preface

This book is a university-level introductory textbook. Physical geodesy is the science of the figure of the earth and of its gravity field. Particular emphasis is put on the interaction between geometry, especially GPS, and modern gravitational techniques. The mathematical tool is potential theory. More about the purpose and application of physical geodesy will be found in the subsequent motivation. For better readability, some repetitions are purposely used. The mathematical apparatus is kept as simple as possible.

The book is divided into 11 chapters, a section of references, and a detailed index which should immediately help in finding certain topics of interest.

The first chapter is an introduction to mathematical potential theory to the extent needed in the present book. More precisely, it is "classical" potential theory as represented, e.g., by the book of Kellogg (1929) (this is our usual mode of quoting references, by name[s] and year). Mathematicians will notice immediately that the presentation, as in most textbooks on theoretical physics, is informal: proofs are frequently omitted or replaced by "heuristic" considerations.

The second chapter introduces the gravity field of the earth, e.g., the force of gravity, level surfaces and plumb lines, the geoid, and coordinates naturally related to them: astronomic latitude and longitude as well as heights above the geoid. A powerful tool are developments in spherical harmonics. The natural reference surface is an ellipsoid of revolution equipped with a "normal" gravity field. This gives us a "Geodetic Reference System" (GRS) or World Geodetic System (WGS). Deviations of the real gravity field quantities from the corresponding reference quantities are small and can be linearized. This makes it possible to treat geodetic problems as relatively simple boundary-value problems of potential theory. A well-known classical solution is Stokes' integral formula.

The third chapter deals with gravity reductions, in particular reductions using the theory of isostasy. A first link to geophysics is established in this way.

The fourth chapter considers the problem of heights, which is more complicated than one would think at the beginning. The first four chapters are an update of the old book by Heiskanen and Moritz (1967), which serves as the template of this book.

The fifth chapter is central in several respects. It is vastly expanded as compared to the former book and has a completely different structure.

The problem of interrelating geometric and physical aspects is met here in all its complexity, from the global (Part I) to the classical local aspects (Part III), the regional "three-dimensional geodesy" in the pre-satellite sense (Part II) forming a transition. One could also say, Part I is integral and Part III is differential. Part I, geocentric and global reference systems, has been made possible only by highly precise geometric satellite methods. The problem of the third dimension, one of the most difficult tasks of geodesy, is formulated and solved here in the most direct and natural way. Part II is an attempt to solve this problem classically in a nondifferential way, but the weak link is the measurement of the zenith distances which are too inaccurate because of atmospheric refraction. The classical way out of this dilemma, still valid today, is the astrogeodetic integration of deflections of the vertical as discussed in Part III.

The sixth chapter is relatively slight, treating the computation of the gravity field up to about 10 km, with a view to application to airborne gravimetry. It is a streamlined version of the old Chapter 6.

The seventh chapter corresponds to the old Chapter 9, but it is greatly expanded to reflect the enormous progress of satellite methods for the determination of the global gravitational field. The main problem has been the gap between this global field at high elevations and the detailed but ill-distributed terrestrial gravity measurements. The new dedicated satellite missions, intended to close this gap, are described.

The eighth chapter, on Molodensky's and related theories, is again considerably expanded, owing to their great conceptual importance. Molodensky was the first to base physical geodesy on boundary-value problems on the physical earth's surface rather than on the geoid. His conceptual frame includes also astrogeodetic methods. Although conceived by Molodensky as a method to avoid the reduction of gravity to sea level, in mountain areas it works best if combined with isostatic and similar gravity reductions which are now familiar as "remove-restore" techniques. A great change with respect to the earlier book is the fact that now we understand Molodensky's theory much better, so that we can treat it now completely with elementary mathematics, without the use of integral equations.

The ninth chapter is a practically unchanged update of the old Chapter 7. The statistical treatment of gravity has undergone an enormous and unexpected increase in importance, theoretical as well as practical, so as to warrant a special new chapter.

The tenth chapter deals with least-squares collocation. This is a great synthesis of a generalization of least-squares prediction of gravity treated in the ninth chapter, with the theory of Hilbert space with kernel functions, and with ideas of gravity reduction in the theory of Molodensky. This synthesis is due to a small publication by Krarup (1969). It has an extremely simple mathematical structure and uses only matrix methods well-suited for electronic computation; still, it permits the combination of virtually all geodetic data types. Thus it has become very popular for numerical computation.

The eleventh chapter illustrates these various methods by computations in Austria, which combines a difficult topography with easily available data.

Internet citations within the text omit the part "http://" if the address contains "www"; therefore, "www.esa.int" means "http://www.esa.int". Usually, internet addresses given in the text are not repeated in the list of references. Therefore, the list of references does not yield a complete picture of the references of which we have been benefiting.

The use of the internet sources caused some troubles for the following reason. When looking for a proper and concise explanation or definition, quite often identical descriptions were found at different locations. So the unsolvable problem arose to figure out the earlier and original source. In these cases, sometimes the decision was made, to avoid a possible conflict of interests, by omitting the citation of the source at all. This means that some phrases or sentences may have been adapted from internet sources. On the other side, as soon as this book is released, it may and will also serve as an input source for several homepages.

For bibliographical references, the most readily accessible or most comprehensive publication of an author on a particular topic is given rather than his first. The list of references does not aim at completeness; some important publications may have been omitted but never on purpose.

The (American) spelling of a word is adopted from Webster's Dictionary of the English Language (third edition, unabridged). Apart from typical differences like the American "leveling" in contrast to the British "levelling", this may lead to other divergences when comparing dictionaries. Webster's Dictionary always combines the negation "non" and the following word without hyphen unless a capital letter follows. Therefore "nongravitational", "nonpropulsed", "nonsimultaneity" and "non-European" are corresponding spellings.

Symbols representing a vector or a matrix are in boldface. The inner or scalar product of two vectors is indicated by a dot "·". The norm of a vector, i.e., its length, is indicated by two double-bars "‖". Vectors not related to matrices are written either as column or as row vectors, whatever is more convenient.

March 2005   B. Hofmann-Wellenhof   H. Moritz

Preface to the second edition

Compared to the original version, this second edition only answers four times the question marks resulting from a wrong label denotation and, for LaTeX2ε experts, some "overfull \hbox" warnings indicating too long lines without finding appropriate linebreaks were eliminated. Furthermore, one misspelling was corrected, some punctuation problems were solved differently, and a few sentences were reformulated or updated. Therefore, it is considered a corrected version and not a revised one. This is a significant difference because it implies that possible advice for improvements from readers or reviewers were not taken into consideration. The reason for this is not a haughty disregard of other ideas but the tightness of the time schedule. To date only a very small number of reviews has been released. This almost coincided with the message of the Springer Publishing Company that the first edition will be sold out shortly. Therefore, we are grateful to those who gave us advice and we further encourage all readers accordingly because their support might help to improve another edition.

April 2006   B. Hofmann-Wellenhof   H. Moritz

Contents

Motivation

As we have already indicated in the Preface, the subject of physical geodesy is the study of the gravity field and the figure of the earth. In former times, the scientifically relevant "figure of the earth" was the *geoid*, which is defined as one of the equipotential surfaces of the earth's gravity potential, of which the (mean) surface of the oceans forms a part. So the gravity field immediately enters into the very definition of "figure of the earth". "Heights above sea level" are heights above the geoid, and thus are both physically and geometrically defined.

Gravity, essentialy caused by the earth's gravitational attraction, has always been determining the life of humankind, from walking on a hilly road to crossing the oceans by ship or by airplane. It has also formed the shape of our planet.

Scientific geodesy started when leading scientists such as Newton recognized that the earth cannot be a sphere but must rather be flattened because of the earth's rotation. Not very much, but probably measurably. This was one of the greatest scientific problems of that time.

Therefore, around 1740, the French Academy of Sciences undertook two expeditions, one under Bouguer to Peru and one under Maupertuis to Lapland. Their purpose was to measure the length of a meridional arc of, say, 1 degree of latitude, one close to the equator and one close to the North Pole. The difference between the two results is a measure of the flattening, which is the deviation (with respect to the sphere) of the earth ellipsoid. These measurements clearly indicated that the global figure of the earth is an ellipsoid of revolution, at least approximately.

The next century was characterized by attempts to define the figure of the earth more precisely. C.F. Gauss (1777–1855), the "princeps mathematicorum", raised geodesy to the rank of a science. He did this by his theory of surfaces – which finally led to General Relativity, cf. Moritz and Hofmann-Wellenhof (1993) – and his adjustment by least squares, the first of all statistical estimation methods. He liked practical geodetic work and measured a triangulation net. Gauss also introduced the geoid as the "mathematical figure of the earth" defined as a level surface of the gravity field.

The geoid deviates from a well-fitting ellipsoid (e.g., the Geodetic Reference System 1980) by less than 100 meters. Geocentric positions nowadays can be determined by GPS to an accuracy of better than 1 decimeter in a purely geometric way. We may define these positions either in terms of geocentric Cartesian coordinates or as ellipsoidal coordinates φ, λ, h.

So the geometry of the earth can be determined largely independently of the gravitational field, thanks to GPS and other satellite techniques. Still, the gravitational field is needed, e.g., for determining the orbits of the satellites themselves.

It is probable that, by the influence of GPS, gravity anomalies Δg will gradually be replaced by gravity disturbances δg. This is taken into account in the present book.

Gravity has become one of the most sought-for and most interrelated data in geophysics, and every increase of accuracy has immediately generated new needs. For instance, the ocean surface as determined by satellite altimetry is not an exact equipotential surface because of small tilts due to ocean currents. Thus, this "ocean topography", measured by comparing the results of satellite altimetry with precise gravimetric geoids determined by combining various methods, provides important boundary conditions for oceanography.

Already Clairaut related the density of the masses inside the earth with internal gravity on the condition of hydrostatic equilibrium, and this question with the classical title "The figure of the earth" can now be reconsidered in the light of satellite data; cf. Moritz (1990).

Geological phenomena in the earth's crust and upper mantle such as isostasy and plate tectonics require an interaction between geodesy, geophysics, and geology.

Polar motion and anomalies in the earth's rotation are largely caused by the ceaseless circulation of the air masses defining weather. Earth rotation is now monitored by laser and GPS, which provides an unexpected link between geodesy and meteorology; cf. Moritz and Mueller (1987).

New measuring techniques related to inertial navigation systems (INS) require an interaction between the geometry and the gravitational field. This has considerable practical consequences, e.g., in tunnel surveying. GPS stops short in front of a tunnel, and INS or conventional surveying methods must take over. Either of them, however, does depend on the gravity field.

The terrestrial measurement of gravity is very time-consuming. Airborne gravimetry has become operational only after the inertial and gravitational forces have become separable by combination with GPS.

Not all of this can be treated in detail in the present introductory book. It is, however, intended as a solid, mathematically oriented and not too difficult treatise on graduate level leading to one's own postgraduate research.

The book by Heiskanen and Moritz (1967) stood at the transition between classical and satellite geodesy. Similarly, the present book stands at the beginning of an era characterized by "sensor integration", data combination, and kinematic and navigational techniques.

1 Fundamentals of potential theory

1.1 Attraction and potential

The purpose in this preparatory chapter is to present the fundamentals of potential theory, including spherical and ellipsoidal harmonics, in sufficient detail to assure a full understanding of the later chapters. Our intent is to explain the meaning of the theorems and formulas, avoiding long derivations that can be found in any textbook on classical (before 1950) potential theory; we recommend Kellogg (1929). A simple rather than completely rigorous presentation is offered in our book.

Nevertheless, the reader might consider this chapter perhaps more difficult and abstract than other parts of the book. Since later practical applications will give concrete meaning to the topics of the present chapter, the reader may wish to read it only cursorily at first and return to it later when necessary.

According to Newton's law of gravitation, two points with masses m_1, m_2, separated by a distance l, attract each other with a force

$$F = G \, \frac{m_1 m_2}{l^2} \, . \tag{1-1}$$

This force is directed along the line connecting the two points; G is Newton's gravitational constant. In SI units (Système International d'unités) based on meter [m], kilogram [kg], and second [s], the gravitational constant has the value

$$G = 6.6742 \cdot 10^{-11} \ \mathrm{m^3 \, kg^{-1} \, s^{-2}} \, . \tag{1-2}$$

The Newtonian gravitational constant G is somewhat of a scandal in measuring physics. It is on the one hand one of the most important physical constants, and at the same time one of the least accurately determined ones. The international authority in this field is the Committee on Data for Science and Technology (CODATA), see under www.codata.org. In July 2002, CODATA recommended the value of G mentioned above, more precisely it assigned the value $G = (6.6742 \pm 0.0010) \cdot 10^{-11} \ \mathrm{m^3 \, kg^{-1} \, s^{-2}}$. The symbol $\pm$ denotes the standard uncertainty, also called standard deviation or standard error. This corresponds to a relative standard uncertainty of $1.5 \cdot 10^{-4}$ or $150 \, \mathrm{ppm}$ which is a deplorably high inaccuracy for such an important constant, see http://physics.nist.gov/cuu/constants. (For other constants we have a relative accuracy of 10^{-7} and better.) For comparison of experimental results see the internet.

Although the masses m_1, m_2 attract each other in a completely symmetrical way, it is convenient to call one of them the attracting mass and the other the attracted mass. For simplicity we set the attracted mass equal to unity and denote the attracting mass by m. The formula

$$F = G \frac{m}{l^2} \qquad (1\text{–}3)$$

expresses the force exerted by the mass m on a unit mass located at P at a distance l from m.

We now introduce a rectangular coordinate system xyz and denote the coordinates of the attracting mass m by ξ, η, ζ and the coordinates of the attracted point P by x, y, z. The force may be represented by a vector $\mathbf{F}$ with magnitude F (Fig. 1.1). The components of $\mathbf{F}$ are given by

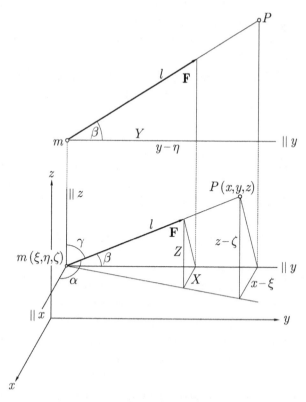

Fig. 1.1. The components of the gravitational force; upper figure shows y-component

$$X = -F \cos \alpha = -\frac{G\,m}{l^2}\frac{x - \xi}{l} = -G\,m\,\frac{x - \xi}{l^3}\,,$$

$$Y = -F \cos \beta = -\frac{G\,m}{l^2}\frac{y - \eta}{l} = -G\,m\,\frac{y - \eta}{l^3}\,,\qquad (1\text{--}4)$$

$$Z = -F \cos \gamma = -\frac{G\,m}{l^2}\frac{z - \zeta}{l} = -G\,m\,\frac{z - \zeta}{l^3}\,,$$

where

$$l = \sqrt{(x - \xi)^2 + (y - \eta)^2 + (z - \zeta)^2}\,. \qquad (1\text{--}5)$$

We next introduce a scalar function

$$V = \frac{G\,m}{l}\,, \qquad (1\text{--}6)$$

called the potential of gravitation. The components X, Y, Z of the gravitational force $\mathbf{F}$ are then given by

$$X = \frac{\partial V}{\partial x}\,, \quad Y = \frac{\partial V}{\partial y}\,, \quad Z = \frac{\partial V}{\partial z}\,, \qquad (1\text{--}7)$$

as can be easily verified by differentiating (1–6), since

$$\frac{\partial}{\partial x}\left(\frac{1}{l}\right) = -\frac{1}{l^2}\frac{\partial l}{\partial x} = -\frac{1}{l^2}\frac{x - \xi}{l} = -\frac{x - \xi}{l^3}\,,\dots\,. \qquad (1\text{--}8)$$

In vector notation, Eq. (1–7) is written

$$\mathbf{F} = [X,\ Y,\ Z] = \operatorname{grad} V\,; \qquad (1\text{--}9)$$

that is, the force vector is the gradient vector of the scalar function V.

It is of basic importance that according to (1–7) the three components of the vector $\mathbf{F}$ can be replaced by a single function V. Especially when we consider the attraction of systems of point masses or of solid bodies, as we do in geodesy, it is much easier to deal with the potential than with the three components of the force. Even in these complicated cases the relations (1–7) are applied; the function V is then simply the sum of the contributions of the respective particles.

Thus, if we have a system of several point masses $m_1, m_2, \dots, m_n$, the potential of the system is the sum of the individual contributions (1–6):

$$V = \frac{G\,m_1}{l_1} + \frac{G\,m_2}{l_2} + \dots + \frac{G\,m_n}{l_n} = G\sum_{i=1}^{n}\frac{m_i}{l_i}\,. \qquad (1\text{--}10)$$

1.2   Potential of a solid body

Let us now assume that point masses are distributed continuously over a volume v (Fig. 1.2) with density

$$\varrho = \frac{dm}{dv}\,,\qquad(1\text{--}11)$$

where dv is an element of volume and dm is an element of mass. Then the sum (1–10) becomes an integral (Newton's integral),

$$V = G \iiint\limits_{v} \frac{dm}{l} = G \iiint\limits_{v} \frac{\varrho}{l}\, dv\,,\qquad(1\text{--}12)$$

where l is the distance between the mass element $dm = \varrho\, dv$ and the attracted point P. Denoting the coordinates of the attracted point P by x, y, z and of the mass element m by ξ, η, ζ, we see that l is again given by (1–5), and we can write explicitly

$$V(x, y, z) = G \iiint\limits_{v} \frac{\varrho(\xi, \eta, \zeta)}{\sqrt{(x - \xi)^2 + (y - \eta)^2 + (z - \zeta)^2}}\, d\xi\, d\eta\, d\zeta\,,\qquad(1\text{--}13)$$

since the element of volume is expressed by

$$dv = d\xi\, d\eta\, d\zeta\,.\qquad(1\text{--}14)$$

This is the reason for the triple integrals in (1–12).

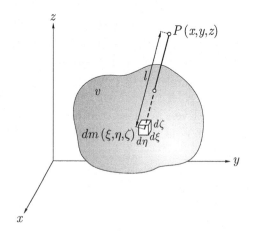

Fig. 1.2. Potential of a solid body

The components of the force of attraction are given by (1–7). For instance,

$$X = \frac{\partial V}{\partial x} = G \frac{\partial}{\partial x} \iiint\limits_{v} \frac{\varrho(\xi, \eta, \zeta)}{l} \, d\xi \, d\eta \, d\zeta$$

$$= G \iiint\limits_{v} \varrho(\xi, \eta, \zeta) \frac{\partial}{\partial x} \left(\frac{1}{l}\right) d\xi \, d\eta \, d\zeta \,.$$

(1–15)

Note that we have interchanged the order of differentiation and integration. Substituting (1–8) into the above expression, we finally obtain

$$X = -G \iiint\limits_{v} \frac{x - \xi}{l^3} \varrho \, dv \,.$$

(1–16)

Analogous expressions result for Y and Z.

The potential V is continuous throughout the whole space and vanishes at infinity like $1/l$ for $l \to \infty$. This can be seen from the fact that for very large distances l the body acts approximately like a point mass, with the result that its attraction is then approximately given by (1–6). Consequently, in celestial mechanics the planets are usually considered as point masses.

The first derivatives of V, that is, the force components, are also continuous throughout space, but not so the second derivatives. At points where the density changes discontinuously, some second derivatives have a discontinuity. This is evident because the potential V may be shown to satisfy *Poisson's equation*

$$\Delta V = -4\pi \, G \varrho \,,$$

(1–17)

where

$$\Delta V = \frac{\partial^2 V}{\partial x^2} + \frac{\partial^2 V}{\partial y^2} + \frac{\partial^2 V}{\partial z^2} \,.$$

(1–18)

The symbol Δ, called the *Laplacian operator*, has the form

$$\frac{\partial^2}{\partial x^2} + \frac{\partial^2}{\partial y^2} + \frac{\partial^2}{\partial z^2} \,.$$

(1–19)

From (1–17) and (1–18) we see that at least one of the second derivatives of V must be discontinuous together with ϱ.

Outside the attracting bodies, in empty space, the density ϱ is zero and (1–17) becomes

$$\Delta V = 0 \,.$$

(1–20)

This is *Laplace's equation*. Its solutions are called *harmonic functions*. Hence, the potential of gravitation is a harmonic function outside the attracting masses but not inside the masses: there it satisfies Poisson's equation.

1.3   Harmonic functions

Earlier we have defined the harmonic functions as solutions of Laplace's equation

$$\Delta V = 0. \tag{1-21}$$

More precisely, a function is called *harmonic in a region* v of space if it satisfies Laplace's equation at every point of v. If the region is the exterior of a certain closed surface S, then it must in addition vanish like $1/l$ for $l \to \infty$. It can be shown that *every harmonic function is analytic* (in the region where it satisfies Laplace's equation); that is, it is continuous and has continuous derivatives of any order and can be developed into a Taylor series.

The simplest harmonic function is the reciprocal distance

$$\frac{1}{l} = \frac{1}{\sqrt{(x-\xi)^2 + (y-\eta)^2 + (z-\zeta)^2}} \tag{1-22}$$

between two points $P(\xi, \eta, \zeta)$ and $P(x, y, z)$. It is the potential of a point mass $m = 1/G$, situated at the point $P(\xi, \eta, \zeta)$; compare (1-5) and (1-6).

It is easy to show that $1/l$ is harmonic. We form the following partial derivatives with respect to x, y, z in the fashion of (1-8):

$$\frac{\partial}{\partial x}\left(\frac{1}{l}\right) = -\frac{x-\xi}{l^3}, \quad \frac{\partial}{\partial y}\left(\frac{1}{l}\right) = -\frac{y-\eta}{l^3}, \quad \frac{\partial}{\partial z}\left(\frac{1}{l}\right) = -\frac{z-\zeta}{l^3};$$

$$\frac{\partial^2}{\partial x^2}\left(\frac{1}{l}\right) = \frac{-l^2 + 3(x-\xi)^2}{l^3}, \quad \frac{\partial^2}{\partial y^2}\left(\frac{1}{l}\right) = \frac{-l^2 + 3(y-\eta)^2}{l^3}, \tag{1-23}$$

$$\frac{\partial^2}{\partial z^2}\left(\frac{1}{l}\right) = \frac{-l^2 + 3(z-\zeta)^2}{l^3}.$$

Adding the last three equations and recalling the definition of Δ, we find

$$\Delta\left(\frac{1}{l}\right) = 0; \tag{1-24}$$

that is, $1/l$ is harmonic.

The point $P(\xi, \eta, \zeta)$, where l is zero and $1/l$ is infinite, is the only point to which we cannot apply the above derivation; $1/l$ is not harmonic at this singular point.

As a matter of fact, the slightly more general potential (1-6) of an arbitrary point mass m is also harmonic except at $P(\xi, \eta, \zeta)$, because (1-24) remains unchanged if both sides are multiplied by Gm.

Not only the potential of a point mass but also any other gravitational potential is harmonic outside the attracting masses. Consider the potential (1–12) of an extended body. Interchanging the order of differentiation and integration, we find from (1–12)

$$\Delta V = G\,\Delta\left[\iiint\limits_v \frac{\varrho}{l}\,dv\right] = G\iiint\limits_v \varrho\,\Delta\left(\frac{1}{l}\right)\,dv = 0\,;\qquad(1\text{–}25)$$

that is, the potential of a solid body is also harmonic at any point $P(x, y, z)$ outside the attracting masses.

If P lies inside the attracting body, the above derivation breaks down, since $1/l$ becomes infinite for that mass element $dm(\xi, \eta, \zeta)$ which coincides with $P(x, y, z)$, and (1–24) does not apply. This is the reason why the potential of a solid body is not harmonic in its interior but instead satisfies Poisson's differential equation (1–17).

1.4   Laplace's equation in spherical coordinates

The most important harmonic functions are the *spherical harmonics*. To find them, we introduce spherical coordinates: r (radius vector; note that this is a standard notation, although it does not represent a vector in the contemporary sense), ϑ (polar distance), λ (geocentric longitude), see Fig. 1.3. Spherical coordinates are related to rectangular coordinates x, y, z by the

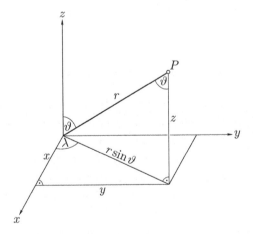

Fig. 1.3. Spherical and rectangular coordinates

equations

$$x = r \sin \vartheta \cos \lambda ,$$
$$y = r \sin \vartheta \sin \lambda ,$$
$$z = r \cos \vartheta ;$$

(1–26)

or inversely by

$$r = \sqrt{x^2 + y^2 + z^2} ,$$
$$\vartheta = \tan^{-1} \frac{\sqrt{x^2 + y^2}}{z} ,$$
$$\lambda = \tan^{-1} \frac{y}{x} .$$

(1–27)

To get Laplace's equation in spherical coordinates, we first determine the element of arc (element of distance) ds in these coordinates. For this purpose we form

$$dx = \frac{\partial x}{\partial r} dr + \frac{\partial x}{\partial \vartheta} d\vartheta + \frac{\partial x}{\partial \lambda} d\lambda ,$$
$$dy = \frac{\partial y}{\partial r} dr + \frac{\partial y}{\partial \vartheta} d\vartheta + \frac{\partial y}{\partial \lambda} d\lambda ,$$
$$dz = \frac{\partial z}{\partial r} dr + \frac{\partial z}{\partial \vartheta} d\vartheta + \frac{\partial z}{\partial \lambda} d\lambda .$$

(1–28)

By differentiating (1–26) and substituting it into the elementary formula

$$ds^2 = dx^2 + dy^2 + dz^2 ,$$

(1–29)

we obtain

$$ds^2 = dr^2 + r^2 \, d\vartheta^2 + r^2 \sin^2\vartheta \, d\lambda^2 .$$

(1–30)

We might have found this well-known formula more simply by geometrical considerations, but the approach used here is more general and can also be applied to ellipsoidal (harmonic) coordinates.

In (1–30) there are no terms with $dr \, d\vartheta$, $dr \, d\lambda$, and $d\vartheta \, d\lambda$. This expresses the evident fact that spherical coordinates are orthogonal: the spheres $r =$ constant, the cones $\vartheta =$ constant, and the planes $\lambda =$ constant intersect each other orthogonally.

The general form of the element of arc in arbitrary orthogonal coordinates q_1, q_2, q_3 is

$$ds^2 = h_1^2 \, dq_1^2 + h_2^2 \, dq_2^2 + h_3^2 \, dq_3^2 .$$

(1–31)

It can be shown that Laplace's operator in these coordinates is

$$\Delta V = \frac{1}{h_1 h_2 h_3} \left[\frac{\partial}{\partial q_1} \left(\frac{h_2 h_3}{h_1} \frac{\partial V}{\partial q_1} \right) + \frac{\partial}{\partial q_2} \left(\frac{h_3 h_1}{h_2} \frac{\partial V}{\partial q_2} \right) + \frac{\partial}{\partial q_3} \left(\frac{h_1 h_2}{h_3} \frac{\partial V}{\partial q_3} \right) \right].$$
(1–32)

For spherical coordinates we have $q_1 = r$, $q_2 = \vartheta$, $q_3 = \lambda$. Comparison of (1–30) and (1–31) shows that

$$h_1 = 1, \quad h_2 = r, \quad h_3 = r \sin \vartheta.$$
(1–33)

Substituting these relations into (1–32) yields

$$\Delta V = \frac{1}{r^2} \frac{\partial}{\partial r} \left(r^2 \frac{\partial V}{\partial r} \right) + \frac{1}{r^2 \sin \vartheta} \frac{\partial}{\partial \vartheta} \left(\sin \vartheta \frac{\partial V}{\partial \vartheta} \right) + \frac{1}{r^2 \sin^2 \vartheta} \frac{\partial^2 V}{\partial \lambda^2}.$$
(1–34)

Performing the differentiations we find

$$\Delta V \equiv \frac{\partial^2 V}{\partial r^2} + \frac{2}{r} \frac{\partial V}{\partial r} + \frac{1}{r^2} \frac{\partial^2 V}{\partial \vartheta^2} + \frac{\cot \vartheta}{r^2} \frac{\partial V}{\partial \vartheta} + \frac{1}{r^2 \sin^2 \vartheta} \frac{\partial^2 V}{\partial \lambda^2} = 0, \quad (1–35)$$

which is *Laplace's equation in spherical coordinates*. An alternative expression is obtained when multiplying both sides by r^2:

$$r^2 \frac{\partial^2 V}{\partial r^2} + 2r \frac{\partial V}{\partial r} + \frac{\partial^2 V}{\partial \vartheta^2} + \cot \vartheta \frac{\partial V}{\partial \vartheta} + \frac{1}{\sin^2 \vartheta} \frac{\partial^2 V}{\partial \lambda^2} = 0.$$
(1–36)

This form will be somewhat more convenient for our subsequent development.

1.5   Spherical harmonics

We attempt to solve Laplace's equation (1–35) or (1–36) by separating the variables r, ϑ, λ using the trial substitution

$$V(r, \vartheta, \lambda) = f(r) \, Y(\vartheta, \lambda),$$
(1–37)

where f is a function of r only and Y is a function of ϑ and λ only. Performing this substitution in (1–36) and dividing by fY, we get

$$\frac{1}{f} \left(r^2 f'' + 2r \, f' \right) = -\frac{1}{Y} \left(\frac{\partial^2 Y}{\partial \vartheta^2} + \cot \vartheta \frac{\partial Y}{\partial \vartheta} + \frac{1}{\sin^2 \vartheta} \frac{\partial^2 Y}{\partial \lambda^2} \right),$$
(1–38)

where the primes denote differentiation with respect to the argument (r, in this case). Since the left-hand side depends only on r and the right-hand side only on ϑ and λ, both sides must be constant. We can therefore separate the equation into two equations:

$$r^2 f''(r) + 2r\, f'(r) - n(n+1)\, f(r) = 0 \qquad (1\text{-}39)$$

and

$$\frac{\partial^2 Y}{\partial \vartheta^2} + \cot\vartheta\, \frac{\partial Y}{\partial \vartheta} + \frac{1}{\sin^2\vartheta}\, \frac{\partial^2 Y}{\partial \lambda^2} + n(n+1)\, Y = 0\,, \qquad (1\text{-}40)$$

where we have denoted the constant by $n(n+1)$.

Solutions of (1–39) are given by the functions

$$f(r) = r^n \quad \text{and} \quad f(r) = r^{-(n+1)}\,; \qquad (1\text{-}41)$$

this should be verified by substitution. Denoting the still unknown solutions of (1–40) by $Y_n(\vartheta,\lambda)$, we see that Laplace's equation (1–35) is solved by the functions

$$V = r^n\, Y_n(\vartheta,\lambda) \quad \text{and} \quad V = \frac{Y_n(\vartheta,\lambda)}{r^{n+1}}\,. \qquad (1\text{-}42)$$

These functions are called *solid spherical harmonics*, whereas the functions $Y_n(\vartheta,\lambda)$ are known as (Laplace's) *surface spherical harmonics*. Both kinds are called *spherical harmonics*; the kind referred to can usually be judged from the context.

Note that n is not an arbitrary constant but must be an integer $0, 1, 2, \ldots$ as we will see later. If a differential equation is linear, and if we know several solutions, then, as is well known, the sum of these solutions is also a solution (this holds for *all* linear equation systems!). Hence, we conclude that

$$V = \sum_{n=0}^{\infty} r^n\, Y_n(\vartheta,\lambda) \quad \text{and} \quad V = \sum_{n=0}^{\infty} \frac{Y_n(\vartheta,\lambda)}{r^{n+1}} \qquad (1\text{-}43)$$

are also solutions of Laplace's equation $\Delta V = 0$; that is, harmonic functions. The important fact is that *every* harmonic function – with certain restrictions – can be expressed in one of the forms (1–43).

1.6 Surface spherical harmonics

Now we have to determine Laplace's surface spherical harmonics $Y_n(\vartheta,\lambda)$. We attempt to solve (1–40) by a new trial substitution

$$Y_n(\vartheta,\lambda) = g(\vartheta)\, h(\lambda)\,, \qquad (1\text{-}44)$$

where the functions g and h individually depend on one variable only. Performing this substitution in (1–40) and multiplying by $\sin^2\vartheta/g\,h$, we find

$$\frac{\sin\vartheta}{g}\left[\sin\vartheta\,g'' + \cos\vartheta\,g' + n(n+1)\sin\vartheta\,g\right] = -\frac{h''}{h}\,, \qquad (1\text{–}45)$$

where the primes denote differentiation with respect to the argument: ϑ in g and λ in h. The left-hand side is a function of ϑ only, and the right-hand side is a function of λ only. Therefore, both sides must again be constant; let the constant be m^2. Thus, the partial differential equation (1–40) splits into two ordinary differential equations for the functions $g(\vartheta)$ and $h(\lambda)$:

$$\sin\vartheta\,g''(\vartheta) + \cos\vartheta\,g'(\vartheta) + \left[n(n+1)\sin\vartheta - \frac{m^2}{\sin\vartheta}\right]g(\vartheta) = 0\,; \qquad (1\text{–}46)$$

$$h''(\lambda) + m^2 h(\lambda) = 0\,. \qquad (1\text{–}47)$$

Solutions of Eq. (1–47) are the functions

$$h(\lambda) = \cos m\lambda \quad \text{and} \quad h(\lambda) = \sin m\lambda\,, \qquad (1\text{–}48)$$

as may be verified by substitution. Equation (1–46), Legendre's differential equation, is more difficult. It can be shown that it has physically meaningful solutions only if n and m are integers $0, 1, 2, \ldots$ and if m is smaller than or equal to n. A solution of (1–46) is the Legendre function $P_{nm}(\cos\vartheta)$, which will be considered in some detail in the next section. Therefore,

$$g(\vartheta) = P_{nm}(\cos\vartheta) \qquad (1\text{–}49)$$

and the functions

$$Y_n(\vartheta, \lambda) = P_{nm}(\cos\vartheta)\cos m\lambda \quad \text{and} \quad Y_n(\vartheta, \lambda) = P_{nm}(\cos\vartheta)\sin m\lambda \quad (1\text{–}50)$$

are solutions of the differential equation (1–40) for Laplace's surface spherical harmonics.

Since these solutions are linear, any linear combination of the solutions (1–50) is also a solution. Such a linear combination has the general form

$$Y_n(\vartheta, \lambda) = \sum_{m=0}^{n}\left[a_{nm}P_{nm}(\cos\vartheta)\cos m\lambda + b_{nm}P_{nm}(\cos\vartheta)\sin m\lambda\right], \qquad (1\text{–}51)$$

where a_{nm} and b_{nm} are arbitrary constants. This is the general expression for the surface spherical harmonics $Y_n(\vartheta, \lambda)$.

Substituting this relation into equations (1–43), we see that

$$V_i(r, \vartheta, \lambda) = \sum_{n=0}^{\infty} r^n \sum_{m=0}^{n} \left[a_{nm} P_{nm}(\cos\vartheta) \cos m\lambda + b_{nm} P_{nm}(\cos\vartheta) \sin m\lambda \right],$$

$$(1\text{–}52)$$

$$V_e(r, \vartheta, \lambda) = \sum_{n=0}^{\infty} \frac{1}{r^{n+1}} \sum_{m=0}^{n} \left[a_{nm} P_{nm}(\cos\vartheta) \cos m\lambda + b_{nm} P_{nm}(\cos\vartheta) \sin m\lambda \right]$$

$$(1\text{–}53)$$

are solutions of Laplace's equation $\Delta V = 0$; that is, harmonic functions. Furthermore, as we have mentioned, they are very general solutions indeed: every function which is harmonic inside a certain sphere can be expanded into a series (1–52), where the subscript i indicates the interior, and every function which is harmonic outside a certain sphere (such as the earth's gravitational potential) can be expanded into a series (1–53), where the subscript e indicates the exterior. Thus, we see how spherical harmonics can be useful in geodesy.

1.7   Legendre's functions

In the preceding section we have introduced Legendre's function $P_{nm}(\cos\vartheta)$ as a solution of Legendre's differential equation (1–46). The subscript n denotes the *degree* and the subscript m the *order* of P_{nm}.

It is convenient to transform Legendre's differential equation (1–46) by the substitution

$$t = \cos\vartheta \,. \tag{1–54}$$

In order to avoid confusion, we use an overbar to denote g as a function of t. Therefore,

$$g(\vartheta) \; = \bar{g}(t) \,,$$

$$g'(\vartheta) = \frac{dg}{d\vartheta} = \frac{dg}{dt}\frac{dt}{d\vartheta} = -\bar{g}'(t)\sin\vartheta \,, \tag{1–55}$$

$$g''(\vartheta) = \bar{g}''(t)\sin^2\vartheta - \bar{g}'(t)\cos\vartheta \,.$$

Inserting these relations into (1–46), dividing by $\sin\vartheta$, and then substituting $\sin^2\vartheta = 1 - t^2$, we get

$$(1 - t^2)\,\bar{g}''(t) - 2t\,\bar{g}'(t) + \left[n(n+1) - \frac{m^2}{1 - t^2} \right] \bar{g}'(t) = 0 \,. \tag{1–56}$$

The Legendre function $\bar{g}(t) = P_{nm}(t)$, which is defined by

$$P_{nm}(t) = \frac{1}{2^n\, n!} (1 - t^2)^{m/2} \frac{d^{n+m}}{dt^{n+m}} (t^2 - 1)^n \,, \tag{1–57}$$

satisfies (1–56). Apart from the factor $(1 - t^2)^{m/2} = \sin^m \vartheta$ and from a constant, the function P_{nm} is the $(n + m)$th derivative of the polynomial $(t^2 - 1)^n$. It can, thus, be evaluated. For instance,

$$P_{11}(t) = \frac{(1 - t^2)^{1/2}}{2 \cdot 1} \frac{d^2}{dt^2}(t^2 - 1) = \frac{1}{2}\sqrt{1 - t^2} \cdot 2 = \sqrt{1 - t^2} = \sin \vartheta . \quad (1\text{–}58)$$

The case $m = 0$ is of particular importance. The functions $P_{n0}(t)$ are often simply denoted by $P_n(t)$. Then (1–57) gives

$$P_n(t) = P_{n0}(t) = \frac{1}{2^n n!} \frac{d^n}{dt^n}(t^2 - 1)^n . \quad (1\text{–}59)$$

Because $m = 0$, there is no square root, that is, no $\sin \vartheta$. Therefore, the $P_n(t)$ are simply polynomials in t. They are called *Legendre's polynomials*. We give the Legendre polynomials for $n = 0$ through $n = 5$:

$$P_0(t) = 1, \qquad P_3(t) = \tfrac{5}{2} t^3 - \tfrac{3}{2} t,$$

$$P_1(t) = t, \qquad P_4(t) = \tfrac{35}{8} t^4 - \tfrac{15}{4} t^2 + \tfrac{3}{8}, \qquad (1\text{–}60)$$

$$P_2(t) = \tfrac{3}{2} t^2 - \tfrac{1}{2}, \qquad P_5(t) = \tfrac{63}{8} t^5 - \tfrac{35}{4} t^3 + \tfrac{15}{8} t.$$

Remember that

$$t = \cos \vartheta . \quad (1\text{–}61)$$

The polynomials may be obtained by means of (1–59) or more simply by the *recursion formula*

$$P_n(t) = -\frac{n - 1}{n} P_{n-2}(t) + \frac{2n - 1}{n} t P_{n-1}(t), \quad (1\text{–}62)$$

by which P_2 can be calculated from P_0 and P_1, P_3 from P_1 and P_2, etc. Graphs of the Legendre polynomials are shown in Fig. 1.4.

The powers of $\cos \vartheta$ can be expressed in terms of the cosines of multiples of ϑ, such as

$$\cos^2 \vartheta = \tfrac{1}{2} \cos 2\vartheta + \tfrac{1}{2}, \quad \cos^3 \vartheta = \tfrac{1}{4} \cos 3\vartheta + \tfrac{3}{4} \cos \vartheta . \quad (1\text{–}63)$$

Therefore, we may also express the $P_n(\cos \vartheta)$ in this way, obtaining

$$P_2(\cos \vartheta) = \tfrac{3}{4} \cos 2\vartheta + \tfrac{1}{4},$$

$$P_3(\cos \vartheta) = \tfrac{5}{8} \cos 3\vartheta + \tfrac{3}{8} \cos \vartheta,$$

$$P_4(\cos \vartheta) = \tfrac{35}{64} \cos 4\vartheta + \tfrac{5}{16} \cos 2\vartheta + \tfrac{9}{64}, \quad (1\text{–}64)$$

$$P_5(\cos \vartheta) = \tfrac{63}{128} \cos 5\vartheta + \tfrac{35}{128} \cos 3\vartheta + \tfrac{15}{64} \cos \vartheta,$$

$$\cdots \qquad = \cdots .$$

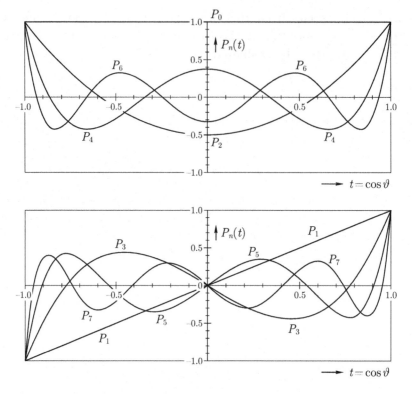

Fig. 1.4. Legendre's polynomials as functions of $t = \cos\vartheta$: n even (top) and n odd (bottom)

If the order m is not zero – that is, for $m = 1, 2, \ldots, n$ – Legendre's functions $P_{nm}(\cos\vartheta)$ are called *associated Legendre functions*. They can be reduced to the Legendre polynomials by means of the equation

$$P_{nm}(t) = (1 - t^2)^{m/2} \frac{d^m P_n(t)}{dt^m}, \qquad (1\text{--}65)$$

which follows from (1–57) and (1–59). Thus, the associated Legendre functions are expressed in terms of the Legendre polynomials of the same degree n. We give some P_{nm}, writing $t = \cos\vartheta$, $\sqrt{1 - t^2} = \sin\vartheta$:

$$P_{11}(\cos\vartheta) = \sin\vartheta\,, \qquad\qquad P_{31}(\cos\vartheta) = \sin\vartheta\left(\tfrac{15}{2}\cos^2\vartheta - \tfrac{3}{2}\right),$$

$$P_{21}(\cos\vartheta) = 3\sin\vartheta\cos\vartheta\,, \quad P_{32}(\cos\vartheta) = 15\sin^2\vartheta\cos\vartheta\,, \qquad (1\text{--}66)$$

$$P_{22}(\cos\vartheta) = 3\sin^2\vartheta\,, \qquad\quad P_{33}(\cos\vartheta) = 15\sin^3\vartheta\,.$$

We also mention an explicit formula for any Legendre function (polynomial or associated function):

$$P_{nm}(t) = 2^{-n}(1-t^2)^{m/2}\sum_{k=0}^{r}(-1)^k\frac{(2n-2k)!}{k!\,(n-k)!\,(n-m-2k)!}\,t^{n-m-2k},$$

(1–67)

where r is the greatest integer $\leq (n-m)/2$; i.e., r is $(n-m)/2$ or $(n-m-1)/2$, whichever is an integer. This formula is convenient for programming.

As this useful formula is seldom found in the literature, we show the derivation, which is quite straightforward. The necessary information on factorials may be obtained from any collection of mathematical formulas. The binomial theorem gives

$$(t^2-1)^n = \sum_{k=0}^{n}(-1)^k\binom{n}{k}t^{2n-2k} = \sum_{k=0}^{n}(-1)^k\frac{n!}{k!\,(n-k)!}\,t^{2n-2k}.$$

(1–68)

Thus, (1–57) becomes

$$P_{nm}(t) = \frac{1}{2^n}(1-t^2)^{m/2}\sum_{k=0}^{n}(-1)^k\frac{1}{k!\,(n-k)!}\frac{d^{n+m}}{dt^{n+m}}(t^{2n-2k}),$$

(1–69)

the quantity $n!$ having been cancelled out. The rth derivative of the power t^s is

$$\frac{d^r}{dt^r}(t^s) = s(s-1)\cdots(s-r+1)\,t^{s-r} = \frac{s!}{(s-r)!}\,t^{s-r}.$$

(1–70)

Setting $r = n+m$ and $s = 2n-2k$, we have

$$\frac{d^{n+m}}{dt^{n+m}}(t^{2n-2k}) = \frac{(2n-2k)!}{(n-m-2k)!}\,t^{n-m-2k}.$$

(1–71)

Inserting this into the above expression for $P_{nm}(t)$ and noting that the lowest possible power of t is either t or $t^0 = 1$, we obtain (1–67).

The surface spherical harmonics are Legendre's functions multiplied by $\cos m\lambda$ or $\sin m\lambda$:

degree 0 $P_0(\cos\vartheta)$;

degree 1 $P_1(\cos\vartheta)$,
    $P_{11}(\cos\vartheta)\cos\lambda,\;\; P_{11}(\cos\vartheta)\sin\lambda$;

degree 2 $P_2(\cos\vartheta)$,
    $P_{21}(\cos\vartheta)\cos\lambda,\;\; P_{21}(\cos\vartheta)\sin\lambda$,
    $P_{22}(\cos\vartheta)\cos 2\lambda,\;\; P_{22}(\cos\vartheta)\sin 2\lambda$;

(1–72)

and so on.

The geometrical representation of these spherical harmonics is useful. The harmonics with $m = 0$ – that is, Legendre's polynomials – are polynomials of degree n in t, so that they have n zeros. These n zeros are all real and situated in the interval $-1 \leq t \leq +1$, that is, $0 \leq \vartheta \leq \pi$ (Fig. 1.4). Therefore, the harmonics with $m = 0$ change their sign n times in this interval; furthermore, they do not depend on λ. Their geometrical representation is therefore similar to Fig. 1.5 a. Since they divide the sphere into zones, they are also called *zonal harmonics*.

The associated Legendre functions change their sign $n - m$ times in the interval $0 \leq \vartheta \leq \pi$. The functions $\cos m\lambda$ and $\sin m\lambda$ have $2m$ zeros in the interval $0 \leq \lambda < 2\pi$, so that the geometrical representation of the harmonics for $m \neq 0$ is similar to that of Fig. 1.5 b. They divide the sphere into compartments in which they are alternately positive and negative, somewhat like a chess board, and are called *tesseral harmonics*. "Tessera" means a square or rectangle, or also a tile. In particular, for $n = m$, they degenerate into functions that divide the sphere into positive and negative sectors, in which case they are called *sectorial harmonics*, see Fig. 1.5 c.

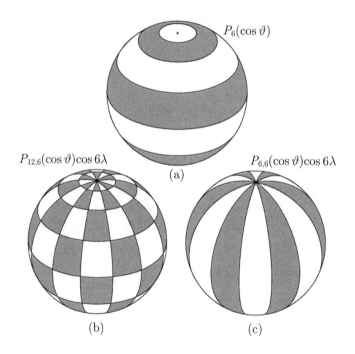

Fig. 1.5. The kinds of spherical harmonics: (a) zonal, (b) tesseral, (c) sectorial

1.8   Legendre's functions of the second kind

The Legendre function $P_{nm}(t)$ is not the only solution of Legendre's differential equation (1–56). There is a completely different function which also satisfies this equation. It is called *Legendre's function of the second kind*, of degree n and order m, and denoted by $Q_{nm}(t)$.

Although the $Q_{nm}(t)$ are functions of a completely different nature, they satisfy relationships very similar to those satisfied by the $P_{nm}(t)$.

The "zonal" functions

$$Q_n(t) \equiv Q_{n0}(t) \tag{1–73}$$

are defined by

$$Q_n(t) = \frac{1}{2} P_n(t) \ln \frac{1+t}{1-t} - \sum_{k=1}^{n} \frac{1}{k} P_{k-1}(t) P_{n-k}(t), \tag{1–74}$$

and the others by

$$Q_{nm}(t) = (1 - t^2)^{m/2} \frac{d^m Q_n(t)}{dt^m}. \tag{1–75}$$

Equation (1–75) is completely analogous to (1–65); furthermore, the functions $Q_n(t)$ satisfy the same recursion formula (1–62) as the functions $P_n(t)$.

If we evaluate the first few Q_n, from (1–74) we find

$$Q_0(t) = \frac{1}{2} \ln \frac{1+t}{1-t} = \tanh^{-1} t,$$

$$Q_1(t) = \frac{t}{2} \ln \frac{1+t}{1-t} - 1 = t \tanh^{-1} t - 1, \tag{1–76}$$

$$Q_2(t) = \left(\frac{3}{4} t^2 - \frac{1}{4}\right) \ln \frac{1+t}{1-t} - \frac{3}{2} t = \left(\frac{3}{2} t^2 - \frac{1}{2}\right) \tanh^{-1} t - \frac{3}{2} t.$$

These formulas and Fig. 1.6 show that the functions Q_{nm} are really quite different from the functions P_{nm}. From the singularity $\pm\infty$ at $t = \pm 1$ (i.e., $\vartheta = 0$ or π), we see that it is impossible to substitute $Q_{nm}(\cos\vartheta)$ for $P_{nm}(\cos\vartheta)$ if ϑ means the polar distance, because harmonic functions must be regular.

However, we will encounter them in the theory of ellipsoidal harmonics (Sect. 1.16), which is applied to the normal gravity field of the earth (Sect. 2.7). For this purpose we need Legendre's functions of the second

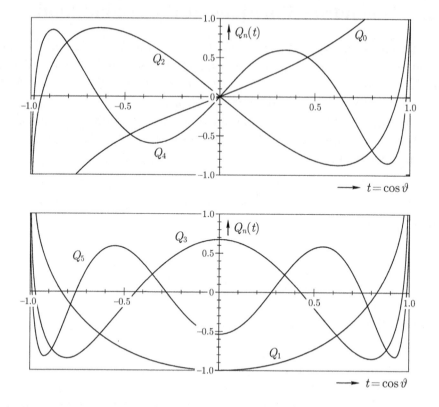

Fig. 1.6. Legendre's functions of the second kind: n even (top) and n odd (bottom)

kind as functions of a complex argument. If the argument z is complex, we must replace the definition (1–74) by

$$Q_n(z) = \frac{1}{2} P_n(z) \ln \frac{z+1}{z-1} - \sum_{k=1}^{n} \frac{1}{k} P_{k-1}(z) \, P_{n-k}(z) \,, \qquad (1\text{–}77)$$

where Legendre's polynomials $P_n(z)$ are defined by the same formulas as in the case of a real argument t. Therefore, the only change as compared to (1–74) is the replacement of

$$\frac{1}{2} \ln \frac{1+t}{1-t} = \tanh^{-1} t \qquad (1\text{–}78)$$

by

$$\frac{1}{2} \ln \frac{z+1}{z-1} = \coth^{-1} z \,. \qquad (1\text{–}79)$$

In particular, we have

$$Q_0(z) = \frac{1}{2} \ln \frac{z+1}{z-1} = \coth^{-1} z \,,$$

$$Q_1(z) = \frac{z}{2} \ln \frac{z+1}{z-1} - 1 = z \coth^{-1} z - 1 \,, \tag{1–80}$$

$$Q_2(z) = \left(\frac{3}{4} z^2 - \frac{1}{4}\right) \ln \frac{z+1}{z-1} - \frac{3}{2} z = \left(\frac{3}{2} z^2 - \frac{1}{2}\right) \coth^{-1} z - \frac{3}{2} z \,.$$

1.9 Expansion theorem and orthogonality relations

In (1–52) and (1–53), we have expanded *harmonic* functions in space into series of *solid* spherical harmonics. In a similar way an *arbitrary* (at least in a very general sense) function $f(\vartheta, \lambda)$ on the surface of the sphere can be expanded into a series of *surface* spherical harmonics:

$$f(\vartheta, \lambda) = \sum_{n=0}^{\infty} Y_n(\vartheta, \lambda) = \sum_{n=0}^{\infty} \sum_{m=0}^{n} [a_{nm} \mathcal{R}_{nm}(\vartheta, \lambda) + b_{nm} \mathcal{S}_{nm}(\vartheta, \lambda)], \quad (1\text{–}81)$$

where we have introduced the abbreviations

$$\mathcal{R}_{nm}(\vartheta, \lambda) = P_{nm}(\cos \vartheta) \cos m\lambda \,,$$
$$\mathcal{S}_{nm}(\vartheta, \lambda) = P_{nm}(\cos \vartheta) \sin m\lambda \,. \tag{1–82}$$

The symbols a_{nm} and b_{nm} are constant coefficients, which we will now determine. Essential for this purpose are the *orthogonality relations*. These remarkable relations mean that the integral over the unit sphere of the product of any two *different* functions $\mathcal{R}_{nm}$ or $\mathcal{S}_{nm}$ is zero:

$$\left. \begin{array}{l} \iint\limits_{\sigma} \mathcal{R}_{nm}(\vartheta, \lambda)\, \mathcal{R}_{sr}(\vartheta, \lambda)\, d\sigma = 0 \\[2mm] \iint\limits_{\sigma} \mathcal{S}_{nm}(\vartheta, \lambda)\, \mathcal{S}_{sr}(\vartheta, \lambda)\, d\sigma = 0 \end{array} \right\} \quad \text{if } s \neq n \text{ or } r \neq m \text{ or both;}$$

$$\iint\limits_{\sigma} \mathcal{R}_{nm}(\vartheta, \lambda)\, \mathcal{S}_{sr}(\vartheta, \lambda)\, d\sigma = 0 \qquad \text{in any case}\,. \tag{1–83}$$

For the product of two *equal* functions $\mathcal{R}_{nm}$ or $\mathcal{S}_{nm}$, we have

$$\iint\limits_{\sigma} [\mathcal{R}_{n0}(\vartheta, \lambda)]^2 \, d\sigma = \frac{4\pi}{2n+1} \,;$$

$$\iint\limits_{\sigma} [\mathcal{R}_{nm}(\vartheta, \lambda)]^2 \, d\sigma = \iint\limits_{\sigma} [\mathcal{S}_{nm}(\vartheta, \lambda)]^2 \, d\sigma = \frac{2\pi}{2n+1} \frac{(n+m)!}{(n-m)!} \quad (m \neq 0)\,.$$

$$\tag{1–84}$$

Note that there is no $\mathcal{S}_{n0}$, since $\sin 0\lambda = 0$. In these formulas we have used the abbreviation

$$\iint_{\sigma} = \int_{\lambda=0}^{2\pi} \int_{\vartheta=0}^{\pi} \tag{1-85}$$

for the integral over the unit sphere. The expression

$$d\sigma = \sin\vartheta\,d\vartheta\,d\lambda \tag{1-86}$$

denotes the surface element of the unit sphere.

Now we turn to the determination of the coefficients a_{nm} and b_{nm} in (1–81). Multiplying both sides of the equation by a certain $\mathcal{R}_{sr}(\vartheta, \lambda)$ and integrating over the unit sphere gives

$$\iint_{\sigma} f(\vartheta, \lambda)\,\mathcal{R}_{sr}(\vartheta, \lambda)\,d\sigma = a_{sr} \iint_{\sigma} [\mathcal{R}_{sr}(\vartheta, \lambda)]^2\,d\sigma\,, \tag{1-87}$$

since in the double integral on the right-hand side all terms except the one with $n = s$, $m = r$ will vanish according to the orthogonality relations (1–83). The integral on the right-hand side has the value given in (1–84), so that a_{sr} is determined. In a similar way we find b_{sr} by multiplying (1–81) by $\mathcal{S}_{sr}(\vartheta, \lambda)$ and integrating over the unit sphere. The result is

$$a_{n0} = \frac{2n+1}{4\pi} \iint_{\sigma} f(\vartheta, \lambda)\,P_n(\cos\vartheta)\,d\sigma\,;$$

$$\left.\begin{aligned}
a_{nm} &= \frac{2n+1}{2\pi}\frac{(n-m)!}{(n+m)!} \iint_{\sigma} f(\vartheta, \lambda)\,\mathcal{R}_{nm}(\vartheta, \lambda)\,d\sigma \\[2em]
b_{nm} &= \frac{2n+1}{2\pi}\frac{(n-m)!}{(n+m)!} \iint_{\sigma} f(\vartheta, \lambda)\,\mathcal{S}_{nm}(\vartheta, \lambda)\,d\sigma
\end{aligned}\right\} \quad (m \neq 0)\,. \tag{1-88}$$

The coefficients a_{nm} and b_{nm} can, thus, be determined by integration.

We note that the Laplace spherical harmonics $Y_n(\vartheta, \lambda)$ in (1–81) may also be found directly by the formula

$$Y_n(\vartheta, \lambda) = \frac{2n+1}{4\pi} \int_{\lambda'=0}^{2\pi} \int_{\vartheta'=0}^{\pi} f(\vartheta', \lambda')\,P_n(\cos\psi)\,\sin\vartheta'\,d\vartheta'\,d\lambda'\,, \tag{1-89}$$

where ψ is the spherical distance between the points P, represented by ϑ, λ, and P', represented by ϑ', λ' (Fig. 1.7), so that

$$\cos\psi = \cos\vartheta\cos\vartheta' + \sin\vartheta\sin\vartheta'\cos(\lambda' - \lambda)\,. \tag{1-90}$$

Later, when being acquainted with Sect. 1.11, Eq. (1–89) may be verified by straightforward computation, substituting $P_n(\cos\psi)$ from the decomposition formula (1–105).

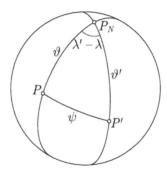

Fig. 1.7. Spherical distance ψ

1.10   Fully normalized spherical harmonics

The formulas of the preceding section for the expansion of a function into a series of surface spherical harmonics are somewhat inconvenient to handle. If we look at equations (1–84) and (1–88), we see that there are different formulas for $m = 0$ and $m \neq 0$; furthermore, the expressions are rather complicated and difficult to remember.

Therefore, it has been proposed that the "conventional" harmonics $\mathcal{R}_{nm}$ and $\mathcal{S}_{nm}$, defined by (1–82) together with (1–57), be replaced by other functions which differ by a constant factor and are easier to handle. We consider here only the *fully normalized* harmonics, which seem to be the most convenient and the most widely used.

The "fully normalized" harmonics are simply "normalized" in the sense of the theory of real functions; we have to use this clumsy expression because the term "normalized spherical harmonics" has already been used for other functions, unfortunately often for some that are not "normalized" at all in the mathematical sense.

We denote the fully normalized harmonics by $\bar{\mathcal{R}}_{nm}$ and $\bar{\mathcal{S}}_{nm}$; they are defined by

$$\bar{\mathcal{R}}_{n0}(\vartheta, \lambda) = \sqrt{2n+1}\,\mathcal{R}_{n0}(\vartheta, \lambda) \equiv \sqrt{2n+1}\,P_n(\cos \vartheta)\,;$$

$$\left. \begin{aligned} \bar{\mathcal{R}}_{nm}(\vartheta, \lambda) &= \sqrt{2(2n+1)\frac{(n-m)!}{(n+m)!}}\,\mathcal{R}_{nm}(\vartheta, \lambda) \\[2ex] \bar{\mathcal{S}}_{nm}(\vartheta, \lambda) &= \sqrt{2(2n+1)\frac{(n-m)!}{(n+m)!}}\,\mathcal{S}_{nm}(\vartheta, \lambda) \end{aligned} \right\} \quad (m \neq 0)\,. \qquad (1\text{–}91)$$

The orthogonality relations (1–83) also apply for these fully normalized harmonics, whereas Eqs. (1–84) are thoroughly simplified: they become

$$\frac{1}{4\pi} \iint_\sigma \bar{\mathcal{R}}_{nm}^2 \, d\sigma = \frac{1}{4\pi} \iint_\sigma \bar{\mathcal{S}}_{nm}^2 \, d\sigma = 1 . \tag{1–92}$$

This means that the *average square of any fully normalized harmonic is unity*, the average being taken over the sphere (the average corresponds to the integral divided by the area 4π). This formula now applies for any m, whether it is zero or not.

If we expand an arbitrary function $f(\vartheta, \lambda)$ into a series of fully normalized harmonics, analogously to (1–81),

$$f(\vartheta, \lambda) = \sum_{n=0}^{\infty} \sum_{m=0}^{n} [\bar{a}_{nm} \bar{\mathcal{R}}_{nm}(\vartheta, \lambda) + \bar{b}_{nm} \bar{\mathcal{S}}_{nm}(\vartheta, \lambda)] , \tag{1–93}$$

then the coefficients $\bar{a}_{nm}, \bar{b}_{nm}$ are simply given by

$$\bar{a}_{nm} = \frac{1}{4\pi} \iint_\sigma f(\vartheta, \lambda) \, \bar{\mathcal{R}}_{nm}(\vartheta, \lambda) \, d\sigma ,$$

$$\bar{b}_{nm} = \frac{1}{4\pi} \iint_\sigma f(\vartheta, \lambda) \, \bar{\mathcal{S}}_{nm}(\vartheta, \lambda) \, d\sigma ; \tag{1–94}$$

that is, the coefficients are the average products of the function and the corresponding harmonic $\bar{\mathcal{R}}_{nm}$ or $\bar{\mathcal{S}}_{nm}$.

The simplicity of formulas (1–92) and (1–94) constitutes the main advantage of the fully normalized spherical harmonics and makes them useful in many respects, even though the functions $\bar{\mathcal{R}}_{nm}$ and $\bar{\mathcal{S}}_{nm}$ in (1–91) are a little more complicated than the conventional $\mathcal{R}_{nm}$ and $\mathcal{S}_{nm}$. We have

$$\bar{\mathcal{R}}_{nm}(\vartheta, \lambda) = \bar{P}_{nm}(\cos \vartheta) \, \cos m\lambda ,$$

$$\bar{\mathcal{S}}_{nm}(\vartheta, \lambda) = \bar{P}_{nm}(\cos \vartheta) \, \sin m\lambda , \tag{1–95}$$

where

$$\bar{P}_{n0}(t) = \sqrt{2n+1}\, 2^{-n} \sum_{k=0}^{r} (-1)^k \frac{(2n-2k)!}{k!\,(n-k)!\,(n-2k)!} t^{n-2k} \tag{1–96}$$

for $m = 0$, and

$$\bar{P}_{nm}(t) = \sqrt{2(2n+1)\frac{(n-m)!}{(n+m)!}}\, 2^{-n} (1-t^2)^{m/2} \cdot$$

$$\sum_{k=0}^{r} (-1)^k \frac{(2n-2k)!}{k!\,(n-k)!\,(n-m-2k)!} t^{n-m-2k} \tag{1–97}$$

for $m \neq 0$. This corresponds to (1–67); here, as in (1–67), r is the greatest integer $\leq (n-m)/2$.

There are relations between the coefficients $\bar{a}_{nm}$ and $\bar{b}_{nm}$ for fully normalized harmonics and the coefficients a_{nm} and b_{nm} for conventional harmonics that are inverse to those in (1–91):

$$
\left.
\begin{aligned}
\bar{a}_{n0} &= \frac{a_{n0}}{\sqrt{2n+1}} \,; \\[2mm]
\bar{a}_{nm} &= \sqrt{\frac{1}{2(2n+1)} \frac{(n+m)!}{(n-m)!}} \, a_{nm} \\[2mm]
\bar{b}_{nm} &= \sqrt{\frac{1}{2(2n+1)} \frac{(n+m)!}{(n-m)!}} \, b_{nm}
\end{aligned}
\right\} \quad (m \neq 0)\,.
$$

(1–98)

1.11 Expansion of the reciprocal distance into zonal harmonics and decomposition formula

The distance l between two points with spherical coordinates

$$P(r, \vartheta, \lambda), \quad P'(r', \vartheta', \lambda') \tag{1–99}$$

is given by

$$l^2 = r^2 + r'^2 - 2r\,r'\cos\psi\,, \tag{1–100}$$

where ψ is the angle between the radius vectors r and r' (Fig. 1.8), so that, from (1–90),

$$\cos\psi = \cos\vartheta\cos\vartheta' + \sin\vartheta\sin\vartheta'\cos(\lambda' - \lambda) \tag{1–101}$$

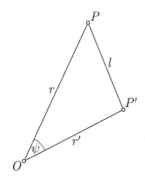

Fig. 1.8. The spatial distance l

results. Assuming $r' < r$, we may write

$$\frac{1}{l} = \frac{1}{\sqrt{r^2 - 2r\,r'\cos\psi + r'^2}} = \frac{1}{r\sqrt{1 - 2\alpha\,u + \alpha^2}}, \tag{1-102}$$

where we have put $\alpha = r'/r$ and $u = \cos\psi$. If $r' < r$, this can be expanded into a power series with respect to α. It is remarkable that the coefficients of α^n are the (conventional) zonal harmonics, or Legendre's polynomials $P_n(u) = P_n(\cos\psi)$:

$$\frac{1}{\sqrt{1 - 2\alpha\,u + \alpha^2}} = \sum_{n=0}^{\infty} \alpha^n\,P_n(u) = P_0(u) + \alpha\,P_1(u) + \alpha^2 P_2(u) + \cdots . \tag{1-103}$$

Hence, we obtain

$$\frac{1}{l} = \sum_{n=0}^{\infty} \frac{r'^n}{r^{n+1}}\,P_n(\cos\psi), \tag{1-104}$$

which is an important formula.

It would still be desirable in this equation to express $P_n(\cos\psi)$ in terms of functions of the spherical coordinates ϑ, λ and ϑ', λ' of which ψ is composed according to (1–90). This is achieved by the *decomposition formula*

$$P_n(\cos\psi) = P_n(\cos\vartheta)\,P_n(\cos\vartheta') +$$

$$2\sum_{m=1}^{n} \frac{(n-m)!}{(n+m)!}\,[\mathcal{R}_{nm}(\vartheta,\lambda)\mathcal{R}_{nm}(\vartheta',\lambda') + \mathcal{S}_{nm}(\vartheta,\lambda)\mathcal{S}_{nm}(\vartheta',\lambda')].$$

$$\tag{1-105}$$

Substituting this into (1–104), we obtain

$$\frac{1}{l} = \sum_{n=0}^{\infty} \left\{ \frac{P_n(\cos\vartheta)}{r^{n+1}}\,r'^n\,P_n(\cos\vartheta') + 2\sum_{m=1}^{n} \frac{(n-m)!}{(n+m)!} \cdot \right.$$

$$\tag{1-106}$$

$$\left. \left[\frac{\mathcal{R}_{nm}(\vartheta,\lambda)}{r^{n+1}}\,r'^n\,\mathcal{R}_{nm}(\vartheta',\lambda') + \frac{\mathcal{S}_{nm}(\vartheta,\lambda)}{r^{n+1}}\,r'^n\,\mathcal{S}_{nm}(\vartheta',\lambda') \right] \right\} .$$

The use of fully normalized harmonics simplifies these formulas. Replacing the conventional harmonics in (1–105) and (1–106) by fully normalized harmonics by means of (1–91), we find

$$P_n(\cos\psi) = \frac{1}{2n+1} \sum_{m=0}^{n} [\bar{\mathcal{R}}_{nm}(\vartheta,\lambda)\bar{\mathcal{R}}_{nm}(\vartheta',\lambda') + \bar{\mathcal{S}}_{nm}(\vartheta,\lambda)\bar{\mathcal{S}}_{nm}(\vartheta',\lambda')] ;$$

$$\tag{1-107}$$

$$\frac{1}{l} = \sum_{n=0}^{\infty} \sum_{m=0}^{n} \frac{1}{2n+1} \cdot$$

$$\left[\frac{\bar{\mathcal{R}}_{nm}(\vartheta, \lambda)}{r^{n+1}} r'^n \bar{\mathcal{R}}_{nm}(\vartheta', \lambda') + \frac{\bar{\mathcal{S}}_{nm}(\vartheta, \lambda)}{r^{n+1}} r'^n \bar{\mathcal{S}}_{nm}(\vartheta', \lambda') \right].$$

(1–108)

The last formula will be fundamental for the expansion of the earth's gravitational field in spherical harmonics.

1.12 Solution of Dirichlet's problem by means of spherical harmonics and Poisson's integral

We define *Dirichlet's problem*, or the *first boundary-value problem of potential theory*, as follows: Given is an arbitrary function on a surface S, to determine is a function V which is harmonic either inside or outside S and which assumes on S the values of the prescribed function.

If the surface S is a sphere, then Dirichlet's problem can be solved by means of spherical harmonics. Let us take first the unit sphere, $r = 1$, and expand the prescribed function, given on the unit sphere and denoted by $V(1, \vartheta, \lambda)$, into a series of surface spherical harmonics (1–81):

$$V(1, \vartheta, \lambda) = \sum_{n=0}^{\infty} Y_n(\vartheta, \lambda),$$

(1–109)

the $Y_n(\vartheta, \lambda)$ being determined by (1–89). (This series converges for very general functions V.) The functions

$$V_{\mathrm{i}}(r, \vartheta, \lambda) = \sum_{n=0}^{\infty} r^n Y_n(\vartheta, \lambda)$$

(1–110)

and

$$V_{\mathrm{e}}(r, \vartheta, \lambda) = \sum_{n=0}^{\infty} \frac{Y_n(\vartheta, \lambda)}{r^{n+1}}$$

(1–111)

assume the given values $V(1, \vartheta, \lambda)$ on the surface $r = 1$. The series (1–109) converges, and we have for $r < 1$

$$r^n Y_n < Y_n$$

(1–112)

and for $r > 1$

$$\frac{Y_n}{r^{n+1}} < Y_n.$$

(1–113)

Hence, the series (1–110) converges for $r \leq 1$, and the series (1–111) converges for $r \geq 1$; furthermore, both series have been found to represent harmonic functions. Therefore, we see that Dirichlet's problem is solved by $V_\mathrm{i}(r, \vartheta, \lambda)$ for the interior of the sphere $r = 1$, and by $V_\mathrm{e}(r, \vartheta, \lambda)$ for its exterior.

For a sphere of arbitrary radius $r = R$, the solution is similar. We expand the given function

$$V(R, \vartheta, \lambda) = \sum_{n=0}^{\infty} Y_n(\vartheta, \lambda) . \tag{1–114}$$

The surface spherical harmonics Y_n are determined by

$$Y_n(\vartheta, \lambda) = \frac{2n + 1}{4\pi} \int_{\lambda'=0}^{2\pi} \int_{\vartheta'=0}^{\pi} V(R, \vartheta', \lambda') P_n(\cos \psi) \sin \vartheta' \, d\vartheta' \, d\lambda' . \tag{1–115}$$

Then the series

$$V_\mathrm{i}(r, \vartheta, \lambda) = \sum_{n=0}^{\infty} \left(\frac{r}{R}\right)^n Y_n(\vartheta, \lambda) \tag{1–116}$$

solves the first boundary-value problem for the interior, and the series

$$V_\mathrm{e}(r, \vartheta, \lambda) = \sum_{n=0}^{\infty} \left(\frac{R}{r}\right)^{n+1} Y_n(\vartheta, \lambda) \tag{1–117}$$

solves it for the exterior of the sphere $r = R$.

Thus, we see that Dirichlet's problem can always be solved for the sphere. It is evident that this is closely related to the possibility of expanding an *arbitrary* function on the sphere into a series of *surface* spherical harmonics and a *harmonic* function in space into a series of *solid* spherical harmonics.

Dirichlet's boundary-value problem can be solved not only for the sphere but also for any sufficiently smooth boundary surface. An example is given in Sect. 1.16.

The solvability of Dirichlet's problem is also essential to Molodensky's problem (Sect. 8.3). See also Kellogg (1929: Chap. XI).

Poisson's integral

A more direct solution is obtained as follows. We consider only the exterior problem, which is of greater interest in geodesy. Substituting $Y_n(\vartheta, \lambda)$ from (1–89) into (1–117), we obtain

$$V_\mathrm{e}(r, \vartheta, \lambda) =$$

$$\sum_{n=0}^{\infty} \left(\frac{R}{r}\right)^{n+1} \frac{2n + 1}{4\pi} \int_{\lambda'=0}^{2\pi} \int_{\vartheta'=0}^{\pi} V(R, \vartheta', \lambda') P_n(\cos \psi) \sin \vartheta' \, d\vartheta' \, d\lambda' .$$

$$\tag{1–118}$$

We can rearrange this as

$$V_e(r, \vartheta, \lambda) = \frac{1}{4\pi} \int_{\lambda'=0}^{2\pi} \int_{\vartheta'=0}^{\pi} V(R, \vartheta', \lambda') \cdot$$

$$\left[\sum_{n=0}^{\infty} (2n+1) \left(\frac{R}{r} \right)^{n+1} P_n(\cos \psi) \right] \sin \vartheta' \, d\vartheta' \, d\lambda' . \qquad (1\text{–}119)$$

The sum in the brackets can be evaluated. We denote the spatial distance between the points $P(r, \vartheta, \lambda)$ and $P'(R, \vartheta', \lambda')$ by l. Then, using (1–104),

$$\frac{1}{l} = \frac{1}{\sqrt{r^2 + R^2 - 2Rr \cos \psi}} = \frac{1}{R} \sum_{n=0}^{\infty} \left(\frac{R}{r} \right)^{n+1} P_n(\cos \psi) \qquad (1\text{–}120)$$

results. Differentiating with respect to r, we get

$$-\frac{r - R \cos \psi}{l^3} = -\frac{1}{R} \sum_{n=0}^{\infty} (n+1) \frac{R^{n+1}}{r^{n+2}} P_n(\cos \psi) . \qquad (1\text{–}121)$$

Multiplying this equation by $-2Rr$, multiplying the expression for $1/l$ by $-R$, and then adding the two equations yields

$$\frac{R(r^2 - R^2)}{l^3} = \sum_{n=0}^{\infty} (2n+1) \left(\frac{R}{r} \right)^{n+1} P_n(\cos \psi) . \qquad (1\text{–}122)$$

The right-hand side is the bracketed expression in (1–119). Substituting the left-hand side, we finally obtain

$$V_e(r, \vartheta, \lambda) = \frac{R(r^2 - R^2)}{4\pi} \int_{\lambda'=0}^{2\pi} \int_{\vartheta'=0}^{\pi} \frac{V(R, \vartheta', \lambda')}{l^3} \sin \vartheta' \, d\vartheta' \, d\lambda' , \qquad (1\text{–}123)$$

where

$$l = \sqrt{r^2 + R^2 - 2Rr \cos \psi} . \qquad (1\text{–}124)$$

This is *Poisson's integral*. It is an explicit solution of Dirichlet's problem for the exterior of the sphere, which has many applications in physical geodesy.

1.13 Other boundary-value problems

There are other similar boundary-value problems. In *Neumann's problem*, or the *second boundary-value problem of potential theory*, the normal derivative $\partial V/\partial n$ is given on the surface S, instead of the function V itself. The normal derivative is the derivative along the outward-directed surface normal n to

S. In the *third boundary-value problem*, a linear combination of V and of its normal derivative

$$h\,V + k\,\frac{\partial V}{\partial n} \qquad (1\text{--}125)$$

is given on S.

For the sphere, the solution of these boundary-value problems is also easily expressed in terms of spherical harmonics. We consider the exterior problems only, because these are of special interest to geodesy.

In *Neumann's problem*, we expand the given values of $\partial V / \partial n$ on the sphere $r = R$ into a series of surface spherical harmonics:

$$\left(\frac{\partial V}{\partial n}\right)_{r=R} = \sum_{n=0}^{\infty} Y_n(\vartheta, \lambda). \qquad (1\text{--}126)$$

The harmonic function which solves Neumann's problem for the exterior of the sphere is then

$$V_{\mathrm{e}}(r, \vartheta, \lambda) = -R \sum_{n=0}^{\infty} \left(\frac{R}{r}\right)^{n+1} \frac{Y_n(\vartheta, \lambda)}{n+1}. \qquad (1\text{--}127)$$

To verify it, we differentiate with respect to r, getting

$$\frac{\partial V_{\mathrm{e}}}{\partial r} = \sum_{n=0}^{\infty} \left(\frac{R}{r}\right)^{n+2} Y_n(\vartheta, \lambda). \qquad (1\text{--}128)$$

Since for the sphere the normal coincides with the radius vector, we have

$$\left(\frac{\partial V}{\partial n}\right)_{r=R} = \left(\frac{\partial V}{\partial r}\right)_{r=R}, \qquad (1\text{--}129)$$

and we see that (1–126) is satisfied.

The *third boundary-value problem* is particularly relevant to physical geodesy, because the determination of the undulations of the geoid from gravity anomalies is just such a problem. To solve the general case, we again expand the function defined by the given boundary values into surface spherical harmonics:

$$h\,V + k\,\frac{\partial V}{\partial n} = \sum_{n=0}^{\infty} Y_n(\vartheta, \lambda). \qquad (1\text{--}130)$$

The harmonic function

$$V_{\mathrm{e}}(r, \vartheta, \lambda) = \sum_{n=0}^{\infty} \left(\frac{R}{r}\right)^{n+1} \frac{Y_n(\vartheta, \lambda)}{h - (k/R)(n+1)} \qquad (1\text{--}131)$$

solves the third boundary-value problem for the exterior of the sphere $r = R$. The straightforward verification is analogous to the case of (1–127).

In the determination of the geoidal undulations, the constants h, k have the values

$$h = -\frac{2}{R}, \quad k = -1, \tag{1–132}$$

so that

$$V_e(r, \vartheta, \lambda) = R \sum_{n=0}^{\infty} \left(\frac{R}{r}\right)^{n+1} \frac{Y_n(\vartheta, \lambda)}{n - 1} \tag{1–133}$$

solves the *boundary-value problem of physical geodesy*.

As we have seen in the preceding section, the first boundary-value problem can also be solved directly by *Poisson's integral*. Similar integral formulas also exist for the second and the third problem. The integral formula that corresponds to (1–133) for the boundary-value problem of physical geodesy is *Stokes' integral*, which will be considered in detail in Chap. 2.

Remark on inverse problems

Boundary-value problems give the potential *outside* the earth, where there are no masses and where the potential, satisfying Laplace's equation, is harmonic. The determination of the potential *inside* the earth is of a quite different character since the earth is filled by masses, and the interior potential satisfies Poisson's rather than Laplace's equation, as we have seen in Sect. 1.2. Unfortunately, the density ϱ inside the earth is generally unknown.

To see the difficulties of the problem, let us consider Newton's integral (1–12). If the interior masses were known, we could easily use this formula to compute the potential inside (and outside) the earth, in a direct and straightforward way. The determination of the potential from the masses is a "direct" problem. The "inverse" problem is to determine the masses from the potential, finding a solution of Newton's integral for the density ϱ, which is essentially more difficult.

In fact, it is impossible to determine uniquely the generating masses from the external potential. This *inverse problem of potential theory* has no unique solution. Such inverse problems occur in geophysical prospecting by gravity measurements: underground masses are inferred from disturbances of the gravity field. To determine the problem more completely, additional information is necessary, which is furnished, for example, by geology or by seismic measurements.

Generally, nowadays we know that many problems in geophysics and other sciences including medicine (e.g., seismic and medical tomography) are inverse problems. We cannot pursue this interesting problem here and refer only to the extensive literature, e.g., the book by Moritz (1995), the internet page www.inas.tugraz.at/forschung/InverseProblems/AngerMoritz.html or Anger and Moritz (2003).

1.14   The radial derivative of a harmonic function

For later application to problems related with the vertical gradient of gravity, we will now derive an integral formula for the derivative along the radius vector r of an arbitrary harmonic function which we denote by V. Such a harmonic function satisfies Poisson's integral (1–123):

$$V(r, \vartheta, \lambda) = \frac{R(r^2 - R^2)}{4\pi} \int_{\lambda'=0}^{2\pi} \int_{\vartheta'=0}^{\pi} \frac{V(R, \vartheta', \lambda')}{l^3} \sin \vartheta' \, d\vartheta' \, d\lambda' . \quad (1\text{–}134)$$

Forming the radial derivative $\partial V / \partial r$, we note that $V(R, \vartheta', \lambda')$ does not depend on r. Thus, we need only to differentiate $(r^2 - R^2)/l^3$, obtaining

$$\frac{\partial V(r, \vartheta, \lambda)}{\partial r} = \frac{R}{4\pi} \int_{\lambda'=0}^{2\pi} \int_{\vartheta'=0}^{\pi} M(r, \psi) \, V(R, \vartheta', \lambda') \sin \vartheta' \, d\vartheta' \, d\lambda' , \quad (1\text{–}135)$$

where

$$M(r, \psi) \equiv \frac{\partial}{\partial r} \left(\frac{r^2 - R^2}{l^3} \right) = \frac{1}{l^5} (5R^2 r - r^3 - R \, r^2 \cos \psi - 3R^3 \cos \psi) . \quad (1\text{–}136)$$

Applying (1–135) to the special harmonic function $V_1(r, \vartheta, \lambda) = R/r$, where

$$\frac{\partial V_1}{\partial r} = -\frac{R}{r^2} \quad \text{and} \quad V_1(R, \vartheta', \lambda') = \frac{R}{R} = 1 , \quad (1\text{–}137)$$

we obtain

$$-\frac{R}{r^2} = \frac{R}{4\pi} \int_{\lambda'=0}^{2\pi} \int_{\vartheta'=0}^{\pi} M(r, \psi) \sin \vartheta' \, d\vartheta' \, d\lambda' . \quad (1\text{–}138)$$

Multiplying both sides of this equation by $V(r, \vartheta, \lambda)$ and subtracting it from (1–135) gives

$$\frac{\partial V}{\partial r} + \frac{R}{r^2} V_P = \frac{R}{4\pi} \int_{\lambda'=0}^{2\pi} \int_{\vartheta'=0}^{\pi} M(r, \psi) \, (V - V_P) \sin \vartheta' \, d\vartheta' \, d\lambda' , \quad (1\text{–}139)$$

where

$$V_P = V(r, \vartheta, \lambda) , \quad V = V(R, \vartheta', \lambda') . \quad (1\text{–}140)$$

In order to find the radial derivative at the surface of the sphere of radius R, we must set $r = R$. Then l becomes (Fig. 1.9)

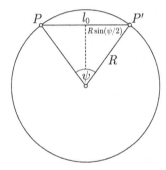

Fig. 1.9. Spatial distance between two points on a sphere

$$l_0 = 2R \sin \frac{\psi}{2} \, , \tag{1-141}$$

and the function M takes the simple form

$$M(R, \psi) = \frac{1}{4R^2 \sin^3 \frac{\psi}{2}} = \frac{2R}{l_0^3} \, . \tag{1-142}$$

For $\psi \to 0$ we have $M(R, \psi) \to \infty$, and we cannot use the original formula (1–135) at the surface of the sphere $r = R$. In the transformed equation (1–139), however, we have $V - V_P \to 0$ for $\psi \to 0$, and the singularity of M for $\psi \to 0$ will be neutralized (provided V is differentiable twice at P). Thus, we obtain the *gradient formula*

$$\frac{\partial V}{\partial r} = -\frac{1}{R} V_P + \frac{R^2}{2\pi} \int_{\lambda'=0}^{2\pi} \int_{\vartheta'=0}^{\pi} \frac{V - V_P}{l_0^3} \sin \vartheta' \, d\vartheta' \, d\lambda' \, . \tag{1-143}$$

This equation expresses $\partial V / \partial r$ on the sphere $r = R$ in terms of V on this sphere; thus, we now have

$$V_P = V(R, \vartheta, \lambda) \, , \quad V = V(R, \vartheta', \lambda') \, . \tag{1-144}$$

Solution in terms of spherical harmonics

We may express V_P as

$$V_P = \sum_{n=0}^{\infty} \left(\frac{R}{r} \right)^{n+1} Y_n(\vartheta, \lambda) \, . \tag{1-145}$$

Differentiation yields

$$\frac{\partial V}{\partial r} = -\sum_{n=0}^{\infty} (n+1) \frac{R^{n+1}}{r^{n+2}} Y_n(\vartheta, \lambda) \, . \tag{1-146}$$

For $r = R$, this becomes

$$\frac{\partial V}{\partial r} = -\frac{1}{R} \sum_{n=0}^{\infty} (n+1) \, Y_n(\vartheta, \lambda) \,. \tag{1–147}$$

This is the equivalent of (1–143) in terms of spherical harmonics. From this equation, we get an interesting by-product. Writing (1–147) as

$$\frac{\partial V}{\partial r} = -\frac{1}{R} V_P - \frac{1}{R} \sum_{n=0}^{\infty} n \, Y_n(\vartheta, \lambda) \tag{1–148}$$

and comparing this with (1–143), we see that

$$\frac{R^2}{2\pi} \int_{\lambda'=0}^{2\pi} \int_{\vartheta'=0}^{\pi} \frac{V - V_P}{l_0^3} \, \sin \vartheta' \, d\vartheta' \, d\lambda' = -\frac{1}{R} \sum_{n=0}^{\infty} n \, Y_n(\vartheta, \lambda) \,. \tag{1–149}$$

This equation is formulated entirely in terms of quantities referred to the spherical surface only. Furthermore, for any function prescribed on the surface of a sphere, one can find a function in space that is harmonic outside the sphere and assumes the values of the function prescribed on it. This is done by solving Dirichlet's exterior problem. From these facts, we conclude that (1–149) holds for any (reasonably) arbitrary function V defined on the surface of a sphere. These developments will be used in Sect. 2.20.

1.15 Laplace's equation in ellipsoidal-harmonic coordinates

Spherical harmonics are most frequently used in geodesy because they are relatively simple and the earth is nearly spherical. Since the earth is more nearly an ellipsoid of revolution, it might be expected that ellipsoidal harmonics, which are defined in a way similar to that of the spherical harmonics, would be even more suitable. The whole matter is a question of mathematical convenience, since both spherical and ellipsoidal harmonics may be used for any attracting body, regardless of its form. As ellipsoidal harmonics are more complicated, however, they are used only in certain special cases which nevertheless are important, namely, in problems involving rigorous computation of normal gravity.

We introduce *ellipsoidal-harmonic coordinates* u, ϑ, λ (Fig. 1.10). In a rectangular system, a point P has the coordinates x, y, z. Now we pass through P the surface of an ellipsoid of revolution whose center is the origin O, whose rotation axis coincides with the z-axis, and whose linear eccentricity has the constant value E. The coordinate u is the semiminor axis of this

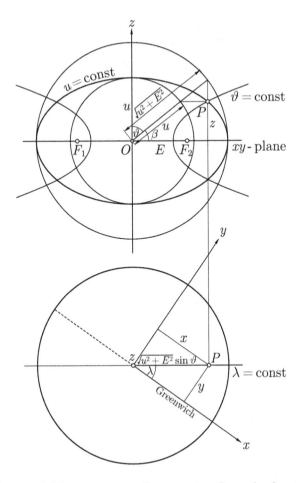

Fig. 1.10. Ellipsoidal-harmonic coordinates: view from the front (top) and view from above (bottom)

ellipsoid, ϑ is the complement of the "reduced latitude" β of P with respect to this ellipsoid (the definition is seen in Fig. 1.10), i.e., $\vartheta = 90° - \beta$, and λ is the geocentric longitude in the usual sense.

It should be carefully noted that in spherical harmonics ϑ is the polar distance, which is nothing but the complement of the *geocentric* latitude, whereas in ellipsoidal-harmonic coordinates ϑ is the complement of the *reduced* latitude denoted by β.

The ellipsoidal-harmonic coordinates u, ϑ, λ are related to x, y, z by

$$x = \sqrt{u^2 + E^2} \sin\vartheta \cos\lambda ,$$

$$y = \sqrt{u^2 + E^2} \sin\vartheta \sin\lambda , \qquad (1\text{–}150)$$

$$z = u \cos\vartheta ,$$

which can be read from Fig. 1.10, considering that $\sqrt{u^2 + E^2}$ is the semi-major axis of the ellipsoid whose surface passes through P. Because of $\vartheta = 90° - \beta$, we may equivalently write

$$x = \sqrt{u^2 + E^2} \cos \beta \cos \lambda\,,$$

$$y = \sqrt{u^2 + E^2} \cos \beta \sin \lambda\,, \qquad (1\text{--}151)$$

$$z = u \sin \beta\,.$$

Taking $u = $ constant, we find

$$\frac{x^2 + y^2}{u^2 + E^2} + \frac{z^2}{u^2} = 1\,, \qquad (1\text{--}152)$$

which represents an ellipsoid of revolution. For $\vartheta = $ constant, we obtain

$$\frac{x^2 + y^2}{E^2 \sin^2\vartheta} - \frac{z^2}{E^2 \cos^2\vartheta} = 1\,, \qquad (1\text{--}153)$$

which represents a hyperboloid of one sheet, and for $\lambda = $ constant, we get the meridian plane

$$y = x \tan \lambda\,. \qquad (1\text{--}154)$$

The constant focal length E, i.e., the distance between the coordinate origin O and one of the focal points F_1 or F_2, which is the same for *all* ellipsoids $u = $ constant, characterizes the coordinate system. For $E = 0$ we have the usual spherical coordinates $u = r$ and ϑ, λ as a limiting case.

To find ds, the element of arc, in ellipsoidal-harmonic coordinates, we proceed in the same way as in spherical coordinates, Eq. (1–30), and obtain

$$ds^2 = \frac{u^2 + E^2 \cos^2\vartheta}{u^2 + E^2}\, du^2 + (u^2 + E^2 \cos^2\vartheta)\, d\vartheta^2 + (u^2 + E^2) \sin^2\vartheta\, d\lambda^2\,. \quad (1\text{--}155)$$

The coordinate system u, ϑ, λ is again orthogonal: the products $du\, d\vartheta$, etc., are missing in the equation above. Setting $u = q_1$, $\vartheta = q_2$, $\lambda = q_3$, we have in (1–31)

$$h_1^2 = \frac{u^2 + E^2 \cos^2\vartheta}{u^2 + E^2}\,, \qquad h_2^2 = u^2 + E^2 \cos^2\vartheta\,, \qquad h_3^2 = (u^2 + E^2) \sin^2\vartheta\,.$$

$$(1\text{--}156)$$

If we substitute these relations into (1–32), we obtain

$$\Delta V = \frac{1}{(u^2 + E^2 \cos^2\vartheta) \sin\vartheta} \left\{ \frac{\partial}{\partial u} \left[(u^2 + E^2) \sin\vartheta\, \frac{\partial V}{\partial u} \right] + \right.$$

$$\left. \frac{\partial}{\partial\vartheta} \left(\sin\vartheta\, \frac{\partial V}{\partial\vartheta} \right) + \frac{\partial}{\partial\lambda} \left[\frac{u^2 + E^2 \cos^2\vartheta}{(u^2 + E^2) \sin\vartheta}\, \frac{\partial V}{\partial\lambda} \right] \right\}\,. \qquad (1\text{--}157)$$

Performing the differentiations and cancelling $\sin \vartheta$, we get

$$\Delta V \equiv \frac{1}{u^2 + E^2 \cos^2\vartheta} \left[(u^2 + E^2) \frac{\partial^2 V}{\partial u^2} + 2u \frac{\partial V}{\partial u} + \frac{\partial^2 V}{\partial \vartheta^2} + \right.$$

$$\left. \cot \vartheta \frac{\partial V}{\partial \vartheta} + \frac{u^2 + E^2 \cos^2\vartheta}{(u^2 + E^2) \sin^2\vartheta} \frac{\partial^2 V}{\partial \lambda^2} \right] = 0 \,, \tag{1-158}$$

which is *Laplace's equation in ellipsoidal-harmonic coordinates*. An alternative expression is obtained by omitting the factor $(u^2 + E^2 \cos^2\vartheta)^{-1}$:

$$(u^2 + E^2) \frac{\partial^2 V}{\partial u^2} + 2u \frac{\partial V}{\partial u} + \frac{\partial^2 V}{\partial \vartheta^2} + \cot \vartheta \frac{\partial V}{\partial \vartheta} + \frac{u^2 + E^2 \cos^2\vartheta}{(u^2 + E^2) \sin^2\vartheta} \frac{\partial^2 V}{\partial \lambda^2} = 0 \,. \tag{1-159}$$

In the limiting case, $E \to 0$, these equations reduce to the spherical expressions (1–35) and (1–36).

1.16   Ellipsoidal harmonics

To solve (1–158) or (1–159), we proceed in a way which is analogous to the method used to solve the corresponding equation (1–36) in spherical coordinates. What we did there may be summarized as follows. By the trial substitution

$$V(r, \vartheta, \lambda) = f(r)\, g(\vartheta)\, h(\lambda) \,, \tag{1-160}$$

we separated the variables r, ϑ, λ, so that the original partial differential equation (1–36) was decomposed into three ordinary differential equations (1–39), (1–46), and (1–47).

In order to solve Laplace's equation in ellipsoidal coordinates (1–159), we correspondingly make the ansatz (trial substitution)

$$V(u, \vartheta, \lambda) = f(u)\, g(\vartheta)\, h(\lambda) \,. \tag{1-161}$$

Substituting and dividing by $f\, g\, h$, we get

$$\frac{1}{f}[(u^2 + E^2)\, f'' + 2u\, f'] + \frac{1}{g}(g'' + g' \cot \vartheta) + \frac{u^2 + E^2 \cos^2\vartheta}{(u^2 + E^2) \sin^2\vartheta} \frac{h''}{h} = 0 \,. \tag{1-162}$$

The variable λ occurs only through the quotient h''/h, which consequently must be constant. One sees this more clearly by writing the equation in the form

$$-\frac{(u^2 + E^2) \sin^2\vartheta}{u^2 + E^2 \cos^2\vartheta} \left\{ \frac{1}{f}[(u^2 + E^2)\, f'' + 2u\, f'] + \frac{1}{g}(g'' + g' \cot \vartheta) \right\} = \frac{h''}{h} \,. \tag{1-163}$$

The left-hand side depends only on u and ϑ, the right-hand side only on λ. The two sides cannot be identically equal unless both are equal to the same constant. Therefore,

$$\frac{h''}{h} = -m^2. \tag{1-164}$$

The factor by which h''/h is to be multiplied, i.e., the inverse of the main factor on the left-hand side of (1–163), can be decomposed as follows:

$$\frac{u^2 + E^2 \cos^2\vartheta}{(u^2 + E^2) \sin^2\vartheta} = \frac{1}{\sin^2\vartheta} - \frac{E^2}{u^2 + E^2}. \tag{1-165}$$

Substituting (1–164) and (1–165) into (1–163) and combining functions of the same variable, we obtain

$$\frac{1}{f}[(u^2 + E^2)\, f'' + 2u\, f'] + \frac{E^2}{u^2 + E^2}\, m^2 = -\frac{1}{g}(g'' + g' \cot\vartheta) + \frac{m^2}{\sin^2\vartheta}. \tag{1-166}$$

The two sides are functions of different independent variables and must therefore be constant. Denoting this constant by $n(n+1)$, we finally get

$$(u^2 + E^2)\, f''(u) + 2u\, f'(u) - \left[n(n+1) - \frac{E^2}{u^2 + E^2}\, m^2\right] f(u) = 0; \tag{1-167}$$

$$\sin\vartheta\, g''(\vartheta) + \cos\vartheta\, g'(\vartheta) + \left[n(n+1)\sin\vartheta - \frac{m^2}{\sin\vartheta}\right] g(\vartheta) = 0; \tag{1-168}$$

$$h''(\lambda) + m^2 h(\lambda) = 0. \tag{1-169}$$

These are the three ordinary differential equations into which the partial differential equation (1–159) is decomposed by the separation of variables (1–161).

The second and third equations are the same as in the spherical case, Eqs. (1–46) and (1–47); the first equation is different. The substitutions

$$\tau = i\,\frac{u}{E} \quad (\text{where } i = \sqrt{-1}) \quad \text{and} \quad t = \cos\vartheta \tag{1-170}$$

transform the first and second equations into

$$(1 - \tau^2)\, \bar{f}''(\tau) - 2\tau\, \bar{f}'(\tau) + \left[n(n+1) - \frac{m^2}{1 - \tau^2}\right] \bar{f}(\tau) = 0,$$

$$(1 - t^2)\, \bar{g}''(t) - 2t\, \bar{g}'(t) + \left[n(n+1) - \frac{m^2}{1 - t^2}\right] \bar{g}(t) = 0, \tag{1-171}$$

where the overbar indicates that the functions f and g are expressed in terms of the new arguments τ and t. From spherical harmonics we are already familiar with the substitution $t = \cos \vartheta$ and the corresponding equation for $\bar{g}(t)$.

Note that $\bar{f}(\tau)$ satisfies formally the same differential equation as $\bar{g}(t)$, namely, Legendre's equation (1–56). As we have seen, this differential equation has two solutions: Legendre's function P_{nm} and Legendre's function of the second kind Q_{nm}. For $\bar{g}(t)$, where $t = \cos \vartheta$, the $Q_{nm}(t)$ are ruled out for obvious reasons, as we have seen in Sect. 1.8. For $\bar{f}(\tau)$, however, both sets of functions $P_{nm}(\tau)$ and $Q_{nm}(\tau)$ are possible solutions; they correspond to the two different solutions $f = r^n$ and $f = r^{-(n+1)}$ in the spherical case. Finally, (1–169) has as before the solutions $\cos m\lambda$ and $\sin m\lambda$.

We summarize all individual solutions:

$$f(u) = P_{nm}\left(i\,\frac{u}{E}\right) \quad \text{or} \quad Q_{nm}\left(i\,\frac{u}{E}\right) ;$$

$$g(\vartheta) = P_{nm}(\cos \vartheta) ; \qquad (1\text{–}172)$$

$$h(\lambda) = \cos m\lambda \quad \text{or} \quad \sin m\lambda .$$

Here n and $m < n$ are integers $0, 1, 2, \ldots$, as before. Hence, the functions

$$V(u, \vartheta, \lambda) = P_{nm}\left(i\,\frac{u}{E}\right) P_{nm}(\cos \vartheta) \begin{Bmatrix} \cos m\lambda \\ \sin m\lambda \end{Bmatrix} ,$$

$$V(u, \vartheta, \lambda) = Q_{nm}\left(i\,\frac{u}{E}\right) P_{nm}(\cos \vartheta) \begin{Bmatrix} \cos m\lambda \\ \sin m\lambda \end{Bmatrix} \qquad (1\text{–}173)$$

are solutions of Laplace's equation $\Delta V = 0$, that is, harmonic functions.

From these functions we may form by linear combination the series

$$V_{\mathrm{i}}(u, \vartheta, \lambda) = \sum_{n=0}^{\infty} \sum_{m=0}^{n} \frac{P_{nm}\left(i\,\dfrac{u}{E}\right)}{P_{nm}\left(i\,\dfrac{b}{E}\right)} \cdot$$

$$[a_{nm} P_{nm}(\cos \vartheta) \cos m\lambda + b_{nm} P_{nm}(\cos \vartheta) \sin m\lambda] ;$$

$$\qquad (1\text{–}174)$$

$$V_{\mathrm{e}}(u, \vartheta, \lambda) = \sum_{n=0}^{\infty} \sum_{m=0}^{n} \frac{Q_{nm}\left(i\,\dfrac{u}{E}\right)}{Q_{nm}\left(i\,\dfrac{b}{E}\right)} \cdot$$

$$[a_{nm} P_{nm}(\cos \vartheta) \cos m\lambda + b_{nm} P_{nm}(\cos \vartheta) \sin m\lambda] .$$

Here b is the semiminor axis of an arbitrary but fixed ellipsoid which may be called the *reference ellipsoid* (Fig. 1.11). The division by $P_{nm}(ib/E)$ or $Q_{nm}(ib/E)$ is possible because they are constants; its purpose is to simplify the expressions and to make the coefficients a_{nm} and b_{nm} real.

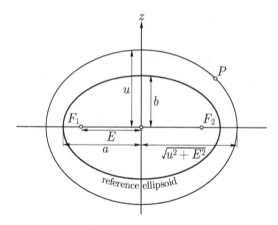

Fig. 1.11. Reference ellipsoid and ellipsoidal-harmonic coordinates

If the eccentricity E reduces to zero, the ellipsoidal-harmonic coordinates u, ϑ, λ become spherical coordinates r, ϑ, λ; the ellipsoid $u = b$ becomes the sphere $r = R$ because then the semiaxes a and b are equal to the radius R; and we find

$$\lim_{E \to 0} \frac{P_{nm}\left(i\frac{u}{E}\right)}{P_{nm}\left(i\frac{b}{E}\right)} = \left(\frac{u}{b}\right)^n = \left(\frac{r}{R}\right)^n, \quad \lim_{E \to 0} \frac{Q_{nm}\left(i\frac{u}{E}\right)}{Q_{nm}\left(i\frac{b}{E}\right)} = \left(\frac{R}{r}\right)^{n+1},$$

$$(1\text{--}175)$$

so that the first series in (1–174) becomes (1–116), and the second series in (1–174) becomes (1–117). Thus, we see that the function $P_{nm}(iu/E)$ corresponds to r^n and $Q_{nm}(iu/E)$ corresponds to $r^{-(n+1)}$ in spherical harmonics.

Hence, the first series in (1–174) is harmonic in the interior of the ellipsoid $u = b$, and the second series in (1–174) is harmonic in its exterior; this case is relevant to geodesy. For $u = b$, the two series are equal:

$$V_{\mathrm{i}}(b, \vartheta, \lambda) = V_{\mathrm{e}}(b, \vartheta, \lambda)$$

$$= \sum_{n=0}^{\infty} \sum_{m=0}^{n} [a_{nm} P_{nm}(\cos \vartheta) \cos m\lambda + b_{nm} P_{nm}(\cos \vartheta) \sin m\lambda].$$

$$(1\text{--}176)$$

Thus, the solution of Dirichlet's boundary-value problem for the ellipsoid of revolution is easy. We expand the function $V(b, \vartheta, \lambda)$, given on the ellipsoid $u = b$, into a series of surface spherical harmonics with the following arguments: $\vartheta = $ complement of reduced latitude, $\lambda = $ geocentric longitude. Then the first series in (1–174) is the solution of the interior problem and the second series in (1–174) is the solution of the exterior Dirichlet problem.

Formula (1–176) shows that not only functions that are defined on the surface of a sphere can be expanded into a series of surface spherical harmonics. Such an expansion is even possible for rather arbitrary functions defined on a convex surface.

A remark on terminology

The ellipsoidal-harmonic coordinates u, ϑ (or β), λ are the generalization of spherical coordinates for the sole use of getting closed solutions of Laplace's equation, in particular, for the gravity field of the reference ellipsoid in Sect. 2.7. The brief name "ellipsoidal coordinates" frequently used for u, β, λ might lead to a confusion with the ellipsoidal coordinates φ, λ, h. In the present book, "ellipsoidal coordinates" will always denote "ellipsoidal geographic coordinates", frequently also called "geodetic coordinates", being represented by φ, λ, h.

2 Gravity field of the earth

2.1 Gravity

The total force acting on a body at rest on the earth's surface is the resultant
of gravitational force and the centrifugal force of the earth's rotation and is
called gravity.

Take a rectangular coordinate system whose origin is at the earth's center
of gravity and whose z-axis coincides with the earth's mean axis of rotation
(Fig. 2.1). The x- and y-axes are so chosen as to obtain a right-handed
coordinate system; otherwise they are arbitrary. For convenience, we may
assume an x-axis which is associated with the mean Greenwich meridian (it
"points" towards the mean Greenwich meridian). Note that we are assuming
in this book that the earth is a solid body rotating with constant speed
around a fixed axis. This is a rather simplified assumption, see Moritz and
Mueller (1987). The centrifugal force f on a unit mass is given by

$$f = \omega^2 p \,, \tag{2-1}$$

where ω is the angular velocity of the earth's rotation and

$$p = \sqrt{x^2 + y^2} \tag{2-2}$$

is the distance from the axis of rotation. The vector $\mathbf{f}$ of this force has the

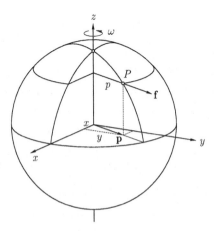

Fig. 2.1. The centrifugal force

direction of the vector

$$\mathbf{p} = [x,\, y,\, 0] \tag{2-3}$$

and is, therefore, given by

$$\mathbf{f} = \omega^2 \mathbf{p} = [\omega^2 x,\, \omega^2 y,\, 0]. \tag{2-4}$$

The centrifugal force can also be derived from a potential

$$\Phi = \frac{1}{2}\,\omega^2(x^2 + y^2), \tag{2-5}$$

so that

$$\mathbf{f} = \operatorname{grad} \Phi \equiv \left[\frac{\partial \Phi}{\partial x},\, \frac{\partial \Phi}{\partial y},\, \frac{\partial \Phi}{\partial z}\right]. \tag{2-6}$$

Substituting (2–5) into (2–6) yields (2–4).

In the introductory remark above, we mentioned that gravity is the resultant of gravitational force and centrifugal force. Accordingly, the potential of gravity, W, is the sum of the potentials of gravitational force, V, cf. (1–12), and centrifugal force, Φ:

$$W = W(x,\, y,\, z) = V + \Phi = G \iiint\limits_{v} \frac{\varrho}{l}\, dv + \frac{1}{2}\,\omega^2(x^2 + y^2), \tag{2-7}$$

where the integration is extended over the earth.

Differentiating (2–5), we find

$$\Delta\Phi \equiv \frac{\partial^2 \Phi}{\partial x^2} + \frac{\partial^2 \Phi}{\partial y^2} + \frac{\partial^2 \Phi}{\partial z^2} = 2\omega^2. \tag{2-8}$$

If we combine this with Poisson's equation (1–17) for V, we get the *generalized Poisson equation* for the gravity potential W:

$$\Delta W = -4\pi\, G\, \varrho + 2\omega^2. \tag{2-9}$$

Since Φ is an analytic function, the discontinuities of W are those of V: some second derivatives have jumps at discontinuities of density.

The gradient vector of W,

$$\mathbf{g} = \operatorname{grad} W \equiv \left[\frac{\partial W}{\partial x},\, \frac{\partial W}{\partial y},\, \frac{\partial W}{\partial z}\right] \tag{2-10}$$

with components

$$g_x = \frac{\partial W}{\partial x} = -G \iiint\limits_v \frac{x - \xi}{l^3} \varrho\, dv + \omega^2 x\,,$$

$$g_y = \frac{\partial W}{\partial y} = -G \iiint\limits_v \frac{y - \eta}{l^3} \varrho\, dv + \omega^2 y\,, \qquad (2\text{--}11)$$

$$g_z = \frac{\partial W}{\partial z} = -G \iiint\limits_v \frac{z - \zeta}{l^3} \varrho\, dv\,,$$

is called the *gravity vector*; it is the total force (gravitational force plus centrifugal force) acting on a unit mass. As a vector, it has *magnitude* and *direction*.

The magnitude g is called gravity in the narrower sense. It has the physical dimension of an acceleration and is measured in gal (1 gal = 1 cm s^{-2}), the unit being named in honor of Galileo Galilei. The numerical value of g is about 978 gal at the equator, and 983 gal at the poles. In geodesy, another unit is often convenient – the milligal, abbreviated mgal (1 mgal = 10^{-3} gal).

In SI units, we have

$$\begin{aligned}
1 \text{ gal} &= 0.01 \text{ m s}^{-2}\,, \\
1 \text{ mgal} &= 10\,\mu\text{m s}^{-2}\,.
\end{aligned} \qquad (2\text{--}12)$$

The direction of the gravity vector is the direction of the *plumb line*, or the vertical; its basic significance for geodetic and astronomical measurements is well known.

In addition to the centrifugal force, another force called the *Coriolis force* acts on a moving body. It is proportional to the velocity with respect to the earth, so that it is zero for bodies resting on the earth. Since in classical geodesy (i.e., not considering navigation) we usually deal with instruments at rest relative to the earth, the Coriolis force plays no role here and need not be considered.

Gravitational and inertial mass

The reader may have noticed that the mass m has been used in two conceptually completely different senses: as *inertial mass* in Newton's law of inertia, *force = mass × acceleration* and as *gravitational mass* in Newton's law of gravitation (1–1). Thus, m in gravitation, which is a "true" force, is the gravitational mass, but m in the centrifugal "force", which is an acceleration, is the inertial mass. The Hungarian physicist Roland Eötvös had shown experimentally already around 1890 that both kinds of masses are equal within 10^{-11}, which is a formidable accuracy. He used the same type of instrument by which experimental physicists have been able to determine the numerical value of the gravitational constant G only to a poor accuracy of about 10^{-4}, as we have seen at the beginning of this book. The coincidence between the inertial and the gravitational mass was far too good to be a physical accident, but, within classical mechanics, it was an inexplicable miracle. It was not before 1915 that Einstein made it one of the pillars of the general theory of relativity!

2.2   Level surfaces and plumb lines

The surfaces
$$W(x,\, y,\, z) = \text{constant}\,, \tag{2–13}$$

on which the potential W is constant, are called *equipotential surfaces* or *level surfaces*.

Differentiating the gravity potential $W = W(x, y, z)$, we find

$$dW = \frac{\partial W}{\partial x}\, dx + \frac{\partial W}{\partial y}\, dy + \frac{\partial W}{\partial z}\, dz\,. \tag{2–14}$$

In vector notation, using the scalar product, this reads

$$dW = \operatorname{grad} W \cdot d\mathbf{x} = \mathbf{g} \cdot d\mathbf{x}\,, \tag{2–15}$$

where

$$d\mathbf{x} = [dx,\, dy,\, dz]\,. \tag{2–16}$$

If the vector $d\mathbf{x}$ is taken along the equipotential surface $W = \text{constant}$, then the potential remains constant and $dW = 0$, so that (2–15) becomes

$$\mathbf{g} \cdot d\mathbf{x} = 0\,. \tag{2–17}$$

If the scalar product of two vectors is zero, then these vectors are orthogonal to each other. This equation therefore expresses the well-known fact that the *gravity vector is orthogonal to the equipotential surface* passing through the same point.

The surface of the oceans, after some slight idealization, is part of a certain level surface. This particular equipotential surface was proposed as the "mathematical figure of the earth" by C.F. Gauss, the "Prince of Mathematicians", and was later termed the *geoid*. This definition has proved highly suitable, and the geoid is still frequently considered by many to be the fundamental surface of physical geodesy. The geoid is thus defined by

$$W = W_0 = \text{constant}\,. \tag{2–18}$$

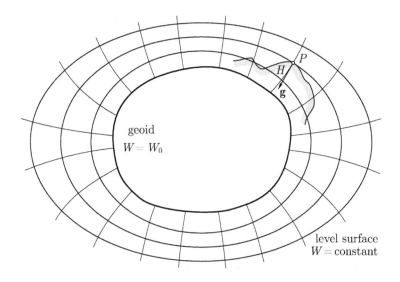

Fig. 2.2. Level surfaces and plumb lines

If we look at equation (2–7) for the gravity potential W, we can see that the equipotential surfaces, expressed by $W(x, y, z) = $ constant, are rather complicated mathematically. The level surfaces that lie completely outside the earth are at least analytical surfaces, although they have no *simple* analytical expression, because the gravity potential W is analytical outside the earth. This is not true of level surfaces that are partly or wholly inside the earth, such as the geoid. They are continuous and "smooth" (i.e., without edges), but they are no longer analytical surfaces; we will see in the next section that the curvature of the interior level surfaces changes discontinuously with the density.

The lines that intersect all equipotential surfaces orthogonally are not exactly straight but slightly curved (Fig. 2.2). They are called *lines of force*, or *plumb lines*. The gravity vector at any point is tangent to the plumb line at that point, hence "direction of the gravity vector", "vertical", and "direction of the plumb line" are synonymous. Sometimes this direction itself is briefly denoted as "plumb line".

As the level surfaces are, so to speak, horizontal everywhere, they share the strong intuitive and physical significance of the horizontal; and they share the geodetic importance of the plumb line because they are orthogonal to it. Thus, we understand why so much attention is paid to the equipotential surfaces.

The height H of a point above sea level (also called the *orthometric height*) is measured along the curved plumb line, starting from the geoid

(Fig. 2.2). If we take the vector dx along the plumb line, in the direction of increasing height H, then its length will be

$$\|d\mathbf{x}\| = dH \qquad (2\text{–}19)$$

and its direction is opposite to the gravity vector $\mathbf{g}$, which points downward, so that the angle between $d\mathbf{x}$ and $\mathbf{g}$ is $180°$. Using the definition of the scalar product (i.e., for two vectors $\mathbf{a}$ and $\mathbf{b}$ it is defined as $\mathbf{a} \cdot \mathbf{b} = \|\mathbf{a}\|\|\mathbf{b}\|\cos\omega$, where ω is the angle between the two vectors), we get

$$\mathbf{g} \cdot d\mathbf{x} = g\,dH\cos 180° = -g\,dH \qquad (2\text{–}20)$$

accordingly, so that Eq. (2–15) becomes

$$dW = -g\,dH\,. \qquad (2\text{–}21)$$

This equation relates the height H to the potential W and will be basic for the theory of height determination (Chap. 4). It shows clearly the inseparable interrelation that characterizes geodesy – the interrelation between the geometrical concepts (H) and the dynamic concepts (W).

Another form of Eq. (2–21) is

$$g = -\frac{\partial W}{\partial H}\,. \qquad (2\text{–}22)$$

It shows that gravity is the negative *vertical gradient* of the potential W, or the negative vertical component of the gradient vector grad W.

Since geodetic measurements (theodolite measurements, leveling, but also satellite techniques etc.) are almost exclusively referred to the system of level surfaces and plumb lines, the geoid plays an essential part. Thus, we see why the aim of physical geodesy has been formulated as the *determination of the level surfaces of the earth's gravity field*. In a still more abstract but equivalent formulation, we may also say that physical geodesy aims at the determination of the potential function $W(x,y,z)$. At a first glance, the reader is probably perplexed about this definition, which is due to Bruns (1878), but its meaning is easily understood: If the potential W is given as a function of the coordinates x, y, z, then we know all level surfaces including the geoid; they are given by the equation

$$W(x,\,y,\,z) = \text{constant}. \qquad (2\text{–}23)$$

2.3   Curvature of level surfaces and plumb lines

The formula for the curvature of a curve $y = f(x)$ is

$$\kappa = \frac{1}{\varrho} = \frac{y''}{(1+y'^2)^{3/2}}\,, \qquad (2\text{–}24)$$

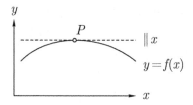

Fig. 2.3. The curvature of a curve

where κ is the curvature, ϱ is the radius of curvature, and

$$y' = \frac{dy}{dx}, \quad y'' = \frac{d^2y}{dx^2}. \tag{2–25}$$

If we use a plane local coordinate system xy in which a parallel to the x-axis is tangent at the point P under consideration (Fig. 2.3), then this implies $y' = 0$ and we get simply

$$\kappa = \frac{1}{\varrho} = \frac{d^2y}{dx^2}. \tag{2–26}$$

Level surfaces

Consider now a point P on a level surface S. Take a local coordinate system xyz with origin at P whose z-axis is vertical, that is, orthogonal to the surface S (Fig. 2.4). We intersect this level surface

$$W(x, y, z) = \text{constant} \tag{2–27}$$

with the xz-plane by setting

$$y = 0. \tag{2–28}$$

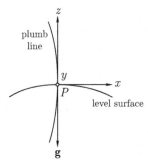

Fig. 2.4. The local coordinate system

Comparing Fig. 2.4 with Fig. 2.3, we see that z now takes the place of y. Therefore, instead of (2–26) we have for the curvature of the intersection of the level surface with the xz-plane:

$$K_1 = \frac{d^2 z}{dx^2} . \qquad (2\text{–}29)$$

If we differentiate $W(x, y, z) = W_0$ with respect to x, considering that y is zero and z is a function of x, we get

$$W_x + W_z \frac{dz}{dx} = 0 ,$$

$$W_{xx} + 2W_{xz} \frac{dz}{dx} + W_{zz} \left(\frac{dz}{dx} \right)^2 + W_z \frac{d^2 z}{dx^2} = 0 , \qquad (2\text{–}30)$$

where the subscripts denote partial differentiation:

$$W_x = \frac{\partial W}{\partial x} , \quad W_{xz} = \frac{\partial^2 W}{\partial x \, \partial z} , \quad \dots . \qquad (2\text{–}31)$$

Since the x-axis is tangent at P, we get $dz/dx = 0$ at P, so that

$$\frac{d^2 z}{dx^2} = -\frac{W_{xx}}{W_z} . \qquad (2\text{–}32)$$

Since the z-axis is vertical, we have, using (2–22),

$$W_z = \frac{\partial W}{\partial z} = \frac{\partial W}{\partial H} = -g . \qquad (2\text{–}33)$$

Therefore, Eq. (2–29) becomes

$$K_1 = \frac{W_{xx}}{g} . \qquad (2\text{–}34)$$

The curvature of the intersection of the level surface with the yz-plane is found by replacing x with y:

$$K_2 = \frac{W_{yy}}{g} . \qquad (2\text{–}35)$$

The mean curvature J of a surface at a point P is defined as the arithmetic mean of the curvatures of the curves in which two mutually perpendicular planes through the surface normal intersect the surface (Fig. 2.5). Hence, we find

$$J = -\frac{1}{2}(K_1 + K_2) = -\frac{W_{xx} + W_{yy}}{2g} . \qquad (2\text{–}36)$$

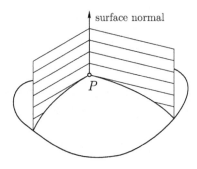

Fig. 2.5. Definition of mean curvature

Here the minus sign is only a convention. This is an expression for the *mean curvature of the level surface.*

From the generalized Poisson equation

$$\Delta W \equiv W_{xx} + W_{yy} + W_{zz} = -4\pi\,G\,\varrho + 2\omega^2\,,\qquad(2\text{--}37)$$

we find

$$-2g\,J + W_{zz} = -4\pi\,G\,\varrho + 2\omega^2\,.\qquad(2\text{--}38)$$

Considering

$$W_z = -g\,,\quad W_{zz} = -\frac{\partial g}{\partial z} = -\frac{\partial g}{\partial H}\,,\qquad(2\text{--}39)$$

we finally obtain

$$\frac{\partial g}{\partial H} = -2g\,J + 4\pi\,G\,\varrho - 2\omega^2\,.\qquad(2\text{--}40)$$

This important equation, relating the *vertical gradient of gravity* $\partial g/\partial H$ to the mean curvature of the level surface, is also due to Bruns (1878). It is another beautiful example of the interrelation between the geometric and dynamic concepts in geodesy.

Plumb lines

The curvature of the plumb line is needed for the reduction of astronomical observations to the geoid. A plumb line may be defined as a curve whose line element vector

$$d\mathbf{x} = [dx,\,dy,\,dz]\qquad(2\text{--}41)$$

has the direction of the gravity vector

$$\mathbf{g} = [W_x,\,W_y,\,W_z]\,;\qquad(2\text{--}42)$$

52 2 Gravity field of the earth

that is, $d\mathbf{x}$ and $\mathbf{g}$ differ only by a proportionality factor. This is best expressed in the form

$$\frac{dx}{W_x} = \frac{dy}{W_y} = \frac{dz}{W_z}.\qquad(2\text{--}43)$$

In the coordinate system of Fig. 2.4, the curvature of the projection of the plumb line onto the xz-plane is given by

$$\kappa_1 = \frac{d^2x}{dz^2};\qquad(2\text{--}44)$$

this is equation (2–26) applied to the present case. Using (2–43), we have

$$\frac{dx}{dz} = \frac{W_x}{W_z}.\qquad(2\text{--}45)$$

We differentiate with respect to z, considering that $y = 0$:

$$\frac{d^2x}{dz^2} = \frac{1}{W_z^2}\left[W_z\left(W_{xz} + W_{xx}\frac{dx}{dz}\right) - W_x\left(W_{zz} + W_{zx}\frac{dx}{dz}\right)\right].\qquad(2\text{--}46)$$

In our particular coordinate system, the gravity vector coincides with the z-axis, so that its x- and y-components are zero:

$$W_x = W_y = 0.\qquad(2\text{--}47)$$

Figure 2.4 shows that we also have

$$\frac{dx}{dz} = 0.\qquad(2\text{--}48)$$

Therefore,

$$\frac{d^2x}{dz^2} = \frac{W_z W_{xz}}{W_z^2} = \frac{W_{xz}}{W_z} = \frac{W_{zx}}{W_z}.\qquad(2\text{--}49)$$

Considering $W_z = -g$, we finally obtain

$$\kappa_1 = \frac{1}{g}\frac{\partial g}{\partial x}\qquad(2\text{--}50)$$

and, similarly,

$$\kappa_2 = \frac{1}{g}\frac{\partial g}{\partial y}.\qquad(2\text{--}51)$$

These are the curvatures of the projections of the plumb line onto the xz- and yz-plane, the z-axis being vertical, that is, coinciding with the gravity vector. The total curvature κ of the plumb line is given, according to differential geometry (essentially Pythagoras' theorem), by

$$\kappa = \sqrt{\kappa_1^2 + \kappa_2^2} = \frac{1}{g}\sqrt{g_x^2 + g_y^2}.\qquad(2\text{--}52)$$

For reducing astronomical observations (Sect. 5.12), we need only the projection curvatures (2–50) and (2–51).

We mention finally that the various formulas for the curvature of level surfaces and plumb lines are equivalent to the single vector equation

$$\operatorname{grad} g = \left(-2g\,J + 4\pi\,G\,\varrho - 2\omega^2\right)\mathbf{n} + g\,\kappa\,\mathbf{n}_1\,, \qquad (2\text{–}53)$$

where $\mathbf{n}$ is the unit vector along the plumb line (its unit tangent vector) and $\mathbf{n}_1$ is the unit vector along the principal normal to the plumb line. This may be easily verified. Using the local xyz-system, we have

$$\begin{aligned}
\mathbf{n} &= [0,\,0,\,1]\,, \\
\mathbf{n}_1 &= [\cos\alpha,\,\sin\alpha,\,0]\,,
\end{aligned} \qquad (2\text{–}54)$$

where α is the angle between the principal normal and the x-axis (Fig. 2.6). The z-component of (2–53) yields Bruns' equation (2–40), and the horizontal components yield

$$\frac{\partial g}{\partial x} = g\,\kappa\,\cos\alpha\,, \quad \frac{\partial g}{\partial y} = g\,\kappa\,\sin\alpha\,. \qquad (2\text{–}55)$$

These are identical to (2–50) and (2–51), since $\kappa_1 = \kappa\cos\alpha$ and $\kappa_2 = \kappa\sin\alpha$, as differential geometry shows. Equation (2–53) is called the *generalized Bruns equation*.

More about the curvature properties and the "inner geometry" of the gravitational field will be found in books by, e.g., Hotine (1969: Chaps. 4–20), Marussi (1985) and Moritz and Hofmann-Wellenhof (1993: Chap. 3).

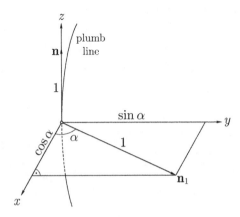

Fig. 2.6. Generalized Bruns equation

2.4   Natural coordinates

The system of level surfaces and plumb lines may be used as a three-dimensional curvilinear coordinate system that is well suited to certain purposes; these coordinates can be measured directly, as opposed to local rectangular coordinates x, y, z. Note, however, that global rectangular coordinates may be measured directly using satellites, see Sect. 5.3.

The direction of the earth's axis of rotation and the position of the equatorial plane (normal to the axis) are well defined astronomically. The *astronomical latitude* Φ of a point P is the angle between the vertical (direction of the plumb line) at P and the equatorial plane, see Fig. 2.7. From this figure, we also see that line PN is parallel to the rotation axis, plane GPF normal to it, that is, parallel to the equatorial plane; $\mathbf{n}$ is the unit vector along the plumb line; plane NPF is the meridian plane of P, and plane NPG is parallel to the meridian plane of Greenwich.

Consider now a straight line through P parallel to the earth's axis of rotation. This parallel and the vertical at P together define the meridian plane of P. The angle between this meridian plane and the meridian plane of Greenwich (or some other fixed plane) is the *astronomical longitude* Λ of P. (Exercise: define Φ and Λ without using the unit sphere. The solution may be found in Sect. 5.9).

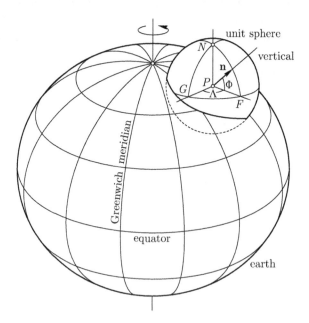

Fig. 2.7. Definition of the astronomical coordinates Φ and Λ of P by means of a unit sphere with center at P

The astronomical coordinates, latitude Φ and longitude Λ, form two of the three spatial coordinates of P. As third coordinate we may take the orthometric height H of P or its potential W. Equivalent to W is the *geopotential number* $C = W_0 - W$, where W_0 is the potential of the geoid. The orthometric height H was defined in Sect. 2.2; see also Fig. 2.2. The relations between W, C, and H are given by the equations

$$W = W_0 - \int_0^H g \, dH = W_0 - C \,,$$

$$C = W_0 - W = \int_0^H g \, dH \,, \qquad (2\text{--}56)$$

$$H = -\int_{W_0}^W \frac{dW}{g} = \int_0^C \frac{dC}{g} \,,$$

which follow from integrating (2–21). The integral is taken along the plumb line of point P, starting from the geoid, where $H = 0$ and $W = W_0$ (see also Fig. 2.8).

The quantities

$$\Phi, \Lambda, W \quad \text{or} \quad \Phi, \Lambda, H \qquad (2\text{--}57)$$

are called *natural coordinates*. They are the real-earth counterparts of the ellipsoidal coordinates. They are related in the following way to the geocentric rectangular coordinates x, y, z of Sect. 2.1. The x-axis is associated with the mean Greenwich meridian; from Fig. 2.7 we read that the unit vector of the vertical $\mathbf{n}$ has the xyz-components

$$\mathbf{n} = [\cos\Phi \cos\Lambda, \ \cos\Phi \sin\Lambda, \ \sin\Phi] \,; \qquad (2\text{--}58)$$

the gravity vector $\mathbf{g}$ is known to be

$$\mathbf{g} = [W_x, \ W_y, \ W_z] \,. \qquad (2\text{--}59)$$

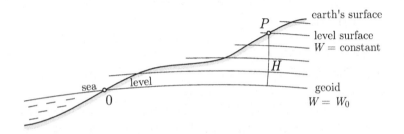

Fig. 2.8. The orthometric height H

On the other hand, since **n** is the unit vector corresponding to **g** but of opposite direction, it is given by

$$\mathbf{n} = -\frac{\mathbf{g}}{\|\mathbf{g}\|} = -\frac{\mathbf{g}}{g}, \tag{2-60}$$

so that

$$\mathbf{g} = -g\,\mathbf{n}. \tag{2-61}$$

This equation, together with (2–58) and (2–59), gives

$$
\begin{aligned}
-W_x &= g \cos \Phi \cos \Lambda, \\
-W_y &= g \cos \Phi \sin \Lambda, \\
-W_z &= g \sin \Phi.
\end{aligned}
\tag{2-62}
$$

Solving for Φ and Λ, we finally obtain

$$
\begin{aligned}
\Phi &= \tan^{-1} \frac{-W_z}{\sqrt{W_x^2 + W_y^2}}, \\
\Lambda &= \tan^{-1} \frac{W_y}{W_x}, \\
W &= W(x, y, z).
\end{aligned}
\tag{2-63}
$$

These three equations relate the natural coordinates Φ, Λ, W to the rectangular coordinates x, y, z, provided the function $W = W(x, y, z)$ is known.

We see that Φ, Λ, H are related to x, y, z in a considerably more complicated way than the spherical coordinates r, ϑ, λ of Sect. 1.4. Note also the conceptual difference between the astronomical longitude Λ and the geocentric longitude λ.

2.5 The potential of the earth in terms of spherical harmonics

Looking at the expression (2–7) for the gravity potential W, we see that the part most difficult to handle is the gravitational potential V, the centrifugal potential being a simple analytic function.

The gravitational potential V can be made more manageable for many purposes if we keep in mind the fact that outside the attracting masses it is a harmonic function and can therefore be expanded into a series of spherical harmonics.

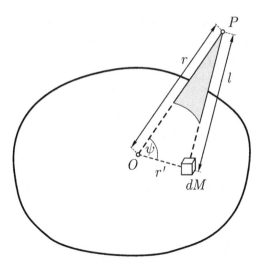

Fig. 2.9. Expansion into spherical harmonics

We now evaluate the coefficients of this series. The gravitational potential V is given by the basic equation (1–12):

$$V = G \iiint\limits_{\text{earth}} \frac{dM}{l},$$ (2–64)

where we now denote the mass element by dM; the integral is extended over the entire earth. Into this integral we substitute the expression (1–104):

$$\frac{1}{l} = \sum_{n=0}^{\infty} \frac{r'^n}{r^{n+1}} P_n(\cos \psi),$$ (2–65)

where the P_n are the conventional Legendre polynomials, r is the radius vector of the fixed point P at which V is to be determined, r' is the radius vector of the variable mass element dM, and ψ is the angle between r and r' (Fig. 2.9).

Since r is a constant with respect to the integration over the earth, it can be taken out of the integral. Thus, we get

$$V = \sum_{n=0}^{\infty} \frac{1}{r^{n+1}} G \iiint\limits_{\text{earth}} r'^n P_n(\cos \psi) \, dM.$$ (2–66)

Writing this in the usual form as a series of solid spherical harmonics,

$$V = \sum_{n=0}^{\infty} \frac{Y_n(\vartheta, \lambda)}{r^{n+1}},$$ (2–67)

we see by comparison that the Laplace surface spherical harmonic $Y_n(\vartheta, \lambda)$ is given by

$$Y_n(\vartheta, \lambda) = G \iiint_{\text{earth}} r'^n P_n(\cos \psi) \, dM \,, \tag{2–68}$$

the dependence on ϑ and λ arises from the angle ψ since

$$\cos \psi = \cos \vartheta \, \cos \vartheta' + \sin \vartheta \, \sin \vartheta' \cos(\lambda' - \lambda) \,. \tag{2–69}$$

The spherical coordinates ϑ, λ have been defined in Sect. 1.4.

A more explicit form is obtained by using the decomposition formula (1–108):

$$\frac{1}{l} = \sum_{n=0}^{\infty} \sum_{m=0}^{n} \frac{1}{2n+1} \left[\frac{\bar{\mathcal{R}}_{nm}(\vartheta, \lambda)}{r^{n+1}} r'^n \bar{\mathcal{R}}_{nm}(\vartheta', \lambda') + \frac{\bar{\mathcal{S}}_{nm}(\vartheta, \lambda)}{r^{n+1}} r'^n \bar{\mathcal{S}}_{nm}(\vartheta', \lambda') \right] .$$
$$\tag{2–70}$$

Substituting this relation into the integral (2–64), we obtain

$$V = \sum_{n=0}^{\infty} \sum_{m=0}^{n} \left[\bar{A}_{nm} \frac{\bar{\mathcal{R}}_{nm}(\vartheta, \lambda)}{r^{n+1}} + \bar{B}_{nm} \frac{\bar{\mathcal{S}}_{nm}(\vartheta, \lambda)}{r^{n+1}} \right], \tag{2–71}$$

where the constant coefficients $\bar{A}_{nm}$ and $\bar{B}_{nm}$ are given by

$$(2n + 1) \, \bar{A}_{nm} = G \iiint_{\text{earth}} r'^n \, \bar{\mathcal{R}}_{nm}(\vartheta', \lambda') \, dM \,,$$

$$(2n + 1) \, \bar{B}_{nm} = G \iiint_{\text{earth}} r'^n \, \bar{\mathcal{S}}_{nm}(\vartheta', \lambda') \, dM \,. \tag{2–72}$$

These formulas are very symmetrical and easy to remember: the coefficient, multiplied by $2n + 1$, of the solid harmonic

$$\frac{\bar{\mathcal{R}}_{nm}(\vartheta, \lambda)}{r^{n+1}} \tag{2–73}$$

is the integral of the solid harmonic

$$r'^n \bar{\mathcal{R}}_{nm}(\vartheta', \lambda') \,. \tag{2–74}$$

An analogous relation results for $\bar{S}_{nm}$.

Note the nice analogy: V is a *sum* and the coefficients are *integrals*!

Since the mass element is

$$dM = \varrho \, dx' \, dy' \, dz' = \varrho \, r'^2 \sin \vartheta' \, dr' \, d\vartheta' \, d\lambda' \,, \tag{2–75}$$

the actual evaluation of the integrals requires that the density ϱ be expressed as a function of r', ϑ', λ'. Although no such expression is available at present, this fact does not diminish the theoretical and practical significance of spherical harmonics, since the coefficients A_{nm}, B_{nm} can be determined from the boundary values of gravity at the earth's surface. This is a boundary-value problem (see Sect. 1.13) and will be elaborated later.

Recalling the relations (1–91) and (1–98) between conventional and fully normalized spherical harmonics, we can also write equations (2–71) and (2–72) in terms of conventional harmonics, readily obtaining

$$V = \sum_{n=0}^{\infty} \sum_{m=0}^{n} \left[A_{nm} \frac{\mathcal{R}_{nm}(\vartheta, \lambda)}{r^{n+1}} + B_{nm} \frac{\mathcal{S}_{nm}(\vartheta, \lambda)}{r^{n+1}} \right], \qquad (2\text{–}76)$$

where

$$A_{n0} = G \iiint\limits_{\text{earth}} r'^{n} P_n(\cos \vartheta') \, dM \; ;$$

$$\left. \begin{aligned} A_{nm} &= 2 \frac{(n-m)!}{(n+m)!} G \iiint\limits_{\text{earth}} r'^{n} \mathcal{R}_{nm}(\vartheta', \lambda') \, dM \\ B_{nm} &= 2 \frac{(n-m)!}{(n+m)!} G \iiint\limits_{\text{earth}} r'^{n} \mathcal{S}_{nm}(\vartheta', \lambda') \, dM \end{aligned} \right\} \quad (m \neq 0). \qquad (2\text{–}77)$$

In connection with satellite dynamics, the potential V is often written in the form

$$V = \frac{GM}{r} \left\{ 1 + \sum_{n=1}^{\infty} \sum_{m=0}^{n} \left(\frac{a}{r} \right)^n \left[C_{nm} \mathcal{R}_{nm}(\vartheta, \lambda) + S_{nm} \mathcal{S}_{nm}(\vartheta, \lambda) \right] \right\}, \quad (2\text{–}78)$$

where a is the equatorial radius of the earth, so that

$$\left. \begin{aligned} A_{nm} &= GM \, a^n \, C_{nm} \\ B_{nm} &= GM \, a^n \, S_{nm} \end{aligned} \right\} \quad (n \neq 0). \qquad (2\text{–}79)$$

Distinguish the coefficient S_{nm} and the function $\mathcal{S}_{nm}$! The coefficient C_{n0} has formerly been denoted by $-J_n$. Note that C is related to cosine and S is related to sine.

The corresponding fully normalized coefficients

$$\bar{C}_{n0} = \frac{1}{\sqrt{2n+1}} \, C_{n0} \, ,$$

$$\left.\begin{array}{l} \bar{C}_{nm} = \sqrt{\dfrac{(n+m)!}{2(2n+1)(n-m)!}} \, C_{nm} \\[3ex] \bar{S}_{nm} = \sqrt{\dfrac{(n+m)!}{2(2n+1)(n-m)!}} \, S_{nm} \end{array}\right\} \quad (m \neq 0) \tag{2–80}$$

are also used.

It is obvious that the nonzonal terms $(m \neq 0)$ would be missing in all these expansions if the earth had complete rotational symmetry, since the terms mentioned depend on the longitude λ. In rotationally symmetrical bodies there is no dependence on λ because all longitudes are equivalent. The tesseral and sectorial harmonics will be small, however, since the deviations from rotational symmetry are slight.

Finally, we discuss the convergence of (2–71), or of the equivalent series expansions, of the earth's potential. This series is an expansion in powers of $1/r$. Therefore, the larger r is, the better the convergence. For smaller r it is not necessarily convergent. For an arbitrary body, the expansion of V into spherical harmonics can be shown to converge always outside the smallest sphere $r = r_0$ that completely encloses the body (Fig. 2.10). Inside this sphere, the series is usually divergent. In certain cases it can converge partly inside the sphere $r = r_0$. If the earth were a homogeneous ellipsoid of about

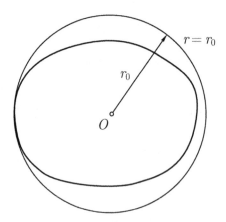

Fig. 2.10. Spherical-harmonic expansion of V converges outside the sphere $r = r_0$

the same dimensions, then the series for V would indeed still converge at the surface of the earth. Owing to the mass irregularities, however, the series of the actual potential V of the earth can be divergent or also convergent at the surface of the earth. *Theoretically,* this makes the use of a harmonic expansion of V at the earth's surface somewhat difficult; *practically,* it is always safe to regard it as convergent. For a detailed discussion see Moritz (1980 a: Sects. 6 and 7) and Sect. 8.6 herein.

It need hardly be pointed out that the spherical-harmonic expansion, always expressing a harmonic function, can represent only the potential outside the attracting masses, never inside.

2.6 Harmonics of lower degree

It is instructive to evaluate the coefficients of the first few spherical harmonics explicitly. For ready reference, we first state some conventional harmonic functions $\mathcal{R}_{nm}$ and $\mathcal{S}_{nm}$, using (1–60), (1–66), and (1–82):

$$
\begin{aligned}
&\mathcal{R}_{00} = 1\,, && \mathcal{S}_{00} = 0\,, \\
&\mathcal{R}_{10} = \cos\vartheta\,, && \mathcal{S}_{10} = 0\,, \\
&\mathcal{R}_{11} = \sin\vartheta\,\cos\lambda\,, && \mathcal{S}_{11} = \sin\vartheta\,\sin\lambda\,, \\
&\mathcal{R}_{20} = \tfrac{3}{2}\cos^2\vartheta - \tfrac{1}{2}\,, && \mathcal{S}_{20} = 0\,, \\
&\mathcal{R}_{21} = 3\sin\vartheta\,\cos\vartheta\,\cos\lambda\,, && \mathcal{S}_{21} = 3\sin\vartheta\,\cos\vartheta\,\sin\lambda\,, \\
&\mathcal{R}_{22} = 3\sin^2\vartheta\,\cos 2\lambda\,, && \mathcal{S}_{22} = 3\sin^2\vartheta\,\sin 2\lambda\,.
\end{aligned}
\tag{2–81}
$$

The corresponding solid harmonics $r^n \mathcal{R}_{nm}$ and $r^n \mathcal{S}_{nm}$ are simply homogeneous polynomials in x, y, z. For instance,

$$
r^2 \mathcal{S}_{22} = 6r^2 \sin^2\vartheta\,\sin\lambda\,\cos\lambda = 6(r\sin\vartheta\,\cos\lambda)(r\sin\vartheta\,\sin\lambda) = 6xy\,. \tag{2–82}
$$

In this way, we find

$$
\begin{aligned}
&\mathcal{R}_{00} = 1\,, && \mathcal{S}_{00} = 0\,, \\
&r\,\mathcal{R}_{10} = z\,, && r\,\mathcal{S}_{10} = 0\,, \\
&r\,\mathcal{R}_{11} = x\,, && r\,\mathcal{S}_{11} = y\,, \\
&r^2\mathcal{R}_{20} = -\tfrac{1}{2}\,x^2 - \tfrac{1}{2}\,y^2 + z^2\,, && r^2\mathcal{S}_{20} = 0\,, \\
&r^2\mathcal{R}_{21} = 3x\,z\,, && r^2\mathcal{S}_{21} = 3y\,z\,, \\
&r^2\mathcal{R}_{22} = 3x^2 - 3y^2\,, && r^2\mathcal{S}_{22} = 6x\,y\,.
\end{aligned}
\tag{2–83}
$$

Substituting these functions into the expression (2–77) for the coefficients A_{nm} and B_{nm} yields for the zero-degree term

$$A_{00} = G \iiint\limits_{\text{earth}} dM = GM \, ;\tag{2–84}$$

that is, the product of the mass of the earth times the gravitational constant. For the first-degree coefficients, we get

$$A_{10} = G \iiint\limits_{\text{earth}} z' \, dM \, , \quad A_{11} = G \iiint\limits_{\text{earth}} x' \, dM \, , \quad B_{11} = G \iiint\limits_{\text{earth}} y' \, dM \, ;$$
$$\tag{2–85}$$

and for the second-degree coefficients

$$A_{20} = \frac{1}{2} G \iiint\limits_{\text{earth}} (-x'^2 - y'^2 + 2z'^2) \, dM \, ,$$

$$A_{21} = G \iiint\limits_{\text{earth}} x' z' \, dM \, , \quad B_{21} = G \iiint\limits_{\text{earth}} y' z' \, dM \, ,\tag{2–86}$$

$$A_{22} = \frac{1}{4} G \iiint\limits_{\text{earth}} (x'^2 - y'^2) \, dM \, , \quad B_{22} = \frac{1}{2} G \iiint\limits_{\text{earth}} x' y' \, dM \, .$$

It is known from mechanics that

$$x_c = \frac{1}{M} \iiint x' \, dM \, , \quad y_c = \frac{1}{M} \iiint y' \, dM \, , \quad z_c = \frac{1}{M} \iiint z' \, dM \tag{2–87}$$

are the rectangular coordinates of the center of gravity (center of mass, geocenter). If the origin of the coordinate system coincides with the center of gravity, then these coordinates and, hence, the integrals (2–85) are zero. *If the origin $r = 0$ is the center of gravity of the earth, then there will be no first-degree terms in the spherical-harmonic expansion of the potential V.* Therefore, this is true for our geocentric coordinate system.

The integrals

$$\iiint x' y' \, dM \, , \quad \iiint y' z' \, dM \, , \quad \iiint z' x' \, dM \tag{2–88}$$

are the *products of inertia*. They are zero if the coordinate axes coincide with the principal axes of inertia. If the z-axis is identical with the mean rotational axis of the earth, which coincides with the axis of maximum inertia, at least the second and third of these products of inertia must vanish. Hence, A_{21} and B_{21} will be zero, but not so B_{22}, which is proportional to the first product of inertia; B_{22} would vanish only if the earth had complete rotational symmetry or if a principal axis of inertia happened to coincide with the Greenwich meridian.

The five harmonics $A_{10}\,\mathcal{R}_{10}$, $A_{11}\,\mathcal{R}_{11}$, $B_{11}\,\mathcal{S}_{11}$, $A_{21}\,\mathcal{R}_{21}$, and $B_{21}\,\mathcal{S}_{21}$ – all first-degree harmonics and those of degree 2 and order 1 – which must, thus, vanish in any spherical-harmonic expansion of the earth's potential, are called *forbidden* or *inadmissible harmonics*.

Introducing the *moments of inertia* with respect to the x-, y-, z-axes by the definitions

$$A = \iiint (y'^2 + z'^2)\, dM\,,$$

$$B = \iiint (z'^2 + x'^2)\, dM\,, \qquad (2\text{--}89)$$

$$C = \iiint (x'^2 + y'^2)\, dM\,,$$

and denoting the xy-product of inertia, which cannot be said to vanish, by

$$D = \iiint x'y'\, dM\,, \qquad (2\text{--}90)$$

we finally have

$$A_{00} = GM\,,$$

$$A_{10} = A_{11} = B_{11} = 0\,,$$

$$A_{20} = G\left[(A+B)/2 - C\right],$$

$$A_{21} = B_{21} = 0\,, \qquad (2\text{--}91)$$

$$A_{22} = \tfrac{1}{4}\,G\,(B-A)\,,$$

$$B_{22} = \tfrac{1}{2}\,G\,D\,.$$

Now let the x- and y-axes actually coincide with the corresponding principal axes of inertia of the earth. This is only theoretically possible, since the principal axes of inertia of the earth are only inaccurately known. Then $B_{22} = 0$; taking into account (2–78) and (2–79), we may write explicitly

$$V = \frac{GM}{r} + \frac{G}{r^3}\left\{\frac{1}{2}\left[C - (A+B)/2\right](1 - 3\cos^2\vartheta) + \right.$$

$$\left. \frac{3}{4}(B-A)\sin^2\vartheta\,\cos 2\lambda\right\} + O(1/r^4)\,. \qquad (2\text{--}92)$$

In rectangular coordinates this assumes the symmetrical form

$$V = \frac{GM}{r} + \frac{G}{2r^5}\Big[(B + C - 2A)\,x^2 + (C + A - 2B)\,y^2 +$$

$$(A + B - 2C)\,z^2\Big] + O(1/r^4)\,, \tag{2-93}$$

which is obtained by taking into account the relations (1–26) between rectangular and spherical coordinates.

Terms of order higher than $1/r^3$ may be neglected for larger distances (say, for the distance to the moon), so that (2–92) or (2–93), omitting the higher-order terms $0(1/r^4)$, are sufficient for many astronomical purposes, cf. Moritz and Mueller (1987). Note that the notation $0(1/r^4)$ means terms of the order of $1/r^4$. For planetary distances even the first term,

$$V = \frac{GM}{r}\,, \tag{2-94}$$

is generally sufficient; it represents the potential of a point mass. Thus, for very large distances, every body acts like a point mass.

Using the form (2–78) of the spherical-harmonic expansion of V, then the coefficients of lower degree are obtained from (2–79) and (2–91). We find

$$C_{10} = C_{11} = S_{11} = 0\,,$$

$$C_{20} = -\frac{C - (A + B)/2}{M\,a^2}\,,$$

$$C_{21} = S_{21} = 0\,, \tag{2-95}$$

$$C_{22} = \frac{B - A}{4M\,a^2}\,,$$

$$S_{22} = \frac{D}{2M\,a^2}\,.$$

The first of these formulas shows that the summation in (2–78) actually begins with $n = 2$; the others relate the coefficients of second degree to the mass and the moments and products of inertia of the earth.

2.7   The gravity field of the level ellipsoid

As a first approximation, the earth is a sphere; as a second approximation, it may be considered an ellipsoid of revolution. Although the earth is not an exact ellipsoid, the gravity field of an ellipsoid is of fundamental practical importance because it is easy to handle mathematically and the deviations of the actual gravity field from the ellipsoidal "normal" field are so small that they can be considered linear. This splitting of the earth's gravity field into a "normal" and a remaining small "disturbing" field considerably simplifies the problem of its determination; the problem could hardly be solved otherwise.

Therefore, we assume that the normal figure of the earth is a level ellipsoid, that is, an ellipsoid of revolution which is an equipotential surface of a normal gravity field. This assumption is necessary because the ellipsoid is to be the normal form of the geoid, which is an equipotential surface of the actual gravity field. Denoting the potential of the normal gravity field by

$$U = U(x, y, z), \qquad (2\text{-}96)$$

we see that the level ellipsoid, being a surface $U = $ constant, exactly corresponds to the geoid, which is defined as a surface $W = $ constant.

The basic point here is that by postulating that the given ellipsoid be an equipotential surface of the normal gravity field, and by prescribing the total mass M, we completely and uniquely determine the normal potential U. The detailed density distribution inside the ellipsoid, which produces the potential U, is quite uninteresting and need not be known at all. In fact, we do not know of any "reasonable" mass distribution for the level ellipsoid (Moritz 1990: Chap. 5). Pizzetti (1894) unsuccessfully used a homogeneous density distribution combined with a surface layer of negative density, which is quite "unnatural".

This determination is possible by Dirichlet's principle (Sect. 1.12): The gravitational potential outside a surface S is completely determined by knowing the geometric shape of S and the value of the potential on S. Originally it was shown only for the gravitational potential V, but it can be applied to the gravity potential

$$U = V + \tfrac{1}{2}\,\omega^2(x^2 + y^2) \qquad (2\text{-}97)$$

as well if the angular velocity ω is given. The proof follows that in Sect. 1.12, with obvious modifications. Hence, the normal potential function $U(x, y, z)$ is completely determined by

1. the shape of the ellipsoid of revolution, that is, its semiaxes a and b,
2. the total mass M, and
3. the angular velocity ω.

The calculation will now be carried out in detail. The given ellipsoid S_0,

$$\frac{x^2 + y^2}{a^2} + \frac{z^2}{b^2} = 1, \qquad (2\text{-}98)$$

is by definition an equipotential surface

$$U(x, y, z) = U_0 \,. \tag{2–99}$$

It is now convenient to introduce the ellipsoidal-harmonic coordinates u, β, λ of Sect. 1.15. The ellipsoid S_0 is taken as the reference ellipsoid $u = b$.

Since $V(u, \beta)$, the gravitational part of the normal potential U, will be harmonic outside the ellipsoid S_0, we use the second equation of the series (1–174). The field V has rotational symmetry and, hence, does not depend on the longitude λ. Therefore, all nonzonal terms, which depend on λ, must be zero, and there remains

$$V(u, \beta) = \sum_{n=0}^{\infty} \frac{Q_n\left(i\dfrac{u}{E}\right)}{Q_n\left(i\dfrac{b}{E}\right)} A_n P_n(\sin \beta) \,, \tag{2–100}$$

where

$$E = \sqrt{a^2 - b^2} \tag{2–101}$$

is the linear eccentricity. The centrifugal potential $\Phi(u, \beta)$ is given by

$$\Phi(u, \beta) = \tfrac{1}{2}\omega^2(u^2 + E^2)\cos^2\beta \,. \tag{2–102}$$

Therefore, the total normal gravity potential may be written

$$U(u, \beta) = \sum_{n=0}^{\infty} \frac{Q_n\left(i\dfrac{u}{E}\right)}{Q_n\left(i\dfrac{b}{E}\right)} A_n P_n(\sin \beta) + \tfrac{1}{2}\omega^2(u^2 + E^2)\cos^2\beta \,. \tag{2–103}$$

On the ellipsoid S_0 we have $u = b$ and $U = U_0$. Hence,

$$\sum_{n=0}^{\infty} A_n P_n(\sin \beta) + \tfrac{1}{2}\omega^2(u^2 + E^2)\cos^2\beta = U_0 \,. \tag{2–104}$$

This equation applies for all points of S_0, that is, for all values of β. Since

$$b^2 + E^2 = a^2 \tag{2–105}$$

and

$$\cos^2\beta = \tfrac{2}{3}\left[1 - P_2(\sin \beta)\right] \,, \tag{2–106}$$

we have

$$\sum_{n=0}^{\infty} A_n P_n(\sin \beta) + \tfrac{1}{3}\omega^2 a^2 - \tfrac{1}{3}\omega^2 a^2 P_2(\sin \beta) - U_0 = 0 \tag{2–107}$$

or

$$\left(A_0 + \tfrac{1}{3}\omega^2 a^2 - U_0\right) P_0(\sin\beta) + A_1 P_1(\sin\beta)$$

$$+ \left(A_2 - \tfrac{1}{3}\omega^2 a^2\right) P_2(\sin\beta) + \sum_{n=3}^{\infty} A_n P_n(\sin\beta) = 0. \qquad (2\text{--}108)$$

This equation applies for all values of β only if the coefficient of every $P_n(\sin\beta)$ is zero. Thus, we get

$$A_0 = U_0 - \tfrac{1}{3}\omega^2 a^2, \qquad A_1 = 0,$$

$$A_2 = \tfrac{1}{3}\omega^2 a^2, \qquad A_3 = A_4 = \ldots = 0. \qquad (2\text{--}109)$$

Substituting these relations into (2–100) gives

$$V(u,\beta) = \left(U_0 - \tfrac{1}{3}\omega^2 a^2\right) \frac{Q_0\!\left(i\dfrac{u}{E}\right)}{Q_0\!\left(i\dfrac{b}{E}\right)} + \tfrac{1}{3}\omega^2 a^2 \frac{Q_2\!\left(i\dfrac{u}{E}\right)}{Q_2\!\left(i\dfrac{b}{E}\right)} P_2(\sin\beta). \quad (2\text{--}110)$$

This formula is basically the solution of Dirichlet's problem for the level ellipsoid, but we can give it more convenient forms. It is a closed formula!

First, we determine the Legendre functions of the second kind, Q_0 and Q_2. As

$$\coth^{-1}(i\,x) = \frac{1}{i}\cot^{-1}x = -i\tan^{-1}\frac{1}{x}, \qquad (2\text{--}111)$$

we find by (1–80) with $z = i\,u/E$:

$$Q_0\!\left(i\,\frac{u}{E}\right) = -i\tan^{-1}\frac{E}{u},$$

$$Q_2\!\left(i\,\frac{u}{E}\right) = \frac{i}{2}\left[\left(1 + 3\frac{u^2}{E^2}\right)\tan^{-1}\frac{E}{u} - 3\frac{u}{E}\right]. \qquad (2\text{--}112)$$

By introducing in (2–112) the abbreviations

$$q = \frac{1}{2}\left[\left(1 + 3\frac{u^2}{E^2}\right)\tan^{-1}\frac{E}{u} - 3\frac{u}{E}\right],$$

$$q_0 = \frac{1}{2}\left[\left(1 + 3\frac{b^2}{E^2}\right)\tan^{-1}\frac{E}{b} - 3\frac{b}{E}\right] \qquad (2\text{--}113)$$

and substituting them in equation (2–110), we obtain

$$V(u,\beta) = \left(U_0 - \tfrac{1}{3}\omega^2 a^2\right)\frac{\tan^{-1}\dfrac{E}{u}}{\tan^{-1}\dfrac{E}{b}} + \tfrac{1}{3}\omega^2 a^2 \frac{q}{q_0} P_2(\sin\beta). \qquad (2\text{--}114)$$

Now we can express U_0 in terms of the mass M. For large values of u, we have

$$\tan^{-1}\frac{E}{u} = \frac{E}{u} + O(1/u^3). \tag{2-115}$$

From the expressions (1–26) for spherical coordinates and from equations (1–151) for ellipsoidal-harmonic coordinates, we find

$$x^2 + y^2 + z^2 = r^2 = u^2 + E^2 \cos^2\beta, \tag{2-116}$$

so that for large values of r we have

$$\frac{1}{u} = \frac{1}{r} + O(1/r^3) \tag{2-117}$$

and

$$\tan^{-1}\frac{E}{u} = \frac{E}{r} + O(1/r^3), \tag{2-118}$$

where $O(x)$ means "small of order x", i.e., small of order $1/r^3$ in our case. For very large distances r, the first term in (2–114) is dominant, so that asymptotically

$$V = \left(U_0 - \tfrac{1}{3}\omega^2 a^2\right) \frac{E}{\tan^{-1}(E/b)} \frac{1}{r} + O(1/r^3). \tag{2-119}$$

We know from Sect. 2.6 that

$$V = \frac{GM}{r} + O(1/r^3). \tag{2-120}$$

Substituting this expression for V into the left-hand side of (2–119) yields

$$\frac{GM}{r} = \left(U_0 - \tfrac{1}{3}\omega^2 a^2\right) \frac{E}{\tan^{-1}(E/b)} \frac{1}{r} + O(1/r^3). \tag{2-121}$$

Now multiply this equation by r and let then $r \to 0$. The result is (rigorously!)

$$GM = \left(U_0 - \tfrac{1}{3}\omega^2 a^2\right) \frac{E}{\tan^{-1}(E/b)}, \tag{2-122}$$

which may be rearranged to

$$U_0 = \frac{GM}{E} \tan^{-1}\frac{E}{b} + \tfrac{1}{3}\omega^2 a^2. \tag{2-123}$$

This is the desired relation between mass M and potential U_0.

Substituting the result for U_0 obtained in (2–123) into (2–114), simplifies the expression for V to

$$V = \frac{GM}{E} \tan^{-1}\frac{E}{u} + \tfrac{1}{3}\omega^2 a^2 \frac{q}{q_0} P_2(\sin\beta). \tag{2-124}$$

Expressing P_2 as

$$P_2(\sin\beta) = \tfrac{3}{2}\sin^2\beta - \tfrac{1}{2} \qquad (2\text{–}125)$$

and, finally, adding the centrifugal potential $\Phi = \omega^2(u^2 + E^2)\cos^2\beta/2$ from (2–102), the normal gravity potential U results as

$$U(u,\beta) = \frac{GM}{E}\tan^{-1}\frac{E}{u} + \tfrac{1}{2}\omega^2 a^2\frac{q}{q_0}\left(\sin^2\beta - \tfrac{1}{3}\right) + \tfrac{1}{2}\omega^2(u^2 + E^2)\cos^2\beta.$$
$$(2\text{–}126)$$

The only constants that occur in this formula are a, b, GM, and ω. This is in complete agreement with Dirichlet's theorem.

2.8   Normal gravity

Referring to the line element in ellipsoidal-harmonic coordinates according to (1–155), replacing ϑ by its complement $90° - \beta$, we get

$$ds^2 = w^2\,du^2 + w^2(u^2 + E^2)\,d\beta^2 + (u^2 + E^2)\cos^2\beta\,d\lambda^2, \qquad (2\text{–}127)$$

where

$$w = \sqrt{\frac{u^2 + E^2\sin^2\beta}{u^2 + E^2}} \qquad (2\text{–}128)$$

has been introduced. Thus, along the coordinate lines we have

$$u = \text{variable},\ \beta = \text{constant},\ \lambda = \text{constant},\quad ds_u = w\,du,$$
$$\beta = \text{variable},\ u = \text{constant},\ \lambda = \text{constant},\quad ds_\beta = w\sqrt{u^2 + E^2}\,d\beta,$$
$$\lambda = \text{variable},\ u = \text{constant},\ \beta = \text{constant},\quad ds_\lambda = \sqrt{u^2 + E^2}\cos\beta\,d\lambda.$$
$$(2\text{–}129)$$

The components of the normal gravity vector

$$\boldsymbol{\gamma} = \operatorname{grad} U \qquad (2\text{–}130)$$

along these coordinate lines are accordingly given by

$$\gamma_u = \frac{\partial U}{\partial s_u} = \frac{1}{w}\frac{\partial U}{\partial u},$$
$$\gamma_\beta = \frac{\partial U}{\partial s_\beta} = \frac{1}{w\sqrt{u^2 + E^2}}\frac{\partial U}{\partial \beta}, \qquad (2\text{–}131)$$
$$\gamma_\lambda = \frac{\partial U}{\partial s_\lambda} = \frac{1}{\sqrt{u^2 + E^2}\cos\beta}\frac{\partial U}{\partial \lambda} = 0.$$

The component γ_λ is zero because U does not contain λ. This is also evident from the rotational symmetry.

Performing the partial differentiations, we find

$$
\begin{aligned}
-w\,\gamma_u &= \frac{GM}{u^2 + E^2} + \frac{\omega^2 a^2 E}{u^2 + E^2}\,\frac{q'}{q_0}\left(\tfrac{1}{2}\sin^2\beta - \tfrac{1}{6}\right) - \omega^2 u\,\cos^2\beta\,, \\
-w\,\gamma_\beta &= \left(-\frac{\omega^2 a^2}{\sqrt{u^2 + E^2}}\,\frac{q}{q_0} + \omega^2\sqrt{u^2 + E^2}\right)\sin\beta\,\cos\beta\,,
\end{aligned}
\tag{2-132}
$$

where we have set

$$
q' = -\frac{u^2 + E^2}{E}\,\frac{dq}{du} = 3\left(1 + \frac{u^2}{E^2}\right)\left(1 - \frac{u}{E}\tan^{-1}\frac{E}{u}\right) - 1\,.
\tag{2-133}
$$

Note that q' does not mean dq/du; this notation has been borrowed from Hirvonen (1960), where q' is the derivative with respect to another independent variable which we are not using here.

For the level ellipsoid S_0 itself, we have $u = b$ and get

$$
\gamma_{\beta,0} = 0\,.
\tag{2-134}
$$

(Note that we will often mark quantities referred to S_0 by the subscript 0.) This is also evident because on S_0 the gravity vector is normal to the level surface S_0. Hence, in addition to the λ-component, the β-component is also zero on the reference ellipsoid $u = b$. Note that the other coordinate ellipsoids $u = $ constant are not equipotential surfaces $U = $ constant, so that the β-component will not in general be zero.

Thus, the total gravity on the ellipsoid S_0, which we simply denote by γ, is given by

$$
\begin{aligned}
\gamma = |\gamma_{u,0}| &= \frac{GM}{a\sqrt{a^2\sin^2\beta + b^2\cos^2\beta}}\cdot \\
&\cdot\left[1 + \frac{\omega^2 a^2 E}{GM}\,\frac{q_0'}{q_0}\left(\tfrac{1}{2}\sin^2\beta - \tfrac{1}{6}\right) - \frac{\omega^2 a^2 b}{GM}\cos^2\beta\right],
\end{aligned}
\tag{2-135}
$$

since on S_0 we get the relations

$$
\begin{aligned}
\sqrt{u^2 + E^2} &= \sqrt{b^2 + E^2} = a\,, \\
w_0 &= \frac{1}{a}\sqrt{b^2 + E^2\sin^2\beta} = \frac{1}{a}\sqrt{a^2\sin^2\beta + b^2\cos^2\beta}\,.
\end{aligned}
\tag{2-136}
$$

Now we introduce the abbreviation

$$
m = \frac{\omega^2 a^2 b}{GM}
\tag{2-137}
$$

and the second eccentricity

$$e' = \frac{E}{b} = \frac{\sqrt{a^2 - b^2}}{b}.$$

(2–138)

The prime on e does not denote differentiation, but merely distinguishes the second eccentricity from the first eccentricity which is defined as $e = E/a$.

Removing the constant terms by noting that

$$1 = \cos^2\beta + \sin^2\beta,$$

(2–139)

we obtain

$$\gamma = \frac{GM}{a\sqrt{a^2 \sin^2\beta + b^2 \cos^2\beta}}$$

$$\cdot \left[\left(1 + \frac{m}{3} \frac{e'q_0'}{q_0} \right) \sin^2\beta + \left(1 - m - \frac{m}{6} \frac{e'q_0'}{q_0} \right) \cos^2\beta \right].$$

(2–140)

At the equator $(\beta = 0)$, we find

$$\gamma_a = \frac{GM}{a\,b} \left(1 - m - \frac{m}{6} \frac{e'q_0'}{q_0} \right);$$

(2–141)

at the poles $(\beta = \pm 90°)$, normal gravity is given by

$$\gamma_b = \frac{GM}{a^2} \left(1 + \frac{m}{3} \frac{e'q_0'}{q_0} \right).$$

(2–142)

Normal gravity at the equator, γ_a, and normal gravity at the pole, γ_b, satisfy the relation

$$\frac{a - b}{a} + \frac{\gamma_b - \gamma_a}{\gamma_a} = \frac{\omega^2 b}{\gamma_a} \left(1 + \frac{e'q_0'}{2q_0} \right),$$

(2–143)

which should be verified by substitution. This is the rigorous form of an important approximate formula published by Clairaut in 1738. It is, therefore, called Clairaut's theorem. Its significance will become clear in Sect. 2.10.

By comparing expression (2–141) for γ_a and expression (2–142) for γ_b with the quantities within parentheses in formula (2–140), we see that γ can be written in the symmetrical form

$$\gamma = \frac{a\,\gamma_b \sin^2\beta + b\,\gamma_a \cos^2\beta}{\sqrt{a^2 \sin^2\beta + b^2 \cos^2\beta}}.$$

(2–144)

We finally introduce the ellipsoidal latitude on the ellipsoid, φ, which is the angle between the normal to the ellipsoid and the equatorial plane (Fig. 2.11). Using the formula from ellipsoidal geometry,

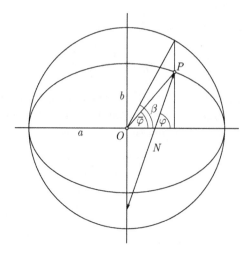

Fig. 2.11. Ellipsoidal latitude φ, geocentric latitude $\bar{\varphi}$, reduced (ellipsoidal-harmonic) latitude β for a point P on the ellipsoid

$$\tan \beta = \frac{b}{a} \tan \varphi \,, \tag{2–145}$$

we obtain

$$\gamma = \frac{a \, \gamma_a \cos^2\varphi + b \, \gamma_b \sin^2\varphi}{\sqrt{a^2 \cos^2\varphi + b^2 \sin^2\varphi}} \,. \tag{2–146}$$

The computation is left as an exercise for the reader. This rigorous formula for normal gravity on the ellipsoid is due to Somigliana from 1929.

We close this section with a short remark on the vertical gradient of gravity at the reference ellipsoid, $\partial\gamma/\partial s_u = \partial\gamma/\partial h$. Bruns' formula (2–40), applied to the normal gravity field with the corresponding ellipsoidal height h and with $\varrho = 0$, yields

$$\frac{\partial\gamma}{\partial h} = -2\gamma J - 2\omega^2 \,. \tag{2–147}$$

The mean curvature of the ellipsoid is given by

$$J = \frac{1}{2} \left(\frac{1}{M} + \frac{1}{N} \right) \,, \tag{2–148}$$

where M and N are the principal radii of curvature: M is the radius in the direction of the meridian, and N is the normal radius of curvature, taken in the direction of the prime vertical. From ellipsoidal geometry, we use the formulas

$$M = \frac{c}{(1 + e'^2 \cos^2\varphi)^{3/2}} \,, \qquad N = \frac{c}{(1 + e'^2 \cos^2\varphi)^{1/2}} \,, \tag{2–149}$$

where

$$c = \frac{a^2}{b} \qquad (2\text{--}150)$$

is the radius of curvature at the pole. The normal radius of curvature, N, admits a simple geometrical interpretation (Fig. 2.11). It is, therefore, also known as the "normal terminated by the minor axis" (Bomford 1962: p. 497).

2.9 Expansion of the normal potential in spherical harmonics

We have found the gravitational potential of the normal figure of the earth in terms of ellipsoidal harmonics in (2–124) as

$$V = \frac{GM}{E} \tan^{-1}\frac{E}{u} + \frac{1}{3}\omega^2 a^2 \frac{q}{q_0} P_2(\sin\beta). \qquad (2\text{--}151)$$

Now we wish to express this equation in terms of spherical coordinates r, ϑ, λ.

We first establish a relation between ellipsoidal-harmonic and spherical coordinates. By comparing the rectangular coordinates in these two systems according to Eqs. (1–26) and (1–151), we get

$$r\sin\vartheta\,\cos\lambda = \sqrt{u^2 + E^2}\,\cos\beta\,\cos\lambda\,,$$

$$r\sin\vartheta\,\sin\lambda = \sqrt{u^2 + E^2}\,\cos\beta\,\sin\lambda\,, \qquad (2\text{--}152)$$

$$r\cos\vartheta = u\sin\beta\,.$$

The longitude λ is the same in both systems. We easily find from these equations

$$\cot\vartheta = \frac{u}{\sqrt{u^2 + E^2}}\tan\beta\,,$$

$$r = \sqrt{u^2 + E^2\cos^2\beta}\,. \qquad (2\text{--}153)$$

The direct transformation of (2–151) by expressing u and β in terms of r and ϑ by means of equations (2–153) is extremely laborious. However, the problem can be solved easily in an indirect way.

We expand $\tan^{-1}(E/u)$ into the well-known power series

$$\tan^{-1}\frac{E}{u} = \frac{E}{u} - \frac{1}{3}\left(\frac{E}{u}\right)^3 + \frac{1}{5}\left(\frac{E}{u}\right)^5 - \dots . \qquad (2\text{--}154)$$

The substitution of this series into the first equation of formula (2–113), i.e.,

$$q = \frac{1}{2}\left[\left(1 + 3\frac{u^2}{E^2}\right)\tan^{-1}\frac{E}{u} - 3\frac{u}{E}\right], \qquad (2\text{--}155)$$

leads, after simple manipulations, to

$$q = 2\left[\frac{1}{3\cdot 5}\left(\frac{E}{u}\right)^3 - \frac{2}{5\cdot 7}\left(\frac{E}{u}\right)^5 + \frac{3}{7\cdot 9}\left(\frac{E}{u}\right)^7 - \cdots\right].$$ (2–156)

More concisely, we have

$$\tan^{-1}\frac{E}{u} = \frac{E}{u} + \sum_{n=1}^{\infty}(-1)^n\frac{1}{2n+1}\left(\frac{E}{u}\right)^{2n+1},$$

$$q = -\sum_{n=1}^{\infty}(-1)^n\frac{2n}{(2n+1)(2n+3)}\left(\frac{E}{u}\right)^{2n+1}.$$ (2–157)

By inserting these relations into (2–151) we obtain

$$V = \frac{GM}{u} + \frac{GM}{E}\sum_{n=1}^{\infty}(-1)^n\frac{1}{2n+1}\left(\frac{E}{u}\right)^{2n+1}$$

$$-\frac{\omega^2 a^2}{3q_0}\sum_{n=1}^{\infty}(-1)^n\frac{2n}{(2n+1)(2n+3)}\left(\frac{E}{u}\right)^{2n+1}P_2(\sin\beta).$$ (2–158)

Introducing m, defined by (2–137), and the second eccentricity $e' = E/b$, we find

$$V = \frac{GM}{u} + \sum_{n=1}^{\infty}(-1)^n\frac{GM}{(2n+1)E}\left(\frac{E}{u}\right)^{2n+1}$$

$$\cdot\left[1 - \frac{m\,e'}{3q_0}\frac{2n}{2n+3}P_2(\sin\beta)\right].$$ (2–159)

We expand the potential V into a series of spherical harmonics. Because of the rotational symmetry, there will be only zonal terms, and because of the symmetry with respect to the equatorial plane, there will be only even zonal harmonics. The zonal harmonics of odd degree change sign for negative latitudes and must, therefore, be absent. Accordingly, the series has the form

$$V = \frac{GM}{r} + A_2\frac{P_2(\cos\vartheta)}{r^3} + A_4\frac{P_4(\cos\vartheta)}{r^5} + \cdots.$$ (2–160)

We next have to determine the coefficients A_2, A_4, ... For this purpose, we consider a point on the axis of rotation, outside the ellipsoid. For this point, we have $\beta = 90°$, $\vartheta = 0°$, and, by (2–153), $u = r$. Then (2–159) becomes

$$V = \frac{GM}{r} + \sum_{n=1}^{\infty}(-1)^n\frac{GM\,E^{2n}}{2n+1}\left(1 - \frac{2n}{2n+3}\frac{m\,e'}{3q_0}\right)\frac{1}{r^{2n+1}},$$ (2–161)

and (2–160) takes the form

$$V = \frac{GM}{r} + \frac{A_2}{r^3} + \frac{A_4}{r^5} + \cdots = \frac{GM}{r} + \sum_{n=1}^{\infty} A_{2n} \frac{1}{r^{2n+1}} \,. \tag{2–162}$$

Here we have used the fact that for all values of n

$$P_n(1) = 1 \tag{2–163}$$

(see also Fig. 1.4). Comparing the coefficients in both expressions for V, we find

$$A_{2n} = (-1)^n \frac{GM\,E^{2n}}{2n+1} \left(1 - \frac{2n}{2n+3} \frac{m\,e'}{3q_0} \right) . \tag{2–164}$$

Equations (2–160) and (2–164) give the desired expression for the potential of the level ellipsoid as a series of spherical harmonics.

The second-degree coefficient A_2 is

$$A_2 = G\,(A - C)\,. \tag{2–165}$$

This follows from (2–91) by using $A = B$ for reasons of symmetry. The C is the moment of inertia with respect to the axis of rotation, and A is the moment of inertia with respect to any axis in the equatorial plane. By letting $n = 1$ in (2–164), we obtain

$$A_2 = -\frac{1}{3} GM\,E^2 \left(1 - \frac{2}{15} \frac{m\,e'}{q_0} \right). \tag{2–166}$$

Comparing this with the preceding Eq. (2–165), we find

$$G\,(C - A) = \frac{1}{3} GM\,E^2 \left(1 - \frac{2}{15} \frac{m\,e'}{q_0} \right). \tag{2–167}$$

Thus, the difference between the principal moments of inertia is expressed in terms of "Stokes' constants" a, b, M, and ω.

It is possible to eliminate q_0 from Eqs. (2–164) and (2–167), obtaining

$$A_{2n} = (-1)^n \frac{3GM\,E^{2n}}{(2n+1)(2n+3)} \left(1 - n + 5n \frac{C - A}{M\,E^2} \right) . \tag{2–168}$$

If we write the potential V in the form

$$V = \frac{GM}{r} \left[1 + C_2 \left(\frac{a}{r} \right)^2 P_2(\cos \vartheta) + C_4 \left(\frac{a}{r} \right)^4 P_4(\cos \vartheta) + \cdots \right]$$

$$= \frac{GM}{r} \left[1 + \sum_{n=1}^{\infty} C_{2n} \left(\frac{a}{r} \right)^{2n} P_{2n}(\cos \vartheta) \right], \tag{2–169}$$

then the C_{2n} are given by

$$C_{2n} = -J_{2n} = (-1)^n \frac{3e^{2n}}{(2n+1)(2n+3)} \left(1 - n + 5n \frac{C-A}{ME^2}\right). \quad (2\text{–}170)$$

Here we have introduced the first eccentricity $e = E/a$. For $n = 1$ this gives the important formula

$$C_{20} = -\frac{C-A}{Ma^2} \quad (2\text{–}171)$$

or, equivalently,

$$J_2 = \frac{C-A}{Ma^2}, \quad (2\text{–}172)$$

which is in agreement with the respective relation in (2–95) when taking into account the rotational symmetry causing $A = B$.

Finally, we note that on eliminating $q_0 = (1/i)\,Q_2(i(b/E))$ by using Eq. (2–167), and U_0 by using Eq. (2–122), we may write the expansion of V in ellipsoidal harmonics, Eq. (2–110), in the form

$$V(u, \beta) = \frac{i}{E}\,GM\,Q_0\left(i\frac{u}{E}\right)$$

$$+ \frac{15i}{2E^3}\,G\left(C - A - \tfrac{1}{3}ME^2\right)Q_2\left(i\frac{u}{E}\right)P_2(\sin\beta). \quad (2\text{–}173)$$

This shows that the coefficients of the ellipsoidal harmonics of degrees zero and two are functions of the mass and of the difference between the two principal moments of inertia. The analogy to the corresponding spherical-harmonic coefficients (2–91) is obvious. *This is a closed formula, not a truncated series!*

2.10   Series expansions for the normal gravity field

Since the earth ellipsoid is very nearly a sphere, the quantities

$$E = \sqrt{a^2 - b^2}, \quad \text{linear eccentricity,}$$

$$e = \frac{E}{a}, \quad \text{first (numerical) eccentricity,}$$

$$e' = \frac{E}{b}, \quad \text{second (numerical) eccentricity,} \quad (2\text{–}174)$$

$$f = \frac{a-b}{a}, \quad \text{flattening,}$$

and similar parameters that characterize the deviation from a sphere are small. Therefore, series expansions in terms of these or similar parameters will be convenient for numerical calculations.

Linear approximation

In order that the readers may find their way through the subsequent practical formulas, we first consider an approximation that is linear in the flattening f. Here we get particularly simple and symmetrical formulas which also exhibit plainly the structure of the higher-order expansions.

It is well known that the radius vector r of an ellipsoid is approximately given by

$$r = a\left(1 - f\sin^2\varphi\right).\qquad(2\text{–}175)$$

As we will see subsequently, normal gravity may, to the same approximation, be written

$$\gamma = \gamma_a\left(1 + f^*\sin^2\varphi\right).\qquad(2\text{–}176)$$

For $\varphi = \pm90°$, at the poles, we have $r = b$ and $\gamma = \gamma_b$. Hence, we may write

$$b = a\left(1 - f\right),\qquad \gamma_b = \gamma_a\left(1 + f^*\right),\qquad(2\text{–}177)$$

and solving for f and f^*, we obtain

$$f = \frac{a - b}{a},$$
$$f^* = \frac{\gamma_b - \gamma_a}{\gamma_a},\qquad(2\text{–}178)$$

so that f is the flattening defined by (2–174), and f^* is an analogous quantity which may be called *gravity flattening*.

To the same approximation, (2–143) becomes

$$f + f^* = \tfrac{5}{2}\,m,\qquad(2\text{–}179)$$

where

$$m \doteq \frac{\omega^2 a}{\gamma_a} = \frac{\text{centrifugal force at equator}}{\text{gravity at equator}}.\qquad(2\text{–}180)$$

This is *Clairaut's theorem* in its original form. It is one of the most striking formulas of physical geodesy: the (geometrical) flattening f in (2–178) can be derived from f^* and m, which are purely dynamical quantities obtained by gravity measurements; that is, *the flattening of the earth can be obtained from gravity measurements*.

Clairaut's formula is only a first approximation and must be improved, first by the inclusion of higher-order ellipsoidal terms in f, and secondly by taking into account the deviation of the earth's gravity field from the normal gravity field. But the principle remains the same.

Second-order expansion

We now expand the closed formulas of the two preceding sections into series in terms of the second numerical eccentricity e' and the flattening f, in general up to and including e'^4 or f^2. Terms of the order of e'^6 or f^3 and higher will usually be neglected.

We start from the series

$$\tan^{-1}\frac{E}{u} = \frac{E}{u} - \frac{1}{3}\left(\frac{E}{u}\right)^3 + \frac{1}{5}\left(\frac{E}{u}\right)^5 - \frac{1}{7}\left(\frac{E}{u}\right)^7 + \cdots ,$$

$$q = 2\left[\frac{1}{3\cdot5}\left(\frac{E}{u}\right)^3 - \frac{2}{5\cdot7}\left(\frac{E}{u}\right)^5 + \frac{3}{7\cdot9}\left(\frac{E}{u}\right)^7 - \cdots\right], \qquad (2\text{--}181)$$

$$q' = 6\left[\frac{1}{3\cdot5}\left(\frac{E}{u}\right)^3 - \frac{1}{5\cdot7}\left(\frac{E}{u}\right)^5 + \frac{1}{7\cdot9}\left(\frac{E}{u}\right)^7 - \cdots\right].$$

The first two series have already been used in the preceding section in (2–154) and (2–156), respectively; the third is obtained by substituting the $\tan^{-1}$ series into the closed formula (2–133) for q'.

On the reference ellipsoid S_0, we have $u = b$ and

$$\frac{E}{u} = \frac{E}{b} = e' , \qquad (2\text{--}182)$$

so that

$$\tan^{-1}e' = e' - \tfrac{1}{3}e'^3 + \tfrac{1}{5}e'^5 \cdots ,$$

$$q_0 = \tfrac{2}{15}e'^3\left(1 - \tfrac{6}{7}e'^2 \cdots\right),$$

$$q_0' = \tfrac{2}{5}e'^2\left(1 - \tfrac{3}{7}e'^2 \cdots\right), \qquad (2\text{--}183)$$

$$\frac{e'\,q_0'}{q_0} = 3\left(1 + \tfrac{3}{7}e'^2 \cdots\right).$$

We also need the series

$$b = \frac{a}{\sqrt{1+e'^2}} = a\left(1 - \tfrac{1}{2}e'^2 + \tfrac{3}{8}e'^4 \cdots\right). \qquad (2\text{--}184)$$

Potential and gravity

By substituting these expressions into the closed formulas (2–123), (2–141), (2–142), and (2–143), we obtain, up to and including the order e'^4, the following relations.

Potential:

$$U_0 = \frac{GM}{b}\left(1 - \tfrac{1}{3}e'^2 + \tfrac{1}{5}e'^4\right) + \tfrac{1}{3}\omega^2 a^2 . \tag{2–185}$$

Gravity at the equator and the pole:

$$\gamma_a = \frac{GM}{a\,b}\left(1 - \tfrac{3}{2}m - \tfrac{3}{14}e'^2 m\right),$$

$$\gamma_b = \frac{GM}{a^2}\left(1 + m + \tfrac{3}{7}e'^2 m\right). \tag{2–186}$$

Clairaut's theorem:

$$f + f^* = \frac{5}{2}\frac{\omega^2 b}{\gamma_a}\left(1 + \frac{9}{35}e'^2\right). \tag{2–187}$$

The ratio $\omega^2 a/\gamma_a$ may be expressed as

$$\frac{\omega^2 a}{\gamma_a} = m + \tfrac{3}{2}m^2 , \tag{2–188}$$

which is a more accurate version of (2–180).

From the first equation of (2–186), we find

$$GM = a\,b\,\gamma_a\left(1 + \tfrac{3}{2}m + \tfrac{3}{14}e'^2 m + \tfrac{9}{4}m^2\right), \tag{2–189}$$

which gives the mass in terms of equatorial gravity. Using this equation, we can express GM in Eq. (2–185) in terms of γ_a, obtaining

$$U_0 = a\,\gamma_a\left(1 - \tfrac{1}{3}e'^2 + \tfrac{11}{6}m + \tfrac{1}{5}e'^4 - \tfrac{2}{7}e'^2 m + \tfrac{11}{4}m^2\right). \tag{2–190}$$

Here we have eliminated $\omega^2 a$ by replacing it with $GM\,m/b$.

Now we can turn to Eq. (2–146) for normal gravity. A simple manipulation yields

$$\gamma = \gamma_a \frac{1 + \frac{b\gamma_b - a\gamma_a}{a\,\gamma_b}\sin^2\varphi}{\sqrt{1 - \frac{a^2 - b^2}{a^2}\sin^2\varphi}} . \tag{2–191}$$

The denominator is expanded into a binomial series:

$$\frac{1}{\sqrt{1 - x}} = 1 + \tfrac{1}{2}x + \tfrac{3}{8}x^2 + \cdots . \tag{2–192}$$

Then the abbreviated series

$$\frac{a^2 - b^2}{a^2} = \frac{e'^2}{1 + e'^2} = e'^2 - e'^4,$$

(2–193)

$$\frac{b\gamma_b - a\gamma_a}{a\gamma_a} = -e'^2 + \tfrac{5}{2}m^2 + e'^4 - \tfrac{13}{7}e'^2m + \tfrac{15}{4}m^2$$

are introduced and we obtain, upon substitution,

$$\gamma = \gamma_a \left[1 + \left(-\tfrac{1}{2}e'^2 + \tfrac{5}{2}m + \tfrac{1}{2}e'4 - \tfrac{13}{7}e'^2m + \tfrac{15}{4}m^2\right)\sin^2\varphi \right.$$

$$\left. + \left(-\tfrac{1}{8}e'^4 + \tfrac{5}{4}e'^2m\right)\sin^4\varphi\right].$$

(2–194)

We may also express these quantities in terms of the flattening f by substituting the equation

$$e'^2 = \frac{1}{(1-f)^2} - 1 = 2f + 3f^2 + \cdots.$$

(2–195)

The flattening f is most commonly used; it offers a slight advantage over the second eccentricity e' in that it is of the same order of magnitude as m: it is not immediately apparent that m^2, e'^2m, and e'^4 are quantities of the same order of magnitude. We obtain

$$GM = ab\gamma_a\left(1 + \tfrac{3}{2}m + \tfrac{3}{7}fm + + \tfrac{9}{4}m^2\right),$$

(2–196)

$$U_0 = a\gamma_a\left(1 - \tfrac{2}{3}f + \tfrac{11}{6}m - \tfrac{1}{5}f^2 - \tfrac{4}{7}fm + \tfrac{11}{4}m^2\right),$$

(2–197)

$$\gamma = \gamma_a\left[1 + \left(-f + \tfrac{5}{2}m + \tfrac{1}{2}f^2 - \tfrac{26}{7}fm + \tfrac{15}{4}m^2\right)\sin^2\varphi\right.$$

$$\left. + \left(-\tfrac{1}{2}f^2 + \tfrac{5}{2}fm\right)\sin^4\varphi\right].$$

(2–198)

The last formula is usually abbreviated as

$$\gamma = \gamma_a\left(1 + f_2\sin^2\varphi + f_4\sin^4\varphi\right),$$

(2–199)

so that we have

$$f_2 = -f + \tfrac{5}{2}m + \tfrac{1}{2}f^2 - \tfrac{26}{7}fm + \tfrac{15}{4}m^2,$$

$$f_4 = -\tfrac{1}{2}f^2 + \tfrac{5}{2}fm.$$

(2–200)

By substituting

$$\sin^4\varphi = \sin^2\varphi - \tfrac{1}{4}\sin^2 2\varphi,$$

(2–201)

we finally obtain

$$\gamma = \gamma_a \left(1 + f^* \sin^2\varphi - \tfrac{1}{4} f_4 \sin^2 2\varphi\right), \tag{2–202}$$

where

$$f^* = \frac{\gamma_b - \gamma_a}{\gamma_a} = f_2 + f_4 \tag{2–203}$$

is the "gravity flattening".

Coefficients of spherical harmonics

Equation (2–167) for the principal moments of inertia yields at once

$$\frac{C - A}{M\,E^2} = \frac{1}{3} - \frac{2}{45}\frac{m\,e'}{q_0}. \tag{2–204}$$

Expanding q_0 by means of (2–183), we find

$$\frac{C - A}{M\,E^2} = \frac{1}{e'^2}\left(\tfrac{1}{3} e'^2 - \tfrac{1}{3} m - \tfrac{2}{7} e'^2 m\right). \tag{2–205}$$

Substituting this into (2–170) yields

$$-C_{20} = J_2 = \frac{C - A}{M\,E^2} = \tfrac{1}{3} e'^2 - \tfrac{1}{3} m - \tfrac{1}{3} e'^4 + \tfrac{1}{21} e'^2 m$$

$$= \tfrac{2}{3} f - \tfrac{1}{3} m - \tfrac{1}{3} f^2 + \tfrac{2}{21} f\,m, \tag{2–206}$$

$$-C_{40} = J_4 = -\tfrac{1}{5} e'^4 + \tfrac{2}{7} e'^2 m = -\tfrac{4}{5} f^2 + \tfrac{4}{7} f\,m. \tag{2–207}$$

The higher C or J, respectively, are already of an order of magnitude that we have neglected.

Gravity above the ellipsoid

Denoting the height above the ellipsoid as ellipsoidal height h, then, in case of a small height, the normal gravity γ_h at this height can be expanded into a series in terms of h:

$$\gamma_h = \gamma + \frac{\partial\gamma}{\partial h} h + \frac{1}{2}\frac{\partial^2\gamma}{\partial h^2} h^2 + \cdots, \tag{2–208}$$

where γ and its derivatives are referred to the ellipsoid, where $h = 0$.

The first derivative $\partial\gamma/\partial h$ may be obtained by applying Bruns' formula (2–147) together with (2–148) to the ellipsoidal height h (instead of H):

$$\frac{\partial\gamma}{\partial h} = -\gamma\left(\frac{1}{M} + \frac{1}{N}\right) - 2\omega^2, \tag{2–209}$$

where M, N are the principal radii of curvature of the ellipsoid, defined by (2–149). Since

$$
\frac{1}{M} = \frac{b}{a^2}\left(1 + e'^2 \cos^2\varphi\right)^{3/2} = \frac{b}{a^2}\left(1 + \tfrac{3}{2} e'^2 \cos^2\varphi \cdots\right),
$$

$$
\frac{1}{N} = \frac{b}{a^2}\left(1 + e'^2 \cos^2\varphi\right)^{1/2} = \frac{b}{a^2}\left(1 + \tfrac{1}{2} e'^2 \cos^2\varphi \cdots\right),
$$

$$(2\text{–}210)$$

we have

$$
\frac{1}{M} + \frac{1}{N} = \frac{b}{a^2}\left(2 + 2e'^2 \cos^2\varphi\right) = \frac{2b}{a^2}\left(1 + 2f \cos^2\varphi\right). \tag{2–211}
$$

Here we have limited ourselves to terms linear in f, since the elevation h is already a small quantity. Thus, we find from (2–209) after simple manipulations:

$$
\frac{\partial\gamma}{\partial h} = -\frac{2\gamma}{a}\left(1 + f + m - 2f \sin^2\varphi\right). \tag{2–212}
$$

The second derivative $\partial^2\gamma/\partial h^2$ may be taken from the spherical approximation, obtained by neglecting e'^2 or f:

$$
\gamma = \frac{GM}{a^2}, \quad \frac{\partial\gamma}{\partial h} = \frac{\partial\gamma}{\partial a} = -\frac{2GM}{a^3}, \quad \frac{\partial^2\gamma}{\partial h^2} = \frac{\partial^2\gamma}{\partial a^2} = \frac{6GM}{a^4}, \tag{2–213}
$$

so that

$$
\frac{\partial^2\gamma}{\partial h^2} = \frac{6\gamma}{a^2}. \tag{2–214}
$$

Thus we obtain

$$
\gamma_h = \gamma\left[1 - \frac{2}{a}\left(1 + f + m - 2f \sin^2\varphi\right) h + \frac{3}{a^2} h^2\right]. \tag{2–215}
$$

Using Eq. (2–198) for γ, we may also write the difference $\gamma_h - \gamma$ in the form

$$
\gamma_h - \gamma = -\frac{2\gamma_a}{a}\left[1 + f + m + \left(-3f + \tfrac{5}{2} m\right)\sin^2\varphi\right] h + \frac{3\gamma_a}{a^2} h^2. \tag{2–216}
$$

The symbol γ_h denotes the normal gravity for a point at latitude φ, situated at height h above the ellipsoid; γ is the gravity at the ellipsoid itself, for the same latitude φ, as given by (2–202) or equivalent formulas.

Second-order series developments for the *inner* gravity field are found in Moritz (1990: Chap. 4); this is the main reason for such a development here, because today one uses the closed formulas wherever possible.

2.11   Reference ellipsoid – numerical values

Some history

The reference ellipsoid and its gravity field are completely determined by four constants. Before the satellite era, one took the following four parameters:

$$
\begin{aligned}
a & \ \dots \ \text{semimajor axis}, \\
f & \ \dots \ \text{flattening}, \\
\gamma_a & \ \dots \ \text{equatorial gravity}, \\
\omega & \ \dots \ \text{angular velocity}.
\end{aligned}
\tag{2–217}
$$

The values best known and most widely used have been those of the *International Ellipsoid*:

$$
\begin{aligned}
a &= 6\,378\,388.000 \ \text{m}, \\
f &= 1/297.000, \\
\gamma_a &= 978.049\,000 \ \text{gal}, \\
\omega &= 0.729\,211\,51 \cdot 10^{-4} \ \text{s}^{-1}.
\end{aligned}
\tag{2–218}
$$

The geometric parameters a and f were determined by Hayford in 1909 from isostatically reduced astrogeodetic data in the United States. They were adopted for the International Ellipsoid by the assembly of the International Association of Geodesy (IAG) at Madrid in 1924. The equatorial gravity value γ_a was computed by Heiskanen (1924, 1928) from isostatically reduced gravity data. The corresponding *international gravity formula*,

$$
\gamma = 978.0490\,(1 + 0.005\,2884 \sin^2\varphi - 0.000\,0059 \sin^2 2\varphi) \ \text{gal}, \tag{2–219}
$$

was adopted by the assembly of IAG at Stockholm in 1930; whose coefficients were computed from the assumed values for a, f, γ_a, ω by Cassinis (1930) using Eqs. (2–200), (2–202), (2–203).

All parameters of the International Ellipsoid and its gravity field can be computed from (2–218) to any desired degree of accuracy, which merely expresses the inner consistency. In this way, we find (rounded values)

$$
\begin{aligned}
b &= 6\,356\,912 \ \text{m}, \\
E &= 522\,976 \ \text{m}, \\
e'^2 &= 0.006\,7682, \\
m &= 0.003\,4499.
\end{aligned}
\tag{2–220}
$$

For the constants in the spherical-harmonic expansion of the normal gravity field, we find the values

$$
\begin{aligned}
-C_{20} = J_2 &= \frac{C - A}{M\,a^2} = 0.001\,0920, \\
-C_{40} = J_4 &= -0.000\,002\,43.
\end{aligned}
\tag{2–221}
$$

The change of normal gravity with elevation is given by the formula (2–216), which for the International Ellipsoid becomes

$$\gamma_h = \gamma - (0.308\,77 - 0.000\,45\,\sin^2\varphi)\,h + 0.000\,072\,h^2\,, \qquad (2\text{–}222)$$

where γ_h and γ are measured in gal, and h is the elevation in kilometer.

Although the International Ellipsoid can no longer be considered the closest approximation of the earth by an ellipsoid, it may still be used as a reference ellipsoid for geodetic purposes. An official change of a reference system must be very carefully considered because a large amount of data may be referred to such a system.

The eastern countries have used the ellipsoid of Krassowsky:

$$\begin{aligned} a &= 6\,378\,245 \text{ m}\,, \\ f &= 1/298.3\,. \end{aligned} \qquad (2\text{–}223)$$

Contemporary data

After the start of Sputnik, the first artificial satellite, in 1957, the International Astronomical Union, in 1964, adopted a new set of constants, among them $a = 6\,378\,160$ m and $f = 1/298.25$. The value of a, which is considerably smaller than that for the International Ellipsoid, incorporates astrogeodetic determinations; the change in the value of J_2, and consequently of f, is due to the results from artificial satellites.

In 1967, these values were taken by the International Union of Geodesy and Geophysics (IUGG) as the *Geodetic Reference System 1967*.

This decision was soon seen to be wrong; especially the value of a was recognized to be too large: now we believe to be on the order of $6\,378\,137$ m, the value of the Geodetic Reference System 1980 (GRS 1980) and, based on it, the World Geodetic System 1984 (WGS 84). More details of these two systems are given below.

Geodetic Reference System 1980 (GRS 1980)

The GRS 1980 has been adopted at the XVII General Assembly of the IUGG in Canberra, December 1979, by Resolution No. 7. Inherently, this resolution *recognizing* that the Geodetic Reference System 1967 adopted at the XIV General Assembly of IUGG, Lucerne, 1967, no longer represents the size, shape, and gravity field of the earth to an accuracy adequate for many geodetic, geophysical, astronomical, and hydrographic applications and *considering* that more appropriate values are now available, *recommends* that the Geodetic Reference System 1967 be replaced by the new Geodetic Reference System 1980 which is also based on the theory of the geocentric equipotential ellipsoid. The four defining parameters of the GRS 1980 are given in

Table 2.1. Defining parameters of the GRS 1980

Parameter and value	Description
$a \quad = 6\,378\,137$ m	semimajor axis of the ellipsoid
$GM = 3\,986\,005 \cdot 10^8$ m^3 s^{-2}	geocentric gravitational constant of the earth (including the atmosphere)
$J_2 \quad = 108\,263 \cdot 10^{-8}$	dynamical form factor of the earth (excluding the permanent tidal deformation)
$\omega \quad = 7\,292\,115 \cdot 10^{-11}$ rad s^{-1}	angular velocity of the earth

Table 2.1. Note that these parameters, as given in the table, are defined as exact! Note also that GM, the "geocentric gravitational constant" of the earth, may also more figuratively be denoted as "product of the (Newtonian) gravitational constant and the earth's mass".

On the basis of these defining parameters and by the computational formulas given in Moritz (1980 b), the geometrical and physical constants of Table 2.2 may be derived.

The GRS 1980 is still (2005) valid as the official reference system of the IUGG and it forms the fundamental basis of the WGS 84.

World Geodetic System 1984 (WGS 84)

As just mentioned, the WGS 84 may be regarded as a descendant of the GRS 1980. Due to its still increasing importance, we consider it appropriate to describe the WGS 84 in some more detail.

Following the National Imagery and Mapping Agency (2000) of the USA, the *definition* of the WGS 84 may be described in the following way. The WGS 84 is a Conventional Terrestrial Reference System (CTRS). The definition of this coordinate system follows the criteria as outlined by the International Earth Rotation Service (IERS). The criteria for this system are the following:

- it is geocentric, the center of mass being defined for the whole earth including oceans and atmosphere;
- its scale is that of the local earth frame, in the meaning of a relativistic theory of gravitation;
- its orientation was initially given by the Bureau International de l'Heure (BIH) orientation of 1984.0;
- its time evolution in orientation will create no residual global rotation with regards to the crust.

Table 2.2. GRS 1980 derived constants

Parameter and value	Description
Geometrical constants	
$b\quad = 6\,356\,752.3141$ m	semiminor axis of the ellipsoid
$E\quad = 521\,854.0097$ m	linear eccentricity
$c\quad = 6\,399\,593.6259$ m	polar radius of curvature
$e^2\quad = 0.006\,694\,380\,022\,90$	first eccentricity squared
$e'^2\quad = 0.006\,739\,496\,775\,48$	second eccentricity squared
$f\quad = 0.003\,352\,810\,681\,18$	flattening
$1/f = 298.257\,222\,101$	reciprocal flattening
Physical constants	
$U_0\ = 62\,636\,860.850$ m^2 s^{-2}	normal potential at the ellipsoid
$J_4\ = -0.000\,002\,370\,912\,22$	spherical-harmonic coefficient
$J_6\ = 0.000\,000\,006\,083\,47$	spherical-harmonic coefficient
$J_8\ = -0.000\,000\,000\,014\,27$	spherical-harmonic coefficient
$m\ = 0.003\,449\,786\,003\,08$	$m = \omega^2 a^2 b/(GM)$
$\gamma_a\ = 9.780\,326\,7715$ m s^{-2}	normal gravity at the equator
$\gamma_b\ = 9.832\,186\,3685$ m s^{-2}	normal gravity at the pole

The WGS 84 is a right-handed, earth-fixed orthogonal coordinate system. The origin and axes are defined in the following way:

- Origin: earth's center of mass.
- Z-axis: the direction of the IERS Reference Pole (IRP); this direction corresponds to the direction of the BIH Conventional Terrestrial Pole (CTP) (epoch 1984.0). In other terms, the Z-axis is, by convention, identical to the mean position of the earth's rotational axis.
- X-axis: intersection of the IERS Reference Meridian (IRM) and the plane passing through the origin and normal to the Z-axis; the IRM is coincident with the BIH Zero Meridian (epoch 1984.0); in other terms, the X-axis is associated with the mean Greenwich meridian.
- Y-axis: this axis completes a right-handed, earth-centered-earth-fixed (ECEF) orthogonal coordinate system.

The WGS 84 origin also serves as the geometric center of the WGS 84 ellipsoid and the Z-axis serves as the rotational axis of this ellipsoid of revolution.

This completes the definition of the WGS 84 as given in National Imagery and Mapping Agency (2000). Note that the *definition* of the WGS 84 CTRS has not changed in any fundamental way.

Reference frames: WGS 84 and ITRF

Now we need the distinction between *definition* and *realization*. When using the term "coordinate system" or "reference system", then this implies the definition only; however, when using the term "coordinate frame", then a realization is implied (Mueller 1985). So far, we have only given a definition of the WGS 84; therefore, we ought to denote this as WGS 84 CTRS. Now we consider a realization and, therefore, use the term "coordinate frame".

Following closely National Imagery and Mapping Agency (2000) and Hofmann-Wellenhof et al. (2001: Sect. 3.2.1), an example of a terrestrial reference frame is – on the basis of the previous definition – the WGS 84 reference frame (often simply denoted as WGS 84 – as we will also do). Associated to this frame is a geocentric ellipsoid of revolution, originally defined by the four parameters (1) semimajor axis a, (2) normalized second degree zonal gravitational coefficient $\bar{C}_{20}$, (3) truncated angular velocity of the earth ω, and (4) earth's gravitational constant G. This frame has been used for GPS since 1987.

Another example for a terrestrial reference frame is the one produced by the IERS and is called International Terrestrial Reference Frame (ITRF) (McCarthy 1996). The definition of the axes is analogous to the WGS 84, i.e., the Z-axis is defined by the IERS Reference Pole (IRP) and the X-axis lies in the IERS Reference Meridian (IRM); however, the realization differs! The ITRF is realized by a number of terrestrial sites where temporal effects (plate tectonics, tidal effects) are also taken into account. Thus, ITRF is regularly updated (almost every year) and the acronym is supplemented by the last two digits of the last year whose data were used in the formation of the frame, e.g., ITRF89, ITRF90, ITRF91, ITRF92, ITRF93, ITRF94, ITRF95, ITRF96, ITRF97, or the full designation of the year, e.g., ITRF2000.

The comparison of the original WGS 84 and ITRF revealed remarkable differences (Malys and Slater 1994):

1. The WGS 84 was established through Doppler observations from the TRANSIT satellite system, while ITRF is based on Satellite Laser Ranging (SLR) and Very Long Baseline Interferometry (VLBI) observations. The accuracy of the TRANSIT reference stations was estimated to be in the range of 1 to 2 meters, while the accuracy of the ITRF reference stations is at the centimeter level.

2. The numerical values for the original defining parameters differ from those in the ITRF. The only significant difference, however, was in the earth's gravitational constant $G_{\mathrm{WGS}} - G_{\mathrm{ITRF}} = 0.582 \cdot 10^8 \, \mathrm{m}^3 \, \mathrm{s}^{-2}$, which resulted in measurable differences in the satellite orbits.

On the basis of this information, the former U.S. Defense Mapping Agency

(DMA) has proposed to replace the value of G in the WGS 84 by the standard IERS value and to refine the coordinates of the GPS tracking stations. The revised WGS 84, valid since January 2, 1994, has been given the designation WGS 84 (G 730), where the 'G' indicates that the respective coordinates used were obtained through GPS and the following number 730 indicates the GPS week number when DMA has implemented the refined system.

In 1996, the U.S. National Imagery and Mapping Agency (NIMA) – the successor of DMA – has implemented a revised version of the frame denoted as WGS 84 (G 873) and being valid since September 29, 1996. The frame is realized by monitor stations with refined coordinates. The associated ellipsoid and its gravity field are now defined by the four parameters a, f, GM, ω, which are slightly different compared to the respective ITRF values, e.g., the current WGS 84 (G 873) frame and the ITRF97 show insignificant systematic differences of less than 2 cm. Hence, they are virtually identical.

Note that the refinements applied to the WGS 84 reference frame have reduced the uncertainties in the coordinates of the frame, the uncertainty of the gravitational model, and the uncertainty of the geoid undulations; however, *they have not changed the WGS 84 coordinate system in the sense of definition!*

More general, the relationship between the WGS 84 and the ITRF is characterized by two statements: (1) WGS 84 and ITRF are consistent; (2) the differences between WGS 84 and ITRF are in the centimeter range worldwide (National Imagery and Mapping Agency 2000).

However, if a transformation between reference frames is required, this is accomplished by a datum transformation (see Sect. 5.7).

Numerical values for the WGS 84 (reference frame)

As mentioned at the very beginning of Sect. 2.11, the reference ellipsoid and its gravity field are completely determined by four constants. The current defining parameters for WGS 84 are listed in Table 2.3.

Table 2.3. Defining parameters of the WGS 84

Parameter and value	Description
$a = 6\,378\,137$ m	semimajor axis of the ellipsoid
$f = 1/298.257\,223\,563$	flattening of the ellipsoid
$GM = 3\,986\,004.418 \cdot 10^8\,\mathrm{m}^3\,\mathrm{s}^{-2}$	geocentric gravitational constant of the earth (including the atmosphere)
$\omega = 7\,292\,115 \cdot 10^{-11}\,\mathrm{rad}\,\mathrm{s}^{-1}$	angular velocity of the earth

Table 2.4. WGS 84 reference ellipsoid derived constants

Parameter and value	Description
Geometrical constants	
$\bar{C}_{20} = -0.484\,166\,774\,985 \cdot 10^{-3}$	normalized second-degree harmonic
$b = 6\,356\,752.3142$ m	semiminor axis of the ellipsoid
$e = 8.181\,919\,084\,2622 \cdot 10^{-2}$	first eccentricity
$e^2 = 6.694\,379\,990\,14 \cdot 10^{-3}$	first eccentricity squared
$e' = 8.209\,443\,794\,9696 \cdot 10^{-2}$	second eccentricity
$e'^2 = 6.739\,496\,742\,28 \cdot 10^{-3}$	second eccentricity squared
$E = 5.218\,540\,084\,2339 \cdot 10^5$	linear eccentricity
$c = 6\,399\,593.6258$ m	polar radius of curvature
$b/a = 0.996\,647\,189\,335$	axis ratio
Physical constants	
$U_0 = 62\,636\,851.7146$ m^2 s^{-2}	normal potential at the ellipsoid
$\gamma_a = 9.780\,325\,3359$ m s^{-2}	normal gravity at the equator
$\gamma_b = 9.832\,184\,9378$ m s^{-2}	normal gravity at the pole
$\bar{\gamma} = 9.797\,643\,2222$ m s^{-2}	mean value of normal gravity
$M = 5.973\,3328 \cdot 10^{24}$ kg	mass of the earth (includes atmosphere)
$m = 0.003\,449\,786\,506\,84$	$m = \omega^2 a^2 b/(GM)$

Some history (even if only some years old) is important here because the parameters selected to originally define the WGS 84 reference ellipsoid were the semimajor axis a, the product of the earth's mass and the gravitational constant GM (also denoted as "geocentric gravitational constant of the earth"), the normalized second-degree zonal gravitational coefficient $\bar{C}_{20}$, and the earth's angular velocity ω. Due to significant refinements of these original defining parameters, the DMA recommended, e.g., a refined value for the GM parameter.

Anyway, a decision was made to retain the original WGS 84 reference ellipsoid values for the semimajor axis $a = 6\,378\,137$ m and for the flattening $f = 1/298.257\,223\,563$. For this reason, the four defining parameters were chosen to be a, f, GM, ω.

Readers who like some confusion may continue right here; otherwise skip this short paragraph. Due to this new choice of the defining parameters, there are in addition two distinct values for the $\bar{C}_{20}$ term, one is dynamically derived and the other geometrically by the defining parameters. The geometric derivation based on the four defining parameters a, f, GM, ω yields $\bar{C}_{20} = -0.484\,166\,774\,985 \cdot 10^{-3}$ which differs from the original value by $7.5015 \cdot 10^{-11}$. For many more details refer to National Imagery and Mapping Agency (2000).

We conclude these considerations by a useful table. Using the four defining parameters, it is possible to derive the more commonly used geometric constants and physical constants (Table 2.4) associated with the WGS 84 reference ellipsoid.

Numerical comparison of GRS 1980 and WGS 84

As mentioned previously, the GRS 1980 is the basis of the WGS 84. However, due to different defining parameters on the one hand and, e.g., a refined value for GM for the WGS 84 on the other hand, numerical differences between the GRS 1980 and the WGS 84 arise. Some of these differences are given in Table 2.5.

Table 2.5. Numerical comparison between GRS 1980 and WGS 84

Parameter	GRS 1980	WGS 84
GM	$3\,986\,005 \cdot 10^8\,\mathrm{m^3\,s^{-2}}$	$3\,986\,004.418 \cdot 10^8\,\mathrm{m^3\,s^{-2}}$
$1/f$	$298.257\,222\,101$	$298.257\,223\,563$
b	$6\,356\,752.3141$ m	$6\,356\,752.3142$ m
e^2	$0.006\,694\,380\,022\,90$	$0.006\,694\,379\,990\,14$
e'^2	$0.006\,739\,496\,775\,48$	$0.006\,739\,496\,742\,28$
E	$521\,854.0097$ m	$521\,854.0084$ m
c	$6\,399\,593.6259$ m	$6\,399\,593.6258$ m
U_0	$62\,636\,860.850\,\mathrm{m^2\,s^{-2}}$	$62\,636\,851.7146\,\mathrm{m^2\,s^{-2}}$
γ_a	$9.780\,326\,7715\,\mathrm{m\,s^{-2}}$	$9.780\,325\,3359\,\mathrm{m\,s^{-2}}$
γ_b	$9.832\,186\,3685\,\mathrm{m\,s^{-2}}$	$9.832\,184\,9378\,\mathrm{m\,s^{-2}}$
m	$0.003\,449\,786\,003\,08$	$0.003\,449\,786\,506\,84$

2.12 Anomalous gravity field, geoidal undulations, and deflections of the vertical

The small difference between the actual gravity potential W and the normal gravity potential U is denoted by T, so that

$$W(x, y, z) = U(x, y, z) + T(x, y, z);$$ (2–224)

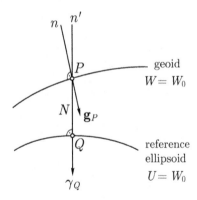

Fig. 2.12. Geoid and reference ellipsoid

T is called the *anomalous potential*, or *disturbing potential*. We compare the geoid

$$W(x, y, z) = W_0 \qquad (2\text{--}225)$$

with a reference ellipsoid

$$U(x, y, z) = W_0 \qquad (2\text{--}226)$$

of the same potential $U_0 = W_0$. A point P of the geoid is projected onto the point Q of the ellipsoid by means of the ellipsoidal normal (Fig. 2.12). The distance PQ between geoid and ellipsoid is called the *geoidal height*, or *geoidal undulation*, and is denoted by N. Unfortunately, there is a conflict of notation here. Denoting both the normal radius of curvature of the ellipsoid and the geoidal height by N is well established in geodetic literature. We continue this practice, as there is little chance of confusion.

Consider now the gravity vector $\mathbf{g}$ at P and the normal gravity vector $\boldsymbol{\gamma}$ at Q. The *gravity anomaly vector* $\Delta\mathbf{g}$ is defined as their difference:

$$\Delta\mathbf{g} = \mathbf{g}_P - \boldsymbol{\gamma}_Q. \qquad (2\text{--}227)$$

A vector is characterized by magnitude and direction. The difference in magnitude is the *gravity anomaly*

$$\Delta g = g_P - \gamma_Q; \qquad (2\text{--}228)$$

the difference in direction is the *deflection of the vertical*.

The deflection of the vertical has two components, a north-south component ξ and an east-west component η (Fig. 2.13). As the direction of the vertical is directly defined by the astronomical coordinates latitude Φ and longitude Λ, the components ξ and η can be expressed by them in a simple way. The actual astronomical coordinates of the geoidal point P, which

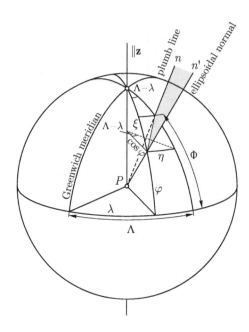

Fig. 2.13. The deflection of the vertical as illustrated by means of a unit sphere with center at P

define the direction of the plumb line n or of the gravity vector $\mathbf{g}$, can be determined by astronomical measurements. The ellipsoidal coordinates (or geodetic coordinates in the sense of geographical coordinates on the ellipsoid) given by the direction of the ellipsoidal normal n' have been denoted by φ and λ – these coordinates should not be confused with the ellipsoidal-harmonic coordinates of Sect. 1.15! It is evident that this λ is identical with the geocentric longitude (and also with the ellipsoidal-harmonic longitude). Thus,

geoidal normal n, astronomical coordinates Φ, Λ ;

ellipsoidal normal n', ellipsoidal coordinates φ, λ .

$$(2\text{--}229)$$

From Fig. 2.13, we read

$$\xi = \Phi - \varphi ,$$
$$\eta = (\Lambda - \lambda) \cos \varphi .$$

$$(2\text{--}230)$$

It is also possible to compare the vectors $\mathbf{g}$ and $\boldsymbol{\gamma}$ at the same point P. Then we get the *gravity disturbance vector*

$$\delta \mathbf{g} = \mathbf{g}_P - \boldsymbol{\gamma}_P .$$

$$(2\text{--}231)$$

Accordingly, the difference in magnitude is the *gravity disturbance*

$$\delta g = g_P - \gamma_P \,. \tag{2-232}$$

The difference in direction – i.e., the deflection of the vertical – is the same as before, since the directions of γ_P and γ_Q coincide virtually.

The gravity disturbance is conceptually even simpler than the gravity anomaly, but it has not been that important in terrestrial geodesy. The significance of the gravity anomaly is that it is given directly: the gravity g is measured on the geoid (or reduced to it), see Chap. 3, and the normal gravity γ is computed for the ellipsoid.

A very important remark

So far, for historical reasons, much more gravity anomalies Δg are available and are being processed than gravity disturbances δg. By GPS, however, the point P is determined rather than Q. *Therefore, in future, we may expect that δg will become more important than Δg.*

However, mirroring the present state of practice of physical geodesy, we continue mainly to work with Δg. Most statements about Δg will also apply for δg, with obvious modifications, such as with Molodensky's corrections (see Chap. 8), and Stokes' formula will be replaced by Koch's formula (see below in this chapter).

Relations

There are several basic mathematical relations between the quantities just defined. Since

$$U_P = U_Q + \left(\frac{\partial U}{\partial n}\right)_Q N = U_Q - \gamma N \,, \tag{2-233}$$

we have

$$W_P = U_P + T_P = U_Q - \gamma N + T_P \,. \tag{2-234}$$

Because

$$W_P = U_Q = W_0 \,, \tag{2-235}$$

we find

$$T = \gamma N \tag{2-236}$$

(where we have omitted the subscript P on the left-hand side) or

$$N = \frac{T}{\gamma} \,. \tag{2-237}$$

This is the famous *Bruns formula*, which relates the geoidal undulation to the disturbing potential.

Next we consider the gravity disturbance. Since

$$\mathbf{g} = \text{grad } W \,,$$
$$\boldsymbol{\gamma} = \text{grad } U \,,$$

$(2\text{–}238)$

the gravity disturbance vector $(2\text{–}231)$ becomes

$$\delta \mathbf{g} = \text{grad } (W - U) = \text{grad } T \equiv \left[\frac{\partial T}{\partial x}, \frac{\partial T}{\partial y}, \frac{\partial T}{\partial z} \right].$$

$(2\text{–}239)$

Then

$$g = -\frac{\partial W}{\partial n} \,, \qquad \gamma = -\frac{\partial U}{\partial n'} \doteq -\frac{\partial U}{\partial n} \,,$$

$(2\text{–}240)$

because the directions of the normals n and n' almost coincide. Therefore, the gravity disturbance is given by

$$\delta g = g_P - \gamma_P = - \left(\frac{\partial W}{\partial n} - \frac{\partial U}{\partial n'} \right) \doteq - \left(\frac{\partial W}{\partial n} - \frac{\partial U}{\partial n} \right)$$

$(2\text{–}241)$

or

$$\delta g = -\frac{\partial T}{\partial n} \,.$$

$(2\text{–}242)$

Since the elevation h is reckoned along the normal, we may also write

$$\delta g = -\frac{\partial T}{\partial h} \,.$$

$(2\text{–}243)$

Comparing $(2\text{–}242)$ with $(2\text{–}239)$, we see that the gravity disturbance δg, besides being the difference in magnitude of the actual and the normal gravity vector, is also the *normal component of the gravity disturbance vector* $\delta \mathbf{g}$.

We now turn to the gravity anomaly Δg. Since

$$\gamma_P = \gamma_Q + \frac{\partial \gamma}{\partial h} N \,,$$

$(2\text{–}244)$

we have

$$-\frac{\partial T}{\partial h} = \delta g = g_P - \gamma_P = g_P - \gamma_Q - \frac{\partial \gamma}{\partial h} N \,.$$

$(2\text{–}245)$

Remembering the definition $(2\text{–}228)$ of the gravity anomaly and taking into account Bruns' formula $(2\text{–}237)$, we find the following equivalent relations:

$$-\frac{\partial T}{\partial h} = \Delta g - \frac{\partial \gamma}{\partial h} N \,,$$

$(2\text{–}246)$

$$\Delta g = -\frac{\partial T}{\partial h} + \frac{\partial \gamma}{\partial h} N \,,$$

$(2\text{–}247)$

$$\Delta g = -\frac{\partial T}{\partial h} + \frac{1}{\gamma}\frac{\partial \gamma}{\partial h}T\,, \qquad (2\text{--}248)$$

$$\delta g = \Delta g - \frac{\partial \gamma}{\partial h}N\,, \qquad (2\text{--}249)$$

$$\delta g = \Delta g - \frac{1}{\gamma}\frac{\partial \gamma}{\partial h}T\,, \qquad (2\text{--}250)$$

relating different quantities of the anomalous gravity field.

Another equivalent form is

$$\frac{\partial T}{\partial h} - \frac{1}{\gamma}\frac{\partial \gamma}{\partial h}T + \Delta g = 0\,. \qquad (2\text{--}251)$$

This expression has been called the *fundamental equation of physical geodesy*, because it relates the measured quantity Δg to the unknown anomalous potential T. In future, the relation

$$\frac{\partial T}{\partial h} + \delta g = 0 \qquad (2\text{--}252)$$

may replace it.

It has the form of a partial differential equation. If Δg were known throughout space, then (2–251) could be discussed and solved as a real partial differential equation. However, since Δg is known only along a surface (the geoid), the fundamental equation (2–251) can be used only as a *boundary condition*, which alone is not sufficient for computing T. Therefore, the name "differential equation of physical geodesy", which is sometimes used for (2–251), is rather misleading.

One usually assumes that there are no masses outside the geoid. This is not really true. But neither do we make observations directly on the geoid; we make them on the physical surface of the earth. In reducing the measured gravity to the geoid, the effect of the masses outside the geoid is removed by computation, so that we can indeed assume that all masses are enclosed by the geoid (see Chaps. 3 and 8).

In this case, since the density ϱ is zero everywhere outside the geoid, the anomalous potential T is harmonic there and satisfies Laplace's equation

$$\Delta T \equiv \frac{\partial^2 T}{\partial x^2} + \frac{\partial^2 T}{\partial y^2} + \frac{\partial^2 T}{\partial z^2} = 0\,. \qquad (2\text{--}253)$$

This is a true partial differential equation and suffices, if supplemented by the boundary condition (2–251), for determining T at every point outside the geoid. If we write the boundary condition in the form

$$-\frac{\partial T}{\partial n} + \frac{1}{\gamma}\frac{\partial \gamma}{\partial n}T = \Delta g\,, \qquad (2\text{--}254)$$

where Δg is assumed to be known at every point of the geoid, then we see that a linear combination of T and $\partial T / \partial n$ is given upon that surface. According to Sect. 1.13, the determination of T is, therefore, a *third boundary-value problem of potential theory*. If it is solved for T, then the geoidal height, which is the most important geometric quantity in physical geodesy, can be computed by Bruns' formula (2–237).

Therefore, we may say that the basic problem of physical geodesy, the determination of the geoid from gravity measurements, is essentially a third boundary-value problem of potential theory.

2.13   Spherical approximation and expansion of the disturbing potential in spherical harmonics

The reference ellipsoid deviates from a sphere only by quantities of the order of the flattening, $f \doteq 3 \cdot 10^{-3}$. Therefore, if we treat the reference ellipsoid as a sphere in equations relating quantities of the anomalous field, this may cause a relative error of the same order. This error is usually permissible in N, T, Δg, δg, etc. For instance, the absolute effect of this relative error on the geoidal height is of the order of $3 \cdot 10^{-3}\, N$; since N hardly exceeds $100\,\mathrm{m}$, this error can usually be expected to be less than $1\,\mathrm{m}$.

As a spherical approximation, we have

$$\gamma = \frac{GM}{r^2}\,,\quad \frac{\partial \gamma}{\partial h} = \frac{\partial \gamma}{\partial r} = -2\,\frac{GM}{r^3}\,,\quad \frac{1}{\gamma}\frac{\partial \gamma}{\partial h} = -\frac{2}{r}\,. \tag{2–255}$$

We introduce a mean radius R of the earth. It is often defined as the radius of a sphere that has the same volume as the earth ellipsoid; from the condition

$$\tfrac{4}{3}\,\pi\,R^3 = \tfrac{4}{3}\,\pi\,a^2\,b\,, \tag{2–256}$$

we get

$$R = \sqrt[3]{a^2\,b}\,. \tag{2–257}$$

In a similar way, we may define a mean value of gravity, γ_0, as normal gravity at latitude $\varphi = 45°$ (Moritz 1980b: p. 403). Numerical values of about

$$R = 6371\ \mathrm{km}\,,\qquad \gamma_0 = 980.6\ \mathrm{gal} \tag{2–258}$$

are usual. Then

$$\frac{1}{\gamma}\frac{\partial \gamma}{\partial h} = -\frac{2}{R}\,,$$

$$\frac{\partial \gamma}{\partial h} = -\frac{2\gamma_0}{R}\,. \tag{2–259}$$

Since the normal to the sphere is the direction of the radius vector r, we have to the same approximation

$$\frac{\partial}{\partial n} = \frac{\partial}{\partial h} = \frac{\partial}{\partial r}.$$
(2–260)

In Bruns' theorem (2–237) we may replace γ by γ_0, and Eqs. (2–246) through (2–250) and (2–251) become

$$-\frac{\partial T}{\partial h} = \Delta g + \frac{2\gamma_0}{R} N,$$
(2–261)

$$\Delta g = -\frac{\partial T}{\partial r} - \frac{2\gamma_0}{R} N,$$
(2–262)

$$\Delta g = -\frac{\partial T}{\partial r} - \frac{2}{R} T,$$
(2–263)

$$\delta g = \Delta g + \frac{2\gamma_0}{R} N,$$
(2–264)

$$\delta g = \Delta g + \frac{2}{R} T,$$
(2–265)

$$\frac{\partial T}{\partial r} + \frac{2}{R} T + \Delta g = 0.$$
(2–266)

The last equation is the spherical approximation of the fundamental boundary condition.

Remark

The meaning of this spherical approximation should be carefully kept in mind. It is used only in equations relating the small quantities T, N, Δg, δg, etc. The reference surface is *never* a sphere in any geometrical sense, but always an ellipsoid. As the flattening f is very small, the ellipsoidal formulas can be expanded into power series in terms of f, and then all terms containing f, f^2, etc., are neglected. In this way one obtains formulas that are rigorously valid for the sphere, but approximately valid for the actual reference ellipsoid as well. However, normal gravity γ in the gravity anomaly $\Delta g = g - \gamma$ must be computed for the ellipsoid to a high degree of accuracy. To speak of a "reference sphere" in space, in any geometric sense, may be highly misleading.

Since the anomalous potential $T = W - U$ is a harmonic function, it can be expanded into a series of spherical harmonics:

$$T(r, \vartheta, \lambda) = \sum_{n=0}^{\infty} \left(\frac{R}{r}\right)^{n+1} T_n(\vartheta, \lambda).$$
(2–267)

$T_n(\vartheta, \lambda)$ is Laplace's surface harmonic of degree n. On the geoid, which as a spherical approximation corresponds to the sphere $r = R$, we have formally

$$T = T(R,\, \vartheta,\, \lambda) = \sum_{n=0}^{\infty} T_n(\vartheta,\, \lambda)\,. \qquad (2\text{--}268)$$

We need not be concerned with questions of convergence here. Differentiating the series (2–267) with respect to r, we find

$$\delta g = -\frac{\partial T}{\partial r} = \frac{1}{r} \sum_{n=0}^{\infty} (n+1) \left(\frac{R}{r}\right)^{n+1} T_n(\vartheta,\, \lambda)\,. \qquad (2\text{--}269)$$

On the geoid, where $r = R$, this becomes

$$\delta g = -\frac{\partial T}{\partial r} = \frac{1}{R} \sum_{n=0}^{\infty} (n+1)\, T_n(\vartheta,\, \lambda)\,. \qquad (2\text{--}270)$$

These series express the gravity disturbance in terms of spherical harmonics.

The equivalent of (2–263) outside the earth is

$$\Delta g = -\frac{\partial T}{\partial r} - \frac{2}{r}\, T\,. \qquad (2\text{--}271)$$

Its exact meaning will be discussed at the end of the following section. The substitution of (2–269) and (2–267) into this equation yields

$$\Delta g = \frac{1}{r} \sum_{n=0}^{\infty} (n-1) \left(\frac{R}{r}\right)^{n+1} T_n(\vartheta,\, \lambda)\,. \qquad (2\text{--}272)$$

On the geoid, this becomes

$$\Delta g = \frac{1}{R} \sum_{n=0}^{\infty} (n-1)\, T_n(\vartheta,\, \lambda)\,. \qquad (2\text{--}273)$$

This is the spherical-harmonic expansion of the gravity anomaly.

Note that even if the anomalous potential T contains a first-degree spherical term $T_1(\vartheta, \lambda)$, it will in the expression for Δg be multiplied by the factor $1 - 1 = 0$, so that Δg can never have a first-degree spherical harmonic – even if T has one.

2.14   Gravity anomalies outside the earth

If a harmonic function H is given at the surface of the earth, then, as a spherical approximation, the values of H outside the earth can be computed by Poisson's integral formula (1–123)

$$H_P = \frac{R}{4\pi} \iint_{\sigma} \frac{r^2 - R^2}{l^3} H \, d\sigma \, . \qquad (2\text{--}274)$$

The symbol $\iint_{\sigma}$ is the usual abbreviation for an integral extended over the whole unit sphere. The meaning of the other notations is read from Fig. 2.14. The value of the harmonic function at the variable surface element $R^2 \, d\sigma$ is denoted simply by H, whereas H_P refers to the fixed point P. Then we get

$$l = \sqrt{r^2 + R^2 - 2Rr \cos \psi} \, . \qquad (2\text{--}275)$$

The harmonic function H can be expanded into a series of spherical harmonics:

$$H = \left(\frac{R}{r}\right) H_0 + \left(\frac{R}{r}\right)^2 H_1 + \sum_{n=2}^{\infty} \left(\frac{R}{r}\right)^{n+1} H_n \, . \qquad (2\text{--}276)$$

By omitting the terms of degrees one and zero, we get a new function

$$H' = H - \left(\frac{R}{r}\right) H_0 - \left(\frac{R}{r}\right)^2 H_1 = \sum_{n=2}^{\infty} \left(\frac{R}{r}\right)^{n+1} H_n \, . \qquad (2\text{--}277)$$

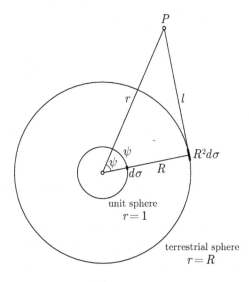

Fig. 2.14. Notations for Poisson's integral and derived formulas

The surface harmonics are given by

$$H_0 = \frac{1}{4\pi} \iint\limits_{\sigma} H \, d\sigma \,, \qquad H_1 = \frac{3}{4\pi} \iint\limits_{\sigma} H \cos \psi \, d\sigma \tag{2–278}$$

according to equation (1–89). Hence, we find from (2–277), on expressing H by Poisson's integral (2–274) and substituting the integrals (2–278) for H_0 and H_1, the basic formula

$$H_P' = \frac{1}{4\pi} \iint\limits_{\sigma} \left(\frac{r^2 - R^2}{l^3} - \frac{1}{r} - \frac{3R}{r^2} \cos \psi \right) H \, d\sigma \,. \tag{2–279}$$

The reason for this modification of Poisson's integral is that the formulas of physical geodesy are simpler if the functions involved do not contain harmonics of degrees zero and one. It is therefore convenient to split off these terms. This is done automatically by the modified Poisson integral (2–279).

We now apply these formulas to the gravity anomalies outside the earth. Equation (2–272) yields at once

$$r \, \Delta g = \sum_{n=0}^{\infty} \left(\frac{R}{r} \right)^{n+1} (n - 1) \, T_n(\vartheta, \lambda) \,. \tag{2–280}$$

Just as $T_n(\vartheta, \lambda)$ is a Laplace surface harmonic, so is $(n-1) \, T_n$. Consequently, $r \, \Delta g$, considered as a function in space, can be expanded into a series of spherical harmonics and *is, therefore, a harmonic function.* Hence, we can apply Poisson's formula to $r \, \Delta g$, getting

$$r \, \Delta g_P = \frac{R}{4\pi} \iint\limits_{\sigma} \left(\frac{r^2 - R^2}{l^3} - \frac{1}{r} - \frac{3R}{r^2} \cos \psi \right) R \, \Delta g \, d\sigma \tag{2–281}$$

or

$$\Delta g_P = \frac{R^2}{4\pi r} \iint\limits_{\sigma} \left(\frac{r^2 - R^2}{l^3} - \frac{1}{r} - \frac{3R}{r^2} \cos \psi \right) \Delta g \, d\sigma \,. \tag{2–282}$$

This is the formula for the computation of gravity anomalies outside the earth from surface gravity anomalies, or for the *upward continuation of gravity anomalies.*

Finally, we discuss the exact meaning of the gravity anomaly δg_P outside the earth. We start with a convenient definition. The level surfaces of the actual gravity potential, the surfaces

$$W = \text{constant} \,, \tag{2–283}$$

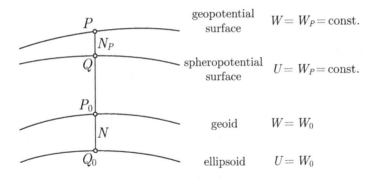

Fig. 2.15. Geopotential and spheropotential surfaces

are often called *geopotential surfaces*; the level surfaces of the normal gravity field, the surfaces

$$U = \text{constant}\,, \tag{2–284}$$

are called *spheropotential surfaces*.

We consider now the point P outside the earth (Fig. 2.15) and denote the geopotential surface passing through it by

$$W = W_P\,. \tag{2–285}$$

There is also a spheropotential surface

$$U = W_P \tag{2–286}$$

of the same constant W_P. The normal plumb line through P intersects this spheropotential surface at the point Q, which is said to correspond to P.

We see that the level surfaces $W = W_P$ and $U = W_P$ are related to each other in exactly the same way as are the geoid $W = W_0$ and the reference ellipsoid $U = W_0$. If, therefore, the gravity anomaly is defined by

$$\Delta g_P = g_P - \gamma_Q\,, \tag{2–287}$$

as in Sect. 2.12, then all derivations and formulas of that section also apply for the present situation, the geopotential surface $W = W_P$ replacing the geoid $W = W_0$, and the spheropotential surface $U = W_P$ replacing the ellipsoid $U = W_0$. This is also the reason why (2–271) applies at P as well as at the geoid.

Note that P in Sect. 2.12 is a point at the geoid, which is denoted by P_0 in Fig. 2.15.

This situation will be taken up again in Chap. 8, in the context of Molodensky's problem.

2.15   Stokes' formula

The basic Eq. (2–271),

$$\Delta g = -\frac{\partial T}{\partial r} - \frac{2}{r} T \, , \qquad (2\text{--}288)$$

can be regarded as a boundary condition only, as long as the gravity anomalies Δg are known only at the surface of the earth. However, by the upward continuation integral (2–282), we are now able to compute the gravity anomalies outside the earth. Thus, our basic equation changes its meaning radically, becoming a real differential equation that can be integrated with respect to r. Note that this is made possible only because T, in addition to the boundary condition, satisfies Laplace's equation $\Delta T = 0$.

Multiplying (2–288) by $-r^2$, we get

$$-r^2 \, \Delta g = r^2 \, \frac{\partial T}{\partial r} + 2r \, T = \frac{\partial}{\partial r} (r^2 \, T) \, . \qquad (2\text{--}289)$$

Integrating the formula

$$\frac{\partial}{\partial r} (r^2 \, T) = -r^2 \, \Delta g(r) \qquad (2\text{--}290)$$

between the limits ∞ and r, we find

$$r^2 \, T \Big|_{\infty}^{r} = - \int_{\infty}^{r} r^2 \, \Delta g(r) \, dr \, , \qquad (2\text{--}291)$$

where $\Delta g(r)$ indicates that Δg is now a function of r, computed from surface gravity anomalies by means of the formula (2–282). Since this formula automatically removes the spherical harmonics of degrees one and zero from $\Delta g(r)$, the anomalous potential T, as computed from $\Delta g(r)$, cannot contain such terms. Thus, we have

$$T = \sum_{n=2}^{\infty} \left(\frac{R}{r} \right)^{n+1} T_n = \frac{R^3}{r^3} T_2 + \frac{R^4}{r^4} T_3 + \cdots . \qquad (2\text{--}292)$$

Therefore,

$$\lim_{r \to \infty} (r^2 \, T) = \lim_{r \to \infty} \left(\frac{R^3}{r} T_2 + \frac{R^4}{r^2} T_3 + \cdots \right) = 0 \, , \qquad (2\text{--}293)$$

so that

$$r^2 \, T \Big|_{\infty}^{r} = r^2 \, T - \lim_{r \to \infty} (r^2 \, T) = r^2 \, T \qquad (2\text{--}294)$$

and

$$r^2 \, T = - \int_{\infty}^{r} r^2 \, \Delta g(r) \, dr \, . \qquad (2\text{--}295)$$

The fact that r is used both as an integration variable and as an upper limit should not cause any difficulty. Substituting the upward continuation integral (2–282), we get

$$r^2 T = \frac{R^2}{4\pi} \int_\infty^r \left[\iint_\sigma \left(-\frac{r^3 - R^2 r}{l^3} + 1 + \frac{3R}{r} \cos \psi \right) \Delta g \, d\sigma \right] dr \,. \quad (2\text{–}296)$$

Interchanging the order of the integrations gives

$$r^2 T = \frac{R^2}{4\pi} \iint_\sigma \left[\int_\infty^r \left(-\frac{r^3 - R^2 r}{l^3} + 1 + \frac{3R}{r} \cos \psi \right) dr \right] \Delta g \, d\sigma \,. \quad (2\text{–}297)$$

The integral in brackets can be evaluated by standard methods. The indefinite integral is

$$\int \left(-\frac{r^3 - R^2 r}{l^3} + 1 + \frac{3R}{r} \cos \psi \right) dr$$

$$= \frac{2r^2}{l} - 3l - 3R \cos \psi \ln(r - R \cos \psi + l) + r + 3R \cos \psi \ln r \,.$$
$$(2\text{–}298)$$

The reader is advised to perform this integration, taking into account (2–275), or at least to check the result by differentiating the right-hand side with respect to r.

For large values of r, we have

$$l = r \left(1 - \frac{R}{r} \cos \psi \cdots \right) = r - R \cos \psi \cdots \quad (2\text{–}299)$$

and, hence, we find that as $r \to \infty$, the right-hand side of the above indefinite integral approaches

$$5R \cos \psi - 3R \cos \psi \ln 2 \,. \quad (2\text{–}300)$$

If we subtract this from the indefinite integral, we get the definite integral, since infinity is its lower limit of integration. Thus,

$$\int_\infty^r \left(-\frac{r^3 - R^2 r}{l^3} + 1 + \frac{3R}{r} \cos \psi \right) dr$$
$$\qquad\qquad\qquad\qquad\qquad\qquad\qquad\qquad\qquad (2\text{–}301)$$
$$= \frac{2r^2}{l} + r - 3l - R \cos \psi \left(5 + 3 \ln \frac{r - R \cos \psi + l}{2r} \right) \,.$$

Hence, we obtain Pizzetti's formula

$$T(r, \vartheta, \lambda) = \frac{R}{4\pi} \iint_\sigma S(r, \psi) \Delta g \, d\sigma \,, \quad (2\text{–}302)$$

where

$$S(r, \psi) = \frac{2R}{l} + \frac{R}{r} - 3\frac{Rl}{r^2} - \frac{R^2}{r^2} \cos\psi \left(5 + 3\ln\frac{r - R\cos\psi + l}{2r}\right). \quad (2\text{--}303)$$

On the geoid itself, we have $r = R$, and denoting $T(R, \vartheta, \lambda)$ simply by T, we find

$$T = \frac{R}{4\pi} \iint_\sigma \Delta g \, S(\psi) \, d\sigma, \quad (2\text{--}304)$$

where

$$S(\psi) = \frac{1}{\sin(\psi/2)} - 6\sin\frac{\psi}{2} + 1 - 5\cos\psi - 3\cos\psi \ln\left(\sin\frac{\psi}{2} + \sin^2\frac{\psi}{2}\right) \quad (2\text{--}305)$$

is obtained from $S(r, \psi)$ by setting

$$r = R \quad \text{and} \quad l = 2R\sin\frac{\psi}{2}. \quad (2\text{--}306)$$

By Bruns' theorem, $N = T/\gamma_0$, we finally get

$$N = \frac{R}{4\pi\,\gamma_0} \iint_\sigma \Delta g \, S(\psi) \, d\sigma. \quad (2\text{--}307)$$

This formula was published by G.G. Stokes in 1849; it is, therefore, called *Stokes' formula*, or *Stokes' integral*. It is by far the most important formula of physical geodesy because it performs *to determine the geoid from gravity data*. Equation (2–304) is also called Stokes' formula, and $S(\psi)$ is known as Stokes' function.

Using formula (2–302), which was derived by Pizzetti (1911) and later on by Vening Meinesz (1928), we can compute the anomalous potential T at any point outside the earth. Dividing T by the normal gravity at the given point P (Bruns' theorem), we obtain the separation N_P between the geopotential surface $W = W_P$ and the corresponding spheropotential surface $U = W_P$, which, outside the earth, takes the place of the geoidal undulation N (see Fig. 2.15 and the explanations at the end of the preceding section).

We mention again that these formulas are based on a spherical approximation; quantities of the order of $3 \cdot 10^{-3} N$ are neglected. This results in an error of probably less than $1\,\mathrm{m}$ in N, which can be neglected for many practical purposes. Sagrebin (1956), Molodenskii et al. (1962: p. 53), Bjerhammar, and Lelgemann have developed higher approximations, which take into account the flattening f of the reference ellipsoid; see Moritz (1980 a: Sect. 39).

We next see from the derivation of Stokes' formula by means of the upward continuation integral (2–282) that it automatically suppresses the harmonic terms of degrees one and zero in T and N. The implications of this will be discussed later. We will see that Stokes' formula in its original form (2–304) and (2–307) only applies for a reference ellipsoid that (1) has the same potential $U_0 = W_0$ as the geoid, (2) encloses a mass that is numerically equal to the earth's mass, and (3) has its center at the center of gravity of the earth. Since the first two conditions are not accurately satisfied by the reference ellipsoids that are in current practical use, and can hardly ever be rigorously fulfilled, Stokes' formula will later be modified for the case of an arbitrary reference ellipsoid.

Finally, T is assumed to be harmonic outside the geoid. This means that the effect of the masses above the geoid must be removed by suitable gravity reductions. This will be discussed in Chaps. 3 and 8.

A bonus application to satellite geodesy

As a somewhat unexpected application, not related to Stokes' formula, we note that Eq. (2–280) can be used to compute gravity anomalies Δg from a satellite-determined spherical-harmonic series of the external gravitational potential V!

2.16   Explicit form of Stokes' integral and Stokes' function in spherical harmonics

We now write Stokes' formula (2–307) more explicitly by introducing suitable coordinate systems on the sphere.

The use of spherical *polar coordinates* with origin at P offers the advantage that the angle ψ, which is the argument of Stokes' function, is one coordinate, the *spherical distance*. The other coordinate is the *azimuth* α, reckoned from north. Their definitions are seen in Fig. 2.16. Denoting by P both a fixed point on the sphere $r = R$ (or in space) and its projection on the unit sphere is common practice and will not cause any trouble.

If P coincides with the north pole, then ψ and α are identical with ϑ and λ. According to Sect. 1.9, the surface element $d\sigma$ is then given by

$$d\sigma = \sin \psi \, d\psi \, d\alpha \,. \tag{2–308}$$

Since all points of the sphere are equivalent, this relation applies for an arbitrary origin P. In the same way, we have

$$\iint\limits_{\sigma} = \int_{\alpha=0}^{2\pi} \int_{\psi=0}^{\pi} \,. \tag{2–309}$$

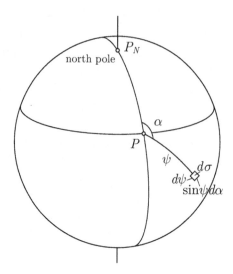

Fig. 2.16. Polar coordinates on the unit sphere

Hence, we find

$$N = \frac{R}{4\pi\,\gamma_0} \int_{\alpha=0}^{2\pi} \int_{\psi=0}^{\pi} \Delta g(\psi, \alpha)\, S(\psi) \sin\psi\, d\psi\, d\alpha \qquad (2\text{--}310)$$

as an explicit form of (2–307). Performing the integration with respect to α first, we obtain

$$N = \frac{R}{2\gamma_0} \int_{\psi=0}^{\pi} \left[\frac{1}{2\pi} \int_{\alpha=0}^{2\pi} \Delta g(\psi, \alpha)\, d\alpha \right] S(\psi) \sin\psi\, d\psi. \qquad (2\text{--}311)$$

The expression in brackets is the average of Δg along a parallel of spherical radius ψ. We denote this average by $\overline{\Delta g}(\psi)$, so that

$$\overline{\Delta g}(\psi) = \frac{1}{2\pi} \int_{\alpha=0}^{2\pi} \Delta g(\psi, \alpha)\, d\alpha. \qquad (2\text{--}312)$$

Thus, Stokes' formula may be written

$$N = \frac{R}{\gamma_0} \int_{\psi=0}^{\pi} \overline{\Delta g}(\psi)\, F(\psi)\, d\psi, \qquad (2\text{--}313)$$

where we have introduced

$$\tfrac{1}{2} S(\psi) \sin\psi = F(\psi). \qquad (2\text{--}314)$$

The functions $S(\psi)$ and $F(\psi)$ are shown in Fig. 2.17. Alternatively, we may

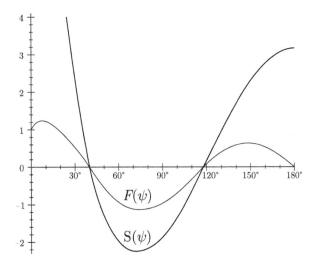

Fig. 2.17. Stokes' functions $S(\psi)$ and $F(\psi)$

use *ellipsoidal coordinates* φ, λ. As a spherical approximation, ϑ is the complement of ellipsoidal latitude:

$$\vartheta = 90° - \varphi \,. \tag{2–315}$$

Hence, we have

$$\iint\limits_{\sigma} d\sigma = \int_{\lambda=0}^{2\pi} \int_{\varphi=-\pi/2}^{\pi/2} \cos\varphi \, d\varphi \, d\lambda \,, \tag{2–316}$$

so that Stokes' formula now becomes

$$N(\varphi, \lambda) = \frac{R}{4\pi\,\gamma_0} \int_{\lambda'=0}^{2\pi} \int_{\varphi'=-\pi/2}^{\pi/2} \Delta g(\varphi', \lambda') \, S(\psi) \cos\varphi' \, d\varphi' \, d\lambda' \,, \tag{2–317}$$

where φ, λ are the ellipsoidal coordinates of the computation point and φ', λ' are the coordinates of the variable surface element $d\sigma$. The spherical distance ψ is expressed as a function of these coordinates by

$$\cos\psi = \sin\varphi \sin\varphi' + \cos\varphi \cos\varphi' \cos(\lambda' - \lambda) \,. \tag{2–318}$$

Stokes' function in terms of spherical harmonics
In Sect. 2.13, Eq. (2–273), we have found

$$\Delta g(\vartheta, \lambda) = \frac{1}{R} \sum_{n=0}^{\infty} (n - 1) \, T_n(\vartheta, \lambda) \,. \tag{2–319}$$

We may also directly express $\Delta g(\vartheta, \lambda)$ as a series of Laplace surface spherical harmonics:

$$\Delta g(\vartheta, \lambda) = \sum_{n=0}^{\infty} \Delta g_n(\vartheta, \lambda) \,. \tag{2-320}$$

Comparing these two series yields

$$\Delta g_n(\vartheta, \lambda) = \frac{n-1}{R} \, T_n(\vartheta, \lambda) \quad \text{or} \quad T_n = \frac{R}{n-1} \Delta g_n \,, \tag{2-321}$$

so that

$$T = \sum_{n=0}^{\infty} T_n = R \sum_{n=0}^{\infty} \frac{\Delta g_n}{n-1} \,. \tag{2-322}$$

This equation shows again that there must not be a first-degree term in the spherical-harmonic expansion of Δg; otherwise the term $\Delta g_n/(n-1)$ would be infinite for $n = 1$. As usual, we now assume that the harmonics of degrees zero and one are missing. Therefore, we start the summation with $n = 2$.

By Eq. (1–89), we may write

$$\Delta g_n = \frac{2n+1}{4\pi} \iint_{\sigma} \Delta g \, P_n(\cos \psi) \, d\sigma \,, \tag{2-323}$$

so that the preceding formula becomes

$$T = \frac{R}{4\pi} \sum_{n=2}^{\infty} \frac{2n+1}{n-1} \iint_{\sigma} \Delta g \, P_n(\cos \psi) \, d\sigma \,. \tag{2-324}$$

By interchanging the order of summation and integration, we get

$$T = \frac{R}{4\pi} \iint_{\sigma} \left[\sum_{n=2}^{\infty} \frac{2n+1}{n-1} P_n(\cos \psi) \right] \Delta g \, d\sigma \,. \tag{2-325}$$

Comparing this with Stokes' formula (2–304), we find the *expression for Stokes' function in terms of Legendre polynomials* (zonal harmonics):

$$S(\psi) = \sum_{n=2}^{\infty} \frac{2n+1}{n-1} P_n(\cos \psi) \,. \tag{2-326}$$

In fact, the analytic expression (2–305) of Stokes' function could have been derived somewhat more simply by direct summation of this series, but we believe that the derivation given in the preceding section is more instructive because it also throws sidelights on important related problems such as the "bonus equation" (2–280).

2.17 Generalization to an arbitrary reference ellipsoid

As we have seen, Stokes' formula, in its original form, suppresses the spherical harmonics of degrees zero and one in the anomalous potential T and is, therefore, strictly valid only if these terms are missing. This fact and the condition $U_0 = W_0$ impose restrictions on the reference ellipsoid and on its normal gravity field that are difficult to fulfil in practice.

Therefore, we generalize Stokes' formula so that it will apply to an arbitrary ellipsoid of reference, which must satisfy only the condition that it is so close to the geoid that the deviations of the geoid from the ellipsoid can be treated as linear.

Consider the anomalous potential T at the surface of the earth. Its expression in surface spherical harmonics is given by

$$T(\vartheta, \lambda) = \sum_{n=0}^{\infty} T_n(\vartheta, \lambda). \tag{2–327}$$

By separating the terms of degrees zero and one, we may write

$$T(\vartheta, \lambda) = T_0 + T_1(\vartheta, \lambda) + T'(\vartheta, \lambda), \tag{2–328}$$

where

$$T'(\vartheta, \lambda) = \sum_{n=2}^{\infty} T_n(\vartheta, \lambda). \tag{2–329}$$

In the general case this function T', rather than T itself, is the quantity given by Stokes' formula. It is equal to T only if T_0 and T_1 are missing. Otherwise, we have to add T_0 and T_1 in order to get the complete function T.

The zero-degree term in the spherical-harmonic expansion of the potential is equal to

$$\frac{GM}{r}, \tag{2–330}$$

where M is the mass. Hence, the zero-degree term of the anomalous potential $T = W - U$ at the surface of the earth, where $r = R$, is given by

$$T_0 = \frac{G\,\delta M}{R}, \tag{2–331}$$

where

$$\delta M = M - M' \tag{2–332}$$

is the difference between the mass M of the earth and the mass M' of the ellipsoid. It would be zero if both masses were equal – but since we do not know the exact mass of the earth, how can we make M' rigorously equal to M?

Subsequently, we will see that the first-degree harmonic can always be assumed to be zero. Under this assumption, we can substitute (2–331) into (2–328) and express T' by the conventional Stokes formula (2–304). Thus we obtain

$$T = \frac{G\,\delta M}{R} + \frac{R}{4\pi} \iint_\sigma \Delta g\, S(\psi)\, d\sigma. \tag{2-333}$$

This is the generalization of Stokes' formula for T. It holds for an arbitrary reference ellipsoid whose center coincides with the center of the earth.

First-degree terms

The coefficients of the first-degree harmonic in the potential W are, according to (2–85) and (2–87), given by

$$GM\,x_c, \quad GM\,y_c, \quad GM\,z_c, \tag{2-334}$$

where x_c, y_c, z_c are the rectangular coordinates of the earth's center of gravity. For the normal potential U, we have the analogous quantities

$$GM'\,x'_c, \quad GM'\,y'_c, \quad GM'\,z'_c. \tag{2-335}$$

As x'_c, y'_c, z'_c are very small in any case, these are practically equal to

$$GM\,x'_c, \quad GM\,y'_c, \quad GM\,z'_c. \tag{2-336}$$

The coefficients of the first-degree harmonic in the anomalous potential $T = W - U$ are, therefore, equal to

$$GM\,(x_c - x'_c), \quad GM\,(y_c - y'_c), \quad GM\,(z_c - z'_c). \tag{2-337}$$

They are zero, and *there is no first-degree harmonic $T_1(\vartheta, \lambda)$ if and only if the center of the reference ellipsoid coincides with the center of gravity of the earth*. This is usually assumed.

In the general case, we find from the first-degree term of (2–76), on putting $r = R$ and using the coefficients (2–85) together with (2–87),

$$T_1(\vartheta, \lambda) = \frac{GM}{R^2} \Big[(z_c - z'_c)\, P_{10}(\cos\vartheta) + (x_c - x'_c)\, P_{11}(\cos\vartheta)\, \cos\lambda$$

$$+ (y_c - y'_c)\, P_{11}(\cos\vartheta)\, \sin\lambda \Big]. \tag{2-338}$$

If the origin of the coordinate system is taken to be the center of the reference ellipsoid, then $x'_c = y'_c = z'_c = 0$. With $P_{10}(\cos\vartheta) = \cos\vartheta$, $P_{11}(\cos\vartheta) = \sin\vartheta$, and $GM/R^2 = \gamma_0$ we then obtain the following expression for the first-degree harmonic of T:

$$T_1(\vartheta, \lambda) = \gamma_0 \left(x_c \sin \vartheta \, \cos \lambda + y_c \sin \vartheta \, \sin \lambda + z_c \, \cos \vartheta \right). \qquad (2\text{--}339)$$

Dividing by γ_0, we find the first-degree harmonic of the geoidal height:

$$N_1(\vartheta, \lambda) = x_c \sin \vartheta \, \cos \lambda + y_c \sin \vartheta \, \sin \lambda + z_c \cos \vartheta \, . \qquad (2\text{--}340)$$

Introducing the vector

$$\mathbf{x}_c = [x_c, \ y_c, \ z_c] \qquad (2\text{--}341)$$

and the unit vector of the direction (ϑ, λ),

$$\mathbf{e} = [\sin \vartheta \cos \lambda, \ \sin \vartheta \sin \lambda, \ \cos \vartheta] \, , \qquad (2\text{--}342)$$

(2–340) may be written as

$$N_1(\vartheta, \lambda) = \mathbf{x}_c \cdot \mathbf{e} \, , \qquad (2\text{--}343)$$

which is interpreted as the projection of the vector $\mathbf{x}_c$ onto the direction (ϑ, λ).

Hence, if the two centers of gravity do not coincide, then we need only add the first-degree terms (2–339) and (2–340) to the generalized Stokes formula (2–333) and to its analogue for N, respectively, in order to get the most general solution for Stokes' problem, the computation of T and N from Δg. Equation (2–273) shows that *any* value of $T_1(\vartheta, \lambda)$ is compatible with a given Δg field because, for $n = 1$, the quantity $(n-1)\, T_1$ is zero and so T_1, whatever be its value, does not at all enter into Δg.

Hence, the most general solution for T and N contains three arbitrary constants x_c, y_c, z_c, which can, thus, be regarded as the constants of integration for Stokes' problem. In actual practice, one always sets $x_c = y_c = z_c = 0$, thus placing the center of the reference ellipsoid at the center of the earth. This constitutes an essential advantage of the gravimetric determination of the geoid over the astrogeodetic method, where the position of the reference ellipsoid with respect to the center of the earth remains unknown.

Zero-degree terms in N and Δg

Let us first extend Bruns' formula (2–237) to an arbitrary reference ellipsoid. Suppose

$$W(x, y, z) = W_0 \, ,$$
$$U(x, y, z) = U_0 \qquad (2\text{--}344)$$

are the equations of the geoid and the ellipsoid, where in general the constants W_0 and U_0 are different. As in Sect. 2.12, we have, using Fig. 2.12, $W_P = U_Q - \gamma N + T$, but now $U_Q = U_0 \neq W_0 = W_P$, so that

$$\gamma N = T - (W_0 - U_0).$$ (2–345)

Denoting the difference between the potentials by

$$\delta W = W_0 - U_0,$$ (2–346)

we obtain the following simple generalization of Bruns' formula:

$$N = \frac{T - \delta W}{\gamma}.$$ (2–347)

We also need the extension of Eqs. (2–246) through (2–250). Those formulas which contain N instead of T are easily seen to hold for an arbitrary reference ellipsoid as well, but the transition from N to T is now effected by means of (2–347). Hence, Eq. (2–247), i.e.,

$$\Delta g = -\frac{\partial T}{\partial h} + \frac{\partial \gamma}{\partial h} N,$$ (2–348)

remains unchanged, but (2–248) becomes

$$\Delta g = -\frac{\partial T}{\partial h} + \frac{1}{\gamma} \frac{\partial \gamma}{\partial h} T - \frac{1}{\gamma} \frac{\partial \gamma}{\partial h} \delta W.$$ (2–349)

Therefore, the fundamental boundary condition is now

$$-\frac{\partial T}{\partial h} + \frac{1}{\gamma} \frac{\partial \gamma}{\partial h} T = \Delta g + \frac{1}{\gamma} \frac{\partial \gamma}{\partial h} \delta W.$$ (2–350)

The spherical approximations of these equations are

$$N = \frac{T - \delta W}{\gamma_0}$$ (2–351)

and

$$\Delta g = -\frac{\partial T}{\partial r} - \frac{2}{R} T + \frac{2}{R} \delta W$$ (2–352)

and

$$-\frac{\partial T}{\partial r} - \frac{2}{R} T = \Delta g - \frac{2}{R} \delta W.$$ (2–353)

Relations between T, N, and Δg

By (2–347), we have

$$T = \gamma_0 \, N + \delta W \, . \tag{2–354}$$

Substituting this into (2–333) and dividing by γ_0, we obtain

$$N = \frac{G \, \delta M}{R \, \gamma_0} - \frac{\delta W}{\gamma_0} + \frac{R}{4\pi \, \gamma_0} \iint\limits_{\sigma} \Delta g \, S(\psi) \, d\sigma \, . \tag{2–355}$$

This is the generalization of Stokes' formula for N. It applies for an arbitrary reference ellipsoid whose center coincides with the center of the earth.

While formula (2–333) for T contains only the effect of a mass difference δM, the formula (2–355) for N contains, in addition, the potential difference δW. These formulas also show clearly that the simple Stokes integrals (2–304) and (2–307) hold only if $\delta M = \delta W = 0$, that is, if the reference ellipsoid has the same potential as the geoid and the same mass as the earth. Otherwise, they give N and T only up to additive constants: putting

$$N_0 = \frac{G \, \delta M}{R \, \gamma_0} - \frac{\delta W}{\gamma_0} \tag{2–356}$$

and taking into account (2–331), we have

$$T = T_0 + \frac{R}{4\pi} \iint\limits_{\sigma} \Delta g \, S(\psi) \, d\sigma \, , \tag{2–357}$$

$$N = N_0 + \frac{R}{4\pi \, \gamma_0} \iint\limits_{\sigma} \Delta g \, S(\psi) \, d\sigma \, . \tag{2–358}$$

Alternative forms of (2–355), which are sometimes useful, are obtained in the following way. Substituting the series (2–268) and (2–270) into (2–352), we get

$$\Delta g(\vartheta, \lambda) = \frac{1}{R} \sum_{n=0}^{\infty} (n-1) \, T_n(\vartheta, \lambda) + \frac{2}{R} \, \delta W \tag{2–359}$$

as the generalization of (2–273). Expanding the function $\Delta g(\vartheta, \lambda)$ into the usual series of Laplace surface spherical harmonics,

$$\Delta g(\vartheta, \lambda) = \sum_{n=0}^{\infty} \Delta g_n(\vartheta, \lambda) \, , \tag{2–360}$$

and comparing the constant terms ($n = 0$) of these two equations, we get

$$-\frac{1}{R} \, T_0 + \frac{2}{R} \, \delta W = \Delta g_0 \, , \tag{2–361}$$

where, by (1–89),

$$\Delta g_0 = \frac{1}{4\pi} \iint_\sigma \Delta g \, d\sigma \, . \tag{2–362}$$

Expressing T_0 by (2–331) in terms of δM, we obtain

$$\Delta g_0 = -\frac{1}{R^2} G \, \delta M + \frac{2}{R} \delta W \, . \tag{2–363}$$

The two equations (2–356) for N_0 and (2–363) for Δg_0 can now be solved for δM and δW:

$$G \, \delta M = R \left(R \, \Delta g_0 + 2\gamma_0 \, N_0 \right) ,$$
$$\delta W = R \, \Delta g_0 + \gamma_0 \, N_0 \, . \tag{2–364}$$

The constant N_0 may be expressed by either of these equations:

$$N_0 = -\frac{R}{2\gamma_0} \Delta g_0 + \frac{G \, \delta M}{2\gamma_0 \, R} \, ,$$
$$N_0 = -\frac{R}{\gamma_0} \Delta g_0 + \frac{\delta W}{\gamma_0} \, . \tag{2–365}$$

A final note

A direct consequence of Eq. (2–356) is that N_0 has an immediate geometrical meaning: if a is the equatorial radius (semimajor axis) of the given reference ellipsoid, then

$$a_E = a + N_0 \tag{2–366}$$

is the equatorial radius of an ellipsoid whose normal potential U_0 is equal to the actual potential W_0 of the geoid, and which encloses the same mass as that of the earth, the flattening f remaining the same. The reason is that for such a new ellipsoid E the new $N_0 = 0$ by (2–356) with $\delta M = 0$ and $\delta W = 0$.

A small *additive* constant N_0 is equivalent to a change of *scale* for a nearly spherical earth. To see this, imagine a nearly spherical orange. Increasing the thickness of the peel of an orange everywhere by 1 mm (say) is equivalent to a similarity transformation (uniform increase of the size) of the orange's surface.

So, the usual Stokes formula, without N_0, gives a global geoid that is determined *only up to the scale* which implicitly is contained in N_0. It is, however, *geocentric*, at least in theory, because it contains no spherical harmonic of first degree, $T_1(\vartheta, \lambda)$. It would be exactly geocentric if the earth were covered uniformly by gravity measurements. The scale was formerly determined astrogeodetically, historically by grade measurements dating back to the 18th century (Clairaut, Maupertuis; see Todhunter [1873]). Today, the scale is furnished by satellites (laser, GPS).

2.18 Gravity disturbances and Koch's formula

It is easy to find *Koch's formula*, which is the alternative of Stokes' formula for gravity disturbances δg. We just indicate the road in its general outlines, leaving the reader to generate a four-lane highway.

Compare equations (2–269) and (2–270) with (2–272) and (2–273). We see that the main difference between gravity disturbances δg and gravity anomalies Δg is the spherical harmonic factor $n+1$ and $n-1$, respectively. The other – very small – difference is that we omit in Δg the terms $n = 0$ and 1 (see comment after (2–273)), which is not necessary in δg.

Using almost literally the development of Sect. 2.14, we get an equation for δg which is the exact equivalent of (2–282) for Δg. Following the integration in Sect. 2.15, we get a formula of form (2–302)

$$T(r, \vartheta, \lambda) = \frac{R}{4\pi} \iint_\sigma K(r, \psi)\, \delta g\, d\sigma\,, \qquad (2\text{–}367)$$

and on the sphere $r = R$ we get a formula of form (2–304) which we call *Koch's formula*:

$$T = \frac{R}{4\pi} \iint_\sigma K(\psi)\, \delta g\, d\sigma\,, \qquad (2\text{–}368)$$

where $K(\psi)$ is the Hotine–Koch function

$$K(\psi) = \frac{1}{\sin(\psi/2)} - \ln\left(1 + \frac{1}{\sin(\psi/2)}\right), \qquad (2\text{–}369)$$

which is very similar to the Stokes function (2–305). By Bruns' theorem, we finally get

$$N = \frac{R}{4\pi\gamma_0} \iint_\sigma K(\psi)\, \delta g\, d\sigma\,. \qquad (2\text{–}370)$$

In absolute analogy with (2–326), we have, simply by replacing $n-1$ by $n+1$ and leaving $n = 0$ as the lower limit of the sum,

$$K(\psi) = \sum_{n=0}^\infty \frac{2n+1}{n+1} P_n(\cos\psi) \qquad (2\text{–}371)$$

as the expression of the Koch–Hotine function *in terms of Legendre polynomials* (zonal harmonics). It is really that simple!

A historical remark

This remark is due to Mrs. M. I. Yurkina, Moscow. Mathematically, the above is the solution of Neumann's problem (the second boundary-value problem of potential theory) for the sphere, cf. Sect. 1.13. It is a classical problem of potential theory, with a history of at least 150 years, similarly to Stokes' formula. "Neumann's problem" is named after the mathematician Carl Neumann, who edited his father's (Franz Neumann) lectures from the 1850s (Neumann 1887: see especially p. 275). The external spherical Neumann problem also occurs in Kellogg (1929: p. 247). It is again found in Hotine (1969: pp. 311, 318).

Their basic significance for modern physical geodesy with a known earth surface was recognized and elaborated by Koch (1971). So the present integral formula should perhaps be called F. Neumann–C. Neumann–Kellogg–Hotine–Koch formula. For brevity, we refer to it as *Koch's formula*.

2.19 Deflections of the vertical and formula of Vening Meinesz

Stokes' formula permits the calculation of the geoidal undulations from gravity anomalies. A similar formula for the computation of the deflections of the vertical from gravity anomalies has been given by Vening Meinesz (1928).

Figure 2.18 shows the intersection of geoid and reference ellipsoid with a vertical plane of arbitrary azimuth. If ε is the component of the deflection of the vertical in this plane, then

$$dN = -\varepsilon\, ds \tag{2–372}$$

or

$$\varepsilon = -\frac{dN}{ds} \; ; \tag{2–373}$$

the minus sign is a convention, its meaning will be explained later.

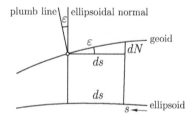

Fig. 2.18. The relation between the geoidal undulation and the deflection of the vertical

In a north-south direction, we have

$$\varepsilon = \xi \quad \text{and} \quad ds = ds_\varphi = R\, d\varphi \, ; \tag{2-374}$$

in an east-west direction,

$$\varepsilon = \eta \quad \text{and} \quad ds = ds_\lambda = R\cos\varphi\, d\lambda \, . \tag{2-375}$$

In the formulas for ds_φ and ds_λ, we have again used the spherical approximation; according to (1–30), the element of arc on the sphere $r = R$ is given by

$$ds^2 = R^2\, d\varphi^2 + R^2 \cos^2\varphi\, d\lambda^2 \, . \tag{2-376}$$

By specializing (2–373), we find

$$\begin{aligned}
\xi &= -\frac{dN}{ds_\varphi} = -\frac{1}{R}\frac{\partial N}{\partial\varphi}\,, \\[2mm]
\eta &= -\frac{dN}{ds_\lambda} = -\frac{1}{R\cos\varphi}\frac{\partial N}{\partial\lambda}\,,
\end{aligned} \tag{2-377}$$

which gives the connection between the geoidal undulation N and the components ξ and η of the deflection of the vertical.

As N is given by Stokes' integral, our problem is to differentiate this formula with respect to φ and λ. For this purpose, we use the form (2–317),

$$N(\varphi,\lambda) = \frac{R}{4\pi\,\gamma_0} \int_{\lambda'=0}^{2\pi} \int_{\varphi'=-\pi/2}^{\pi/2} \Delta g(\varphi',\lambda')\, S(\psi)\cos\varphi'\, d\varphi'\, d\lambda'\,, \tag{2-378}$$

where ψ is defined in (2–318) as a function of φ, λ and φ', λ'.

The integral on the right-hand side of this formula depends on φ and λ only through ψ in $S(\psi)$. Therefore, by differentiating under the integral sign,

$$\frac{\partial N}{\partial\varphi} = \frac{R}{4\pi\,\gamma_0} \int_{\lambda'=0}^{2\pi} \int_{\varphi'=-\pi/2}^{\pi/2} \Delta g(\varphi',\lambda')\, \frac{\partial S(\psi)}{\partial\varphi}\cos\varphi'\, d\varphi'\, d\lambda' \tag{2-379}$$

is obtained and a similar formula for $\partial N/\partial\lambda$. Here we have

$$\frac{\partial S(\psi)}{\partial\varphi} = \frac{dS(\psi)}{d\psi}\frac{\partial\psi}{\partial\varphi}\,, \qquad \frac{\partial S(\psi)}{\partial\lambda} = \frac{dS(\psi)}{d\psi}\frac{\partial\psi}{\partial\lambda}\,. \tag{2-380}$$

Differentiating (2–318) with respect to φ and λ, we obtain

$$\begin{aligned}
-\sin\psi\,\frac{\partial\psi}{\partial\varphi} &= \cos\varphi\sin\varphi' - \sin\varphi\cos\varphi'\,\cos(\lambda'-\lambda)\,, \\[2mm]
-\sin\psi\,\frac{\partial\psi}{\partial\lambda} &= \cos\varphi\cos\varphi'\,\sin(\lambda'-\lambda)\,.
\end{aligned} \tag{2-381}$$

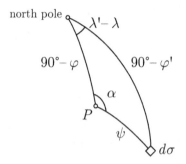

Fig. 2.19. Relation between geographical and polar coordinates on the sphere

We now introduce the azimuth α, as shown in Fig. 2.16. From the spherical triangle of Fig. 2.19 we get, using well-known formulas of spherical trigonometry,

$$
\begin{aligned}
\sin \psi \, \cos \alpha &= \cos \varphi \, \sin \varphi' - \sin \varphi \, \cos \varphi' \, \cos(\lambda' - \lambda) \,, \\
\sin \psi \, \sin \alpha &= \cos \varphi' \, \sin(\lambda' - \lambda) \,.
\end{aligned}
\tag{2–382}
$$

Substituting these relations into the preceding equations, we find the simple expressions

$$
\frac{\partial \psi}{\partial \varphi} = -\cos \alpha \,, \qquad \frac{\partial \psi}{\partial \lambda} = -\cos \varphi \, \sin \alpha \,,
\tag{2–383}
$$

so that

$$
\frac{\partial S(\psi)}{\partial \varphi} = -\frac{dS(\psi)}{d\psi} \cos \alpha \,, \qquad \frac{\partial S(\psi)}{\partial \lambda} = -\frac{dS(\psi)}{d\psi} \cos \varphi \, \sin \alpha \,.
\tag{2–384}
$$

These are substituted into (2–379) and the corresponding formula for $\partial N/\partial \lambda$ and from equations (2–377) we finally obtain

$$
\xi(\varphi, \lambda) = \frac{1}{4\pi \, \gamma_0} \int_{\lambda'=0}^{2\pi} \int_{\varphi'=-\pi/2}^{\pi/2} \Delta g(\varphi', \lambda') \frac{dS(\psi)}{d\psi} \cos \alpha \, \cos \varphi' \, d\varphi' \, d\lambda' \,,
$$

$$
\eta(\varphi, \lambda) = \frac{1}{4\pi \, \gamma_0} \int_{\lambda'=0}^{2\pi} \int_{\varphi'=-\pi/2}^{\pi/2} \Delta g(\varphi', \lambda') \frac{dS(\psi)}{d\psi} \sin \alpha \, \cos \varphi' \, d\varphi' \, d\lambda'
\tag{2–385}
$$

or, written in the usual abbreviated form,

$$
\xi = \frac{1}{4\pi \, \gamma_0} \iint_\sigma \Delta g \, \frac{dS(\psi)}{d\psi} \cos \alpha \, d\sigma \,,
$$

$$
\eta = \frac{1}{4\pi \, \gamma_0} \iint_\sigma \Delta g \, \frac{dS(\psi)}{d\psi} \sin \alpha \, d\sigma \,.
\tag{2–386}
$$

These are the *formulas of Vening Meinesz*. Differentiating Stokes' function $S(\psi)$, Eq. (2–305), with respect to ψ, we obtain *Vening Meinesz' function*

$$
\frac{dS(\psi)}{d\psi} = -\frac{\cos(\psi/2)}{2\sin^2(\psi/2)} + 8\sin\psi - 6\cos(\psi/2) - 3\frac{1 - \sin(\psi/2)}{\sin\psi}
$$

$$
+ 3\sin\psi \, \ln\left[\sin(\psi/2) + \sin^2(\psi/2)\right].
$$

(2–387)

This can be readily verified by using the elementary trigonometric identities. The azimuth α is given by the formula

$$
\tan\alpha = \frac{\cos\varphi' \sin(\lambda' - \lambda)}{\cos\varphi \sin\varphi' - \sin\varphi \cos\varphi' \cos(\lambda' - \lambda)},
$$

(2–388)

which is an immediate consequence of (2–382).

The form (2–385) is an expression of (2–386) in terms of ellipsoidal coordinates φ and λ. As with Stokes' formula (Sect. 2.15), we may also use an expression in terms of spherical polar coordinates ψ and α:

$$
\xi = \frac{1}{4\pi\,\gamma_0} \int_{\alpha=0}^{2\pi} \int_{\psi=0}^{\pi} \Delta g(\psi,\alpha) \, \cos\alpha \, \frac{dS(\psi)}{d\psi} \, \sin\psi \, d\psi \, d\alpha,
$$

$$
\eta = \frac{1}{4\pi\,\gamma_0} \int_{\alpha=0}^{2\pi} \int_{\psi=0}^{\pi} \Delta g(\psi,\alpha) \, \sin\alpha \, \frac{dS(\psi)}{d\psi} \, \sin\psi \, d\psi \, d\alpha.
$$

(2–389)

The reader can easily verify that these equations give the deflection components ξ and η with the correct sign corresponding to the definition (2–230); see also Fig. 2.13. This is the reason why we introduced the minus sign in (2–373).

We note that the formula of Vening Meinesz is valid as it stands for an arbitrary reference ellipsoid, whereas Stokes' formula had to be modified by adding a constant N_0. If we differentiate the modified Stokes formula with respect to φ and λ, to get Vening Meinesz' formula, then this constant N_0 drops out and we get Eqs. (2–386).

For the practical application of Stokes' and Vening Meinesz' formulas and problems, the reader is referred to Sect. 2.21 and to Chap. 3.

2.20 The vertical gradient of gravity

Bruns' formula (2–40), with $\varrho = 0$,

$$
\frac{\partial g}{\partial H} = -2g\,J - 2\omega^2,
$$

(2–390)

cannot be directly applied to determine the gradient $\partial g/\partial H$ because the mean curvature J of the level surfaces is unknown. Therefore, we proceed in the usual way by splitting $\partial g/\partial H$ into a normal and an anomalous part:

$$\frac{\partial g}{\partial H} = \frac{\partial \gamma}{\partial H} + \frac{\partial \Delta g}{\partial H}. \tag{2–391}$$

The normal gradient $\partial \gamma/\partial H$ is given by (2–147) and (2–148). The anomalous part, $\partial \Delta g/\partial H \doteq \partial \Delta g/\partial r$, will be considered now.

Expression in terms of Δg

Equation (2–272) may be written as (note that $r\,\Delta g$ is harmonic and the factor must be 1 for $r = R$)

$$\Delta g(r, \vartheta, \lambda) = \sum_{n=0}^{\infty} \left(\frac{R}{r}\right)^{n+2} \Delta g_n(\vartheta, \lambda). \tag{2–392}$$

By differentiating with respect to r and setting $r = R$, we obtain at sea level:

$$\frac{\partial \Delta g}{\partial r} = -\frac{1}{R} \sum_{n=0}^{\infty} (n+2)\,\Delta g_n = -\frac{1}{R} \sum_{n=0}^{\infty} n\,\Delta g_n - \frac{2}{R}\,\Delta g. \tag{2–393}$$

Now we can apply (1–149), setting $V = \Delta g$ and $Y_n = \Delta g_n$. The result is

$$\frac{\partial \Delta g}{\partial r} = \frac{R^2}{2\pi} \iint_{\sigma} \frac{\Delta g - \Delta g_P}{l_0^3} \, d\sigma - \frac{2}{R}\,\Delta g_P. \tag{2–394}$$

In this equation, Δg_P is referred to the fixed point P at which $\partial \Delta g/\partial r$ is to be computed; l_0 is the spatial distance between the fixed point P and the variable surface element $R^2\, d\sigma$, expressed in terms of the angular distance ψ by

$$l_0 = 2R \sin \frac{\psi}{2}. \tag{2–395}$$

Compare Fig. 1.9 of Sect. 1.14; the element $R^2\, d\sigma$ is at the point P'.

The important integral formula (2–394) expresses the vertical gradient of the gravity anomaly in terms of the gravity anomaly itself. Since the integrand decreases very rapidly with increasing distance l_0, it is sufficient in this formula to extend the integration only over the immediate neighborhood of the point P, as opposed to Stokes' and Vening Meinesz' formulas, where the integration must include the whole earth, if a sufficient accuracy is to be obtained.

Expression in terms of N

By differentiating equation (2–271),

$$\Delta g = -\frac{\partial T}{\partial r} - \frac{2}{r}\,T\,,\qquad(2\text{–}396)$$

with respect to r, we get

$$\frac{\partial \Delta g}{\partial r} = -\frac{\partial^2 T}{\partial r^2} - \frac{2}{r}\frac{\partial T}{\partial r} + \frac{2}{r^2}\,T\,.\qquad(2\text{–}397)$$

To this formula we add Laplace's equation $\Delta T = 0$, which in spherical coordinates has the form

$$\frac{\partial^2 T}{\partial r^2} + \frac{2}{r}\frac{\partial T}{\partial r} - \frac{\tan\varphi}{r^2}\frac{\partial T}{\partial \varphi} + \frac{1}{r^2}\frac{\partial^2 T}{\partial \varphi^2} + \frac{1}{r^2\cos^2\varphi}\frac{\partial^2 T}{\partial \lambda^2} = 0\,;\qquad(2\text{–}398)$$

see Eq. (1–35), modify by replacing V by T and substitute $\vartheta = 90 - \varphi$. The result, on setting $r = R$, is

$$\frac{\partial \Delta g}{\partial r} = \frac{2}{R^2}\,T - \frac{\tan\varphi}{R^2}\frac{\partial T}{\partial \varphi} + \frac{1}{R^2}\frac{\partial^2 T}{\partial \varphi^2} + \frac{1}{R^2\cos^2\varphi}\frac{\partial^2 T}{\partial \lambda^2}\,.\qquad(2\text{–}399)$$

Since $T = \gamma_0\,N$, we may also write

$$\frac{\partial \Delta g}{\partial r} = \frac{2\gamma_0}{R^2}\,N - \frac{\gamma_0}{R^2}\tan\varphi\,\frac{\partial N}{\partial \varphi} + \frac{\gamma_0}{R^2}\frac{\partial^2 N}{\partial \varphi^2} + \frac{\gamma_0}{R^2\cos^2\varphi}\frac{\partial^2 N}{\partial \lambda^2}\,,\qquad(2\text{–}400)$$

where γ_0 is a global mean value as usual. This equation expresses the vertical gradient of the gravity anomaly in terms of the geoidal undulation N and its first and second horizontal derivatives. It can be evaluated by numerical differentiation, using a map of the function N. However, it is less suited for practical application than (2–394) because it requires an extremely accurate and detailed local geoidal map, which is hardly ever available; inaccuracies of N are greatly amplified by forming the second derivatives.

Expression in terms of ξ and η

From equations (2–377), we find

$$\frac{\partial N}{\partial \varphi} = -R\xi\,,\qquad \frac{\partial N}{\partial \lambda} = -R\eta\,\cos\varphi\,,\qquad(2\text{–}401)$$

so that

$$\frac{\partial^2 N}{\partial \varphi^2} = -R\frac{\partial \xi}{\partial \varphi}\,,\qquad \frac{\partial^2 N}{\partial \lambda^2} = -R\frac{\partial \eta}{\partial \lambda}\,\cos\varphi\,.\qquad(2\text{–}402)$$

Substituting these relations into (2–400) yields

$$\frac{\partial \Delta g}{\partial r} = \frac{2\gamma_0}{R^2} N + \frac{\gamma_0}{R} \xi \tan \varphi - \frac{\gamma_0}{R} \frac{\partial \xi}{\partial \varphi} - \frac{\gamma_0}{R \cos \varphi} \frac{\partial \eta}{\partial \lambda}. \tag{2–403}$$

Introducing local rectangular coordinates x, y in the tangent plane, we have

$$R \, d\varphi = ds_\varphi = dx \,,$$
$$\tag{2–404}$$
$$R \cos \varphi \, d\lambda = ds_\lambda = dy \,,$$

so that (2–403) becomes

$$\frac{\partial \Delta g}{\partial r} = \frac{2\gamma_0}{R^2} N + \frac{\gamma_0}{R} \xi \tan \varphi - \gamma_0 \left(\frac{\partial \xi}{\partial x} + \frac{\partial \eta}{\partial y} \right). \tag{2–405}$$

The first two terms on the right-hand side can be shown to be very small in comparison to the third term; hence, to a sufficient accuracy

$$\frac{\partial \Delta g}{\partial r} = -\gamma_0 \left(\frac{\partial \xi}{\partial x} + \frac{\partial \eta}{\partial y} \right) \tag{2–406}$$

may be used. These beautiful formulas express the vertical gradient of the gravity anomaly in terms of the horizontal derivatives of the deflection of the vertical. They can again be evaluated by means of numerical differentiation if a map of ξ and η is available. They are somewhat better suited for practical application than (2–400) because only first derivatives are required.

2.21   Practical evaluation of the integral formulas

Integral formulas such as Stokes' and Vening Meinesz' integrals must be evaluated approximately by summations. The surface elements $d\sigma$ are replaced by small but finite compartments q, which are obtained by suitably subdividing the surface of the earth. Two different methods of subdivision are used:

1. Templates (Fig. 2.20). The subdivision is achieved by concentric circles and their radii. The template is placed on a gravity map of the same scale so that the center of the template coincides with the computation point P on the map. The natural coordinates for this purpose are *polar coordinates* ψ, α with origin at P.

2. Grid lines (Fig. 2.21). The subdivision is achieved by the grid lines of some fixed coordinate system, in particular of *ellipsoidal coordinates* φ, λ. They form rectangular *blocks* – for example, of $10' \times 10'$ or $1° \times 1°$. These blocks are also called squares, although they are usually not squares as defined in plane geometry.

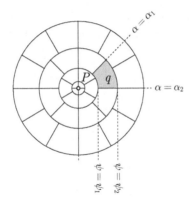

Fig. 2.20. A template

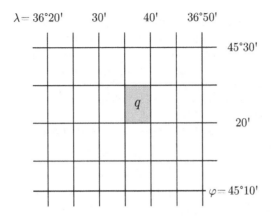

Fig. 2.21. Blocks formed by a grid of ellipsoidal coordinates

The template method is wonderfully easy to understand and to use for theoretical considerations, but completely old-fashioned. Only the gridline method has survived in the computer world.

As a simple and instructive example illustrating the principles of numerical integration consider Stokes' formula

$$N = \frac{R}{4\pi\,\gamma_0} \iint\limits_{\sigma} \Delta g\, S(\psi)\, d\sigma \tag{2–407}$$

with its explicit forms (2–310) for the template method and (2–317) for the method that uses fixed blocks.

For each compartment q_k, the gravity anomalies are replaced by their average value $\overline{\Delta g}_k$ in this compartment. Hence, the above equation becomes

$$N = \frac{R}{4\pi\,\gamma_0} \sum_k \iint\limits_{q_k} \overline{\Delta g}_k\, S(\psi)\, d\sigma = \frac{R}{4\pi\,\gamma_0} \sum_k \overline{\Delta g}_k \iint\limits_{q_k} S(\psi)\, d\sigma \quad (2\text{--}408)$$

or

$$N = \sum_k c_k\, \overline{\Delta g}_k\,, \quad (2\text{--}409)$$

where the coefficients

$$c_k = \frac{R}{4\pi\,\gamma_0} \iint\limits_{q_k} S(\psi)\, d\sigma \quad (2\text{--}410)$$

are obtained by integration over the compartment q_k; they do not depend on Δg.

If the integrand – in our case, Stokes' function $S(\psi)$ – is reasonably constant over the compartment q_k, it may be replaced by its value $S(\psi_k)$ at the center of q_k. Then we have

$$c_k = \frac{R}{4\pi\,\gamma_0} S(\psi) \iint\limits_{q_k} d\sigma = \frac{S(\psi_k)}{4\pi\,\gamma_0\,R} \iint\limits_{q_k} R^2\, d\sigma\,. \quad (2\text{--}411)$$

The final integral is simply the area A_k of the compartment and we obtain

$$c_k = \frac{A_k\, S(\psi_k)}{4\pi\,\gamma_0\,R}\,. \quad (2\text{--}412)$$

The advantage of the template method is its great flexibility. The influence of the compartments near the computation point P is greater than that of the distant ones, and the integrand changes faster in the neighborhood of P. Therefore, a finer subdivision is necessary around P. This can easily be provided by templates. Yet, the method is completely old-fashioned and thus obsolete.

The advantage of the fixed system of blocks formed by a grid of ellipsoidal coordinates lies in the fact that their mean gravity anomalies are needed for many different purposes. These mean anomalies of standard-sized blocks, once they have been determined, can be easily stored and processed by a computer. Also, the same subdivision is used for all computation points, whereas the compartments defined by a template change when the template is moved to the next computation point. The flexibility of the method of standard blocks is limited; however, one may use smaller blocks ($5' \times 5'$, for example) in the neighborhood of P and larger ones ($1° \times 1°$, for example) farther away. With current electronic computation, this method is the only one used in practice. The theoretical usefulness of polar coordinates will be shown now.

Effect of the neighborhood

This issue is interesting and instructive. In the innermost zone, even the template method may pose difficulties if the integrand becomes infinite as $\psi \to 0$. This happens with Stokes' formula, since

$$S(\psi) \doteq \frac{2}{\psi} \tag{2-413}$$

for small ψ. This can be seen from the definition (2–305), because the first term is predominant and is, for small ψ, given by

$$\frac{1}{\sin(\psi/2)} \doteq \frac{1}{(\psi/2)} = \frac{2}{\psi}. \tag{2-414}$$

Vening Meinesz' function becomes infinite as well, since to the same approximation,

$$\frac{dS(\psi)}{d\psi} \doteq -\frac{2}{\psi^2}. \tag{2-415}$$

In the gradient formula (2–394), the integrand

$$\frac{1}{l_0^3} \doteq \frac{1}{R^3 \psi^3} \tag{2-416}$$

behaves in a similar way.

Therefore, it may be convenient to split off the effect of this innermost zone, which will be assumed to be a circle of radius ψ_0 around the computation point. For instance, Stokes' integral becomes in this way

$$N = N_i + N_e, \tag{2-417}$$

where

$$N_i = \frac{R}{4\pi\,\gamma_0} \int_{\alpha=0}^{2\pi} \int_{\psi=0}^{\psi_0} \Delta g\, S(\psi)\, d\sigma,$$

$$N_e = \frac{R}{4\pi\,\gamma_0} \int_{\alpha=0}^{2\pi} \int_{\psi=\psi_0}^{\pi} \Delta g\, S(\psi)\, d\sigma. \tag{2-418}$$

The radius ψ_0 of the inner zone corresponds to a linear distance of a few kilometers. Within this distance, we may treat the sphere as a plane, using polar coordinates s, α, where

$$s \doteq R\psi \doteq R\sin\psi \doteq 2R\sin\frac{\psi}{2}, \tag{2-419}$$

so that the element of area becomes

$$R^2\, d\sigma = s\, ds\, d\alpha. \tag{2-420}$$

It is consistent with this approximation to use (2–413) through (2–416), putting

$$S(\psi) \doteq \frac{2R}{s}, \qquad \frac{dS}{d\psi} \doteq -\frac{2R^2}{s^2}, \qquad \frac{1}{l_0^3} \doteq \frac{1}{s^3}. \qquad (2\text{–}421)$$

In Stokes' and Vening Meinesz' functions as well, the relative error of these approximations is about 1% for $s = 10$ km, and about 3% for $s = 30$ km. In $1/l_0^3$ it is even less. Hence, the effect of this inner zone on our integral formulas becomes

$$N_i = \frac{1}{2\pi \, \gamma_0} \int_{\alpha=0}^{2\pi} \int_{s=0}^{s_0} \frac{\Delta g}{s} \, s \, ds \, d\alpha, \qquad (2\text{–}422)$$

$$\xi_i = -\frac{1}{2\pi \, \gamma_0} \int_{\alpha=0}^{2\pi} \int_{s=0}^{s_0} \frac{\Delta g}{s^2} \cos \alpha \, s \, ds \, d\alpha,$$

$$\eta_i = -\frac{1}{2\pi \, \gamma_0} \int_{\alpha=0}^{2\pi} \int_{s=0}^{s_0} \frac{\Delta g}{s^2} \sin \alpha \, s \, ds \, d\alpha, \qquad (2\text{–}423)$$

$$\left(\frac{\partial \Delta g}{\partial H}\right)_i = \frac{1}{2\pi} \int_{\alpha=0}^{2\pi} \int_{s=0}^{s_0} \frac{\Delta g - \Delta g_P}{s^3} \, s \, ds \, d\alpha. \qquad (2\text{–}424)$$

In order to evaluate these integrals, we expand Δg into a Taylor series at the computation point P:

$$\Delta g = \Delta g_P + x \, g_x + y \, g_y + \frac{1}{2!} \left(x^2 g_{xx} + 2x \, y \, g_{xy} + y^2 g_{yy}\right) + \cdots. \qquad (2\text{–}425)$$

The rectangular coordinates x, y are defined by

$$x = s \cos \alpha, \qquad y = s \sin \alpha, \qquad (2\text{–}426)$$

so that the x-axis points north. We further have

$$g_x = \left(\frac{\partial \Delta g}{\partial x}\right)_P, \qquad g_{xx} = \left(\frac{\partial^2 \Delta g}{\partial x^2}\right)_P, \qquad \text{etc.} \qquad (2\text{–}427)$$

This Taylor series may also be written as

$$\Delta g = \Delta g_P + s \left(g_x \cos \alpha + g_y \sin \alpha\right)$$

$$+ \frac{s^2}{2} \left(g_{xx} \cos^2 \alpha + 2g_{xy} \cos \alpha \sin \alpha + g_{yy} \sin^2 \alpha\right) + \cdots. \qquad (2\text{–}428)$$

Inserting this into the above integrals, we can easily evaluate them. Performing the integration with respect to α first and noting that

$$\int_0^{2\pi} d\alpha = 2\pi \,,$$

$$\int_0^{2\pi} \sin\alpha \, d\alpha = \int_0^{2\pi} \cos\alpha \, d\alpha = \int_0^{2\pi} \sin\alpha \cos\alpha \, d\alpha = 0 \,, \tag{2-429}$$

$$\int_0^{2\pi} \sin^2\alpha \, d\alpha = \int_0^{2\pi} \cos^2\alpha \, d\alpha = \pi \,,$$

we find

$$N_i = \frac{1}{\gamma_0} \int_0^{s_0} \left[\Delta g_P + \frac{s^2}{4}(g_{xx} + g_{yy}) + \cdots \right] ds \,, \tag{2-430}$$

$$\xi_i = -\frac{1}{2\gamma_0} \int_0^{s_0} (g_x + \cdots) \, ds \,,$$

$$\eta_i = -\frac{1}{2\gamma_0} \int_0^{s_0} (g_y + \cdots) \, ds \,, \tag{2-431}$$

$$\left(\frac{\partial \Delta g}{\partial H} \right)_i = \frac{1}{4} \int_{s=0}^{s_0} (g_{xx} + g_{yy} + \cdots) \, ds \,. \tag{2-432}$$

We now perform the integration over s, retaining only the lowest nonvanishing terms. The result is

$$N_i = \frac{s_0}{\gamma_0} \Delta g_P \,, \tag{2-433}$$

$$\xi_i = -\frac{s_0}{2\gamma_0} g_x \,, \quad \eta_i = -\frac{s_0}{2\gamma_0} g_y \,, \tag{2-434}$$

$$\left(\frac{\partial \Delta g}{\partial H} \right)_i = \frac{s_0}{4} (g_{xx} + g_{yy}) \,. \tag{2-435}$$

We see that the effect of the innermost circular zone on Stokes' formula depends, to a first approximation, on the value of Δg at P; the effect on Vening Meinesz' formula depends on the first horizontal derivatives of Δg; and the effect on the vertical gradient depends on the second horizontal derivatives.

Note that the contribution of the innermost zone to the total deflection of the vertical has the same direction as the line of steepest inclination of the "gravity anomaly surface", because the plane vector

$$\vartheta = [\xi_i, \, \eta_i] \tag{2-436}$$

is proportional to the horizontal gradient of Δg,

$$\text{grad } \Delta g = [g_x, \, g_y] \,. \tag{2-437}$$

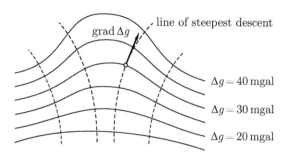

Fig. 2.22. Lines of constant Δg and lines of steepest descent

The direction of grad Δg defines the line of steepest descent (Fig. 2.22). The values of g_x and g_y can be obtained from a gravity map. They are the inclinations of north-south and east-west profiles through P. Values for g_{xx} and g_{yy} may be found by fitting a polynomial in x and y of second degree to the gravity anomaly function in the neighborhood of P.

A remark on accuracy

Deflections of the vertical ξ, η, if combined with astronomical observations of astronomical latitude Φ and astronomical longitude Λ, furnish positions on the reference ellipsoid, expressed by ellipsoidal coordinates

$$\varphi = \Phi - \xi\,,$$
$$\lambda = \Lambda - \eta \sec \varphi\,, \qquad\qquad (2\text{–}438)$$

just as vertical position is obtained by

$$h = H + N\,. \qquad\qquad (2\text{–}439)$$

Unfortunately, to get the same precision for horizontal as for vertical position, is much more difficult, keeping in mind the relation $1'' \cong 30\,\mathrm{m}$ on the earth's surface. So to get an accuracy of $1\,\mathrm{m}$, which is not too difficult with Stokes' formula, means an accuracy better than $0.03''$ in both Φ and ξ (analogously to Λ and η), which is almost impossible to achieve practically.

3 Gravity reduction

3.1 Introduction

Gravity g measured on the physical surface of the earth must be distinguished from normal gravity γ referring to the surface of the ellipsoid. To refer g to sea level, a reduction is necessary. Since there are masses above sea level, the reduction methods differ depending on the way how to deal with these topographic masses. Gravity reduction is essentially the same for gravity anomalies Δg and gravity disturbances δg.

Gravity reduction serves as a tool for three main purposes:

- determination of the geoid,
- interpolation and extrapolation of gravity,
- investigation of the earth's crust.

Only the first two purposes are of a direct geodetic nature. The third is of interest to theoretical geophysicists and geologists, who study the general structure of the crust, and to exploration geophysicists.

The use of Stokes' formula for the determination of the geoid requires that the gravity anomalies Δg represent boundary values at the geoid. This implies two conditions: first, gravity g must refer to the geoid; second, there must be no masses outside the geoid (Sect. 2.12). Hence, figuratively speaking, gravity reduction consists of the following steps:

1. the topographic masses outside the geoid are completely removed or shifted below sea level;
2. then the gravity station is lowered from the earth's surface (point P) to the geoid (point P_0, see Fig. 3.1).

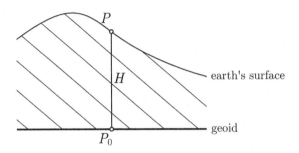

Fig. 3.1. Gravity reduction

The first step requires knowledge of the density of the topographic masses, which is somewhat problematic.

By such a reduction procedure certain irregularities in gravity due to differences in height of the stations are removed so that interpolation and even extrapolation to unobserved areas become easier (Sect. 9.7).

3.2 Auxiliary formulas

Let us compute the potential U and the vertical attraction A of a homogeneous circular cylinder of radius a and height b at a point P situated on its axis at a height c above its base (Fig. 3.2).

P outside cylinder

Assume first that P is above the cylinder, $c > b$. Then the potential is given by the general formula (1–12),

$$U = G \iiint \frac{\varrho}{l}\, dv \,. \tag{3–1}$$

Introducing polar coordinates s, α in the xy-plane by

$$x = s \cos\alpha \,, \quad y = s \sin\alpha \,, \tag{3–2}$$

we have

$$l = \sqrt{s^2 + (c - z)^2} \tag{3–3}$$

and

$$dv = dx\, dy\, dz = s\, ds\, d\alpha\, dz \,. \tag{3–4}$$

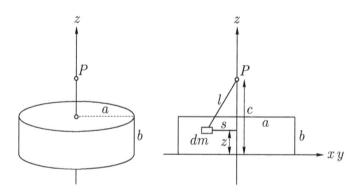

Fig. 3.2. Potential and attraction of a circular cylinder on an external point

Hence, we find, with the density $\varrho = $ constant,

$$U = G\varrho \int_{\alpha=0}^{2\pi} \int_{s=0}^{a} \int_{z=0}^{b} \frac{s \, ds \, dz \, d\alpha}{\sqrt{s^2 + (c-z)^2}}$$

$$= 2\pi \, G\varrho \int_{s=0}^{a} \int_{z=0}^{b} \frac{s \, ds \, dz}{\sqrt{s^2 + (c-z)^2}} \, . \tag{3-5}$$

The integration with respect to s yields

$$\int_{s=0}^{a} \frac{s \, ds}{\sqrt{s^2 + (c-z)^2}} = \sqrt{s^2 + (c-z)^2} \,\Big|_0^a$$

$$= \sqrt{a^2 + (c-z)^2} - c + z \, , \tag{3-6}$$

so that we have

$$U = 2\pi \, G\varrho \int_0^b \left[-c + z + \sqrt{a^2 + (c-z)^2} \right] dz \, . \tag{3-7}$$

The indefinite integral is $2\pi \, G \varrho$ times

$$\tfrac{1}{2}(c-z)^2 - \tfrac{1}{2}(c-z)\sqrt{a^2 + (c-z)^2} - \tfrac{1}{2}a^2 \ln\left[c - z + \sqrt{a^2 + (c-z)^2} \right], \tag{3-8}$$

as may be verified by differentiation. Hence, U finally becomes

$$U_e = \pi \, G \varrho \left\{ (c-b)^2 - c^2 - (c-b)\sqrt{a^2 + (c-b)^2} + c\sqrt{a^2 + c^2} \right.$$

$$\left. - a^2 \ln\left[c - b + \sqrt{a^2 + (c-b)^2} \right] + a^2 \ln\left[c + \sqrt{a^2 + c^2} \right] \right\}, \tag{3-9}$$

where the subscript e denotes that P is external to the cylinder.

The vertical attraction A is the negative derivative of U with respect to the height c [see Eq. (2–22)]:

$$A = -\frac{\partial U}{\partial c} \, . \tag{3-10}$$

Differentiating (3–9), we obtain

$$A_e = 2\pi \, G \varrho \left[b + \sqrt{a^2 + (c-b)^2} - \sqrt{a^2 + c^2} \right]. \tag{3-11}$$

P on cylinder

In this case we have $c = b$, and Eqs. (3–9) and (3–11) become

$$U_0 = \pi G \varrho \left[-b^2 + b\sqrt{a^2 + b^2} + a^2 \ln \frac{b + \sqrt{a^2 + b^2}}{a} \right],\qquad (3\text{–}12)$$

$$A_0 = 2\pi G \varrho \left[a + b - \sqrt{a^2 + b^2} \right].\qquad (3\text{–}13)$$

P inside cylinder

We assume that P is now inside the cylinder, $c < b$. By the plane $z = c$ we separate the cylinder into two parts, 1 and 2 (Fig. 3.3), and compute U as the sum of the contributions of these two parts:

$$U_i = U_1 + U_2,\qquad (3\text{–}14)$$

where the subscript i denotes that P is now inside the cylinder. The term U_1 is given by (3–12) with b replaced by c, and U_2 by the same formula with b replaced by $b - c$. Their sum is

$$U_i = \pi G \varrho \left[-c^2 - (b-c)^2 + c\sqrt{a^2 + c^2} + (b-c)\sqrt{a^2 + (b-c)^2} \right.$$
$$\left. + a^2 \ln \frac{c + \sqrt{a^2 + c^2}}{a} + a^2 \ln \frac{b - c + \sqrt{a^2 + (b-c)^2}}{a} \right].\qquad (3\text{–}15)$$

It is easily seen that the attraction is the difference $A_1 - A_2$:

$$A_i = 2\pi G \varrho \left[2c - b - \sqrt{a^2 + c^2} + \sqrt{a^2 + (b-c)^2} \right];\qquad (3\text{–}16)$$

this formula may also be obtained by differentiating (3–15) according to (3–10).

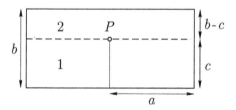

Fig. 3.3. Potential and attraction on an internal point

Circular disk

Let the thickness b of the cylinder go to zero such that the product

$$\kappa = b\,\varrho \qquad (3\text{--}17)$$

remains finite. The quantity κ may then be considered as the surface density with which matter is concentrated on the surface of a circle of radius a. We need potential and attraction for an exterior point. By setting

$$\varrho = \frac{\kappa}{b} \qquad (3\text{--}18)$$

in (3–9) and (3–11) and then letting $b \to 0$, we get by well-known methods of the calculus

$$U_{\mathrm{e}}^{0} = 2\pi\,G\,\kappa\left[\sqrt{a^2 + c^2} - c\right],$$

$$A_{\mathrm{e}}^{0} = 2\pi\,G\,\kappa\left(1 - \frac{c}{\sqrt{a^2 + c^2}}\right). \qquad (3\text{--}19)$$

Sectors and compartments

For a sector of radius a and angle

$$\alpha = \frac{2\pi}{n}, \qquad (3\text{--}20)$$

we must divide the above formulas by n. For a compartment subtending the same angle and bounded by the radii a_1 and a_2 (Fig. 3.4), we get, in an obvious notation,

$$\Delta U = \frac{1}{n}\left[U(a_2) - U(a_1)\right],$$

$$\Delta A = \frac{1}{n}\left[A(a_2) - A(a_1)\right]. \qquad (3\text{--}21)$$

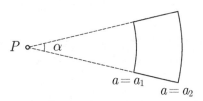

Fig. 3.4. Template compartment

Since A_e and A_i differ only by a constant, this constant drops out in the second equation of (3–21), and we obtain from (3–11) and (3–16)

$$\Delta A_e = \Delta A_i = \frac{2\pi}{n} G \varrho \left[\sqrt{a_2^2 + (c-b)^2} - \sqrt{a_1^2 + (c-b)^2} \right. \tag{3–22}$$
$$\left. - \sqrt{a_2^2 + c^2} + \sqrt{a_1^2 + c^2} \right].$$

On the other hand, $\Delta U_e \neq \Delta U_i$.

Note that we have for didactic reasons purposely used the compartments corresponding to polar coordinates (Fig. 3.4) because they are so simple and instructive, but also still useful for many purposes. For practical computation, rectangular blocks (see Fig. 2.21) are almost exclusively used. For conceptual purposes, however, the polar coordinate template remains invaluable; cf. Sect. 2.21.

3.3   Free-air reduction

For a theoretically correct reduction of gravity to the geoid, we need $\partial g / \partial H$, the vertical gradient of gravity. If g is the observed value at the surface of the earth, then the value g_0 at the geoid may be obtained as a Taylor expansion:

$$g_0 = g - \frac{\partial g}{\partial H} H \cdots, \tag{3–23}$$

where H is the height between P, the gravity station above the geoid, and P_0, the corresponding point on the geoid (Fig. 3.1). Suppose there are no masses above the geoid and neglecting all terms but the linear one, we have

$$g_0 = g + F, \tag{3–24}$$

where

$$F = -\frac{\partial g}{\partial H} H \tag{3–25}$$

is the *free-air reduction* to the geoid. Note that the assumption of no masses above the geoid may be interpreted in the sense that such masses have been mathematically removed beforehand, so that this reduction is indeed carried out "in free air".

For many practical purposes it is sufficient to use instead of $\partial g / \partial H$ the normal gradient of gravity (associated with the ellipsoidal height h) $\partial \gamma / \partial h$, obtaining

$$F \doteq -\frac{\partial \gamma}{\partial h} H \doteq +0.3086 \, H \, \text{[mgal]} \tag{3–26}$$

for H in meters.

3.4 Bouguer reduction

The objective of the Bouguer reduction of gravity is the complete removal
of the topographic masses, that is, the masses outside the geoid.

The Bouguer plate

Assume the area around the gravity station P to be completely flat and
horizontal (Fig. 3.5), and let the masses between the geoid and the earth's
surface have a constant density ϱ. Then the attraction A of this so-called
Bouguer plate is obtained by letting $a \to \infty$ in (3–13), since the plate,
considered plane, may be regarded as a circular cylinder of thickness $b = H$
and infinite radius. By well-known rules of the calculus, we obtain

$$A_B = 2\pi \, G \, \varrho \, H \qquad (3\text{--}27)$$

as the attraction of an infinite Bouguer plate. With standard density $\varrho = 2.67 \text{ g cm}^{-3}$ this becomes

$$A_B = 0.1119 \, H \text{ [mgal]} \qquad (3\text{--}28)$$

for H in meters.

Removing the plate is equivalent to subtracting its attraction (3–27) from
the observed gravity. This is called *incomplete Bouguer reduction*. Note that
this is the usual "plane" Bouguer plate; for a truly "spherical" Bouguer plate
we would have 4π instead of 2π (Moritz 1990: p. 235).

To continue and complete our gravity reduction, we must now apply the
free-air reduction F as given in (3–26). This combined process of removing
the topographic masses and applying the free-air reduction is called *complete
Bouguer reduction*. Its result is Bouguer gravity at the geoid:

$$g_B = g - A_B + F . \qquad (3\text{--}29)$$

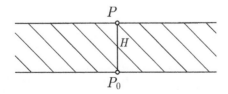

Fig. 3.5. Bouguer plate

With the assumed numerical values, we have

gravity measured at P g
minus Bouguer plate $-0.1119\,H$
plus free-air reduction $+0.3086\,H$ (3–30)

Bouguer gravity at P_0 $g_B = g + 0.1967\,H$.

Since g_B now refers to the geoid, we obtain genuine gravity anomalies in the sense of Sect. 2.12 by subtracting normal gravity γ referred to the ellipsoid:

$$\Delta g_B = g_B - \gamma\,. \qquad (3\text{–}31)$$

They are called *Bouguer anomalies*.

Terrain correction

This simple procedure is refined by taking into account the deviation of the actual topography from the Bouguer plate of P (Fig. 3.6). This is called *terrain correction* or *topographic correction*. At A the mass surplus Δm_+, which attracts upward, is removed, causing g at P to increase. At B the mass deficiency Δm_- is made up, causing g at P to increase again. *The terrain correction is always positive.*

The practical determination of the terrain correction A_t is carried out by means of a template, similar to that shown in Fig. 2.20, using (3–22) and adding the effects of the individual compartments:

$$A_t = \sum \Delta A\,. \qquad (3\text{–}32)$$

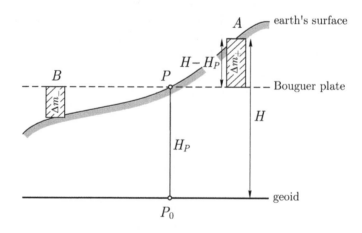

Fig. 3.6. Terrain correction

Again, we can use a template in polar coordinates (Fig. 2.20) for theoretical considerations or a rectangular grid (Fig. 2.21) for numerical computations. For a surplus mass Δm_+, $H > H_p$, we have

$$b = H - H_P, \quad c = 0; \tag{3–33}$$

and for a mass deficiency Δm_-, $H < H_P$,

$$b = c = H_P - H. \tag{3–34}$$

By adding the terrain correction A_t to (3–29), the *refined Bouguer gravity*

$$g_B = g - A_B + A_t + F \tag{3–35}$$

is obtained. The Bouguer reduction and the corresponding Bouguer anomalies Δg_B are called *refined* or *simple*, depending on whether the terrain correction has been applied or not.

In practice it is convenient to separate the Bouguer reduction into the effect of a Bouguer plate and the terrain correction, because the amount of the latter is usually much less. Even for mountains 3000 m in height, the terrain correction is only of the order of 50 mgal (Heiskanen and Vening Meinesz 1958: p. 154).

Unified procedure

It is also possible to compute the total effect of the topographic masses,

$$A_T = A_B - A_t, \tag{3–36}$$

in one step by using columns with base at sea level (Fig. 3.7), again sub-

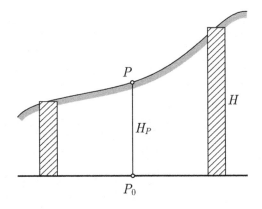

Fig. 3.7. Bouguer reduction

dividing the terrain by means of a template. Note the difference between A_T, the attraction of the topographic masses, and the terrain correction A_t! Then

$$A_T = \sum \Delta A, \qquad (3\text{--}37)$$

where we now have $b = H$, $c = H_P$. Use (3–13) with $b = H_P$ for the innermost circle.

Instead of (3–35), we now have

$$g_B = g - A_T + F. \qquad (3\text{--}38)$$

The Bouguer reduction may be still further refined by the consideration of density anomalies, anomalies in the free-air gradient of gravity (Sect. 2.20), and spherical effects. More computational formulas may be found in Jung (1961: Sect. 6.4).

3.5   Poincaré and Prey reduction

Suppose we need the gravity g' inside the earth. Since g' cannot be measured, it must be computed from the surface gravity. This is done by reducing the measured values of gravity according to the method of Poincaré and Prey.

We denote the point at which g' is to be computed by Q, so that $g' = g_Q$. Let P be the corresponding surface point so that P and Q are situated on the same plumb line (Fig. 3.8). Gravity at P, denoted by g_P, is measured.

The direct way of computing g_Q would be to use the formula

$$g_Q = g_P - \int_Q^P \frac{\partial g}{\partial H}\, dH, \qquad (3\text{--}39)$$

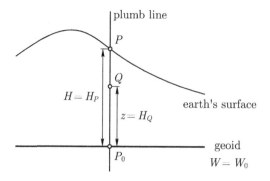

Fig. 3.8. Prey reduction

provided that the actual gravity gradient $\partial g/\partial H$ inside the earth were known. It can be obtained by Bruns' formula (2–40),

$$\frac{\partial g}{\partial H} = -2g\,J + 4\pi\,G\,\varrho - 2\omega^2\,, \tag{3–40}$$

if the mean curvature J of the geopotential surfaces and the density ϱ are known between P and Q.

The normal free-air gradient is given by (2–147):

$$\frac{\partial \gamma}{\partial h} = -2\gamma\,J_0 - 2\omega^2\,, \tag{3–41}$$

where J_0 is the mean curvature of the spheropotential surfaces. If the approximation

$$g\,J \doteq \gamma\,J_0 \tag{3–42}$$

is sufficient, then we get from (3–40) and (3–41)

$$\frac{\partial g}{\partial H} = \frac{\partial \gamma}{\partial h} + 4\pi\,G\,\varrho\,. \tag{3–43}$$

Numerically, neglecting the variation of $\partial \gamma/\partial h$ with latitude, we find for the density $\varrho = 2.67$ g cm^{-3} and (truncated) $G = 6.67 \cdot 10^{-11}$ m^3 kg^{-1} s^{-2}

$$\frac{\partial g}{\partial H} = -0.3086 + 0.2238 = -0.0848 \text{ gal km}^{-1}\,, \tag{3–44}$$

so that (3–39) becomes

$$g_Q = g_P + 0.0848\,(H_P - H_Q) \tag{3–45}$$

with g in gal and H in km. This simple formula, although being rather crude, is often applied in practice.

The accurate way to compute g_Q would be to use (3–39) and (3–40) with the actual mean curvature J of the geopotential surfaces, but this would require knowledge of the detailed shape of these surfaces far beyond what is attainable today.

Another way of computing g_Q, which is more practicable at present, is the following. It is similar to the usual reduction of gravity to sea level (see Sect. 3.4) and consists of three steps:

1. Remove all masses above the geopotential surface $W = W_Q$, which contains Q, and subtract their attraction from g at P.

2. Since the gravity station P is now "in free air", apply the free-air reduction, thus moving the gravity station from P to Q.

3. Restore the removed masses to their former position, and add algebraically their attraction to g at Q.

The purpose of this slightly complicated but logically clear procedure is that in step 2 the *free-air* gradient can be used. If we *here* replace the actual free-air gradient by the normal gradient $\partial\gamma/\partial h$, then the error will presumably be smaller than in using (3–43).

Note that the free-air gradient can also be accurately computed alternatively by (2–394); the gravity anomalies Δg in this formula are the gravity anomalies obtained after performing step 2, that is, Bouguer anomalies referred to the lower point Q.

The effect of the masses above Q (steps 1 and 3) may be computed, e.g., by means of some kind of template or computer procedure for numerical three-dimensional integration. If the terrain correction is neglected and only the infinite Bouguer plate between P and Q of the normal density $\varrho = 2.67 \text{ g cm}^{-3}$ is taken into account, then we obtain with the steps numbered as above:

gravity measured at P	g_P
1. remove Bouguer plate	$-\,0.1119\,(H_P - H_Q)$
2. free-air reduction from P to Q	$+\,0.3086\,(H_P - H_Q)$
3. restore Bouguer plate	$-\,0.1119\,(H_P - H_Q)$
together: gravity at Q	$g_Q = g_P + 0.0848\,(H_P - H_Q)\,.$

$$(3\text{–}46)$$

This is the same as (3–45), which is, thus, confirmed independently. We see now that the use of (3–43) or (3–45) amounts to replacing the terrain with a Bouguer plate.

Finally, we note that the reduction of Poincaré and Prey, abbreviated as *Prey reduction*, yields the actual gravity which would be measured inside the earth if this were possible. Its purpose is, thus, completely different from the purpose of the other gravity reductions which give boundary values at the geoid.

It cannot be directly used for the determination of the geoid but is needed to obtain orthometric heights as will be discussed in Sect. 4.3. Actual gravity g_0 at a geoidal point P_0 is related to Bouguer gravity g_B, Eq. (3–38), by

$$g_0 = g_B - A_{T, P_0}\,. \qquad (3\text{–}47)$$

It is obtained by subtracting from g_B the attraction A_{T, P_0} of the topographic masses on P_0, which corresponds to restoring the topography after the free-air reduction of Bouguer gravity from P to P_0.

3.6 Isostatic reduction

3.6.1 Isostasy

One might be inclined to assume that the topographic masses are simply superposed on an essentially homogeneous crust. If this were the case, the Bouguer reduction would remove the main irregularities of the gravity field so that the Bouguer anomalies would be very small and would fluctuate randomly around zero. However, just the opposite is true. Bouguer anomalies in mountainous areas are systematically negative and may attain large values, increasing in magnitude on the average by 100 mgal per 1000 m of elevation. The only explanation possible is that there is some kind of mass deficiency under the mountains. This means that the topographic masses are *compensated* in some way.

There is a similar effect for the deflections of the vertical. The actual deflections are smaller than the visible topographic masses would suggest. In the middle of the nineteenth century, J.H. Pratt observed such an effect in the Himalayas. At one station in this area he computed a value of $28''$ for the deflection of the vertical from the attraction of the visible masses of the mountains. The value obtained through astrogeodetic measurements was only $5''$. Again, some kind of compensation is needed to account for this discrepancy.

Two different theories for such a compensation were developed at almost exactly the same time, by J.H. Pratt in 1854 and 1859 and by G.B. Airy in 1855. According to Pratt, the mountains have risen from the underground somewhat like a fermenting dough. According to Airy, the mountains are floating on a fluid lava of higher density (somewhat like an iceberg floating on water), so that the higher the mountain, the deeper it sinks.

Pratt–Hayford system

This system of compensation was outlined by Pratt and put into a mathematical form by J.F. Hayford, who used it systematically for geodetic purposes.

The principle is illustrated in Fig. 3.9. Underneath the level of compensation there is uniform density. Above, the mass of each column of the same cross section is equal. Let D be the depth of the level of compensation, reckoned from sea level, and let ϱ_0 be the density of a column of height D. Then the density ϱ of a column of height $D + H$ (H representing the height of the topography) satisfies the equation

$$(D + H)\,\varrho = D\,\varrho_0\,, \qquad (3\text{--}48)$$

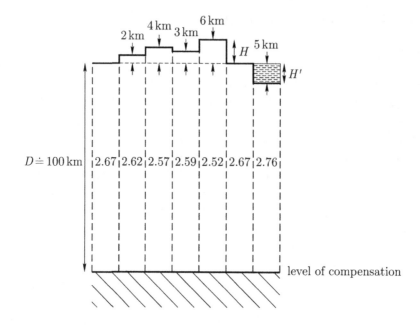

Fig. 3.9. Pratt–Hayford isostasy model

which expresses the condition of equal mass. It may be assumed that

$$\varrho_0 = 2.67 \text{ g cm}^{-3}. \tag{3-49}$$

According to (3–48), the actual density ϱ is slightly smaller than this normal value ϱ_0. Consequently, there is a mass deficiency which, according to (3–48), is given by

$$\Delta\varrho = \varrho_0 - \varrho = \frac{h}{D+H}\varrho_0. \tag{3-50}$$

In the oceans, the condition of equal mass is expressed as

$$(D - H')\varrho + H'\varrho_w = D\varrho_0, \tag{3-51}$$

where

$$\varrho_w = 1.027 \text{ g cm}^{-3} \tag{3-52}$$

is the density and H' the depth of the ocean. Hence, there is a mass surplus of a suboceanic column given by

$$\varrho - \varrho_0 = \frac{H'}{D - H'}(\varrho_0 - \varrho_w). \tag{3-53}$$

As a matter of fact, this model of compensation is idealized and schematic. It can be only approximately fulfilled in nature. Values of the depth of compensation around

$$D = 100 \text{ km} \tag{3-54}$$

are assumed.

For a spheroidal earth, the columns will converge slightly towards its center, and other refinements may be introduced. We may postulate either equality of mass or equality of pressure; each postulate leads to somewhat different spherical refinements. It may be mentioned that for computational reasons Hayford used still another, slightly different model; for instance, he reckoned the depth of compensation D from the earth's surface instead of from sea level.

Airy–Heiskanen system

Airy proposed this model, and Heiskanen gave it a precise formulation for geodetic purposes and applied it extensively. Figure 3.10 illustrates the principle. The mountains of constant density

$$\varrho_0 = 2.67 \text{ g cm}^{-3} \tag{3-55}$$

float on a denser underlayer of constant density

$$\varrho_1 = 3.27 \text{ g cm}^{-3} . \tag{3-56}$$

The higher they are, the deeper they sink. Thus, root formations exist under mountains, and "antiroots" under the oceans.

We denote the density difference $\varrho_1 - \varrho_0$ by $\Delta\varrho$. On the basis of assumed numerical values, we have

$$\Delta\varrho = \varrho_1 - \varrho_0 = 0.6 \text{ g cm}^{-3} . \tag{3-57}$$

Denoting the height of the topography by H and the thickness of the corresponding root by t (Fig. 3.10), then the condition of floating equilibrium is

$$t \, \Delta\varrho = H \, \varrho_0 , \tag{3-58}$$

so that

$$t = \frac{\varrho_0}{\Delta\varrho} H = 4.45 \, H \tag{3-59}$$

results. For the oceans, the corresponding condition is

$$t' \, \Delta\varrho = H' \, (\varrho_0 - \varrho_w) , \tag{3-60}$$

where H' and ϱ_w are defined as above and t' is the thickness of the antiroot (Fig. 3.10), so that we get

$$t' = \frac{\varrho_0 - \varrho_w}{\varrho_1 - \varrho_0} H' = 2.73 \, H' \tag{3-61}$$

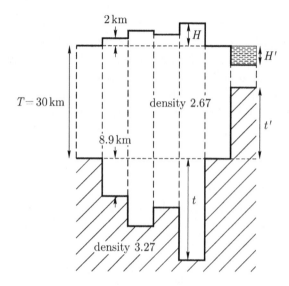

Fig. 3.10. Airy–Heiskanen isostasy model

for the numerical values assumed.

Again, spherical corrections must be applied to these formulas for higher accuracy, and the formulations in terms of equal mass and equal pressure lead to slightly different results.

The normal thickness of the earth's crust is denoted by T (Fig. 3.10); values of around

$$T = 30\,\text{km} \tag{3–62}$$

are assumed. The crustal thickness under mountains is then

$$T + H + t \tag{3–63}$$

and under the oceans it is

$$T - H' - t' . \tag{3–64}$$

Vening Meinesz regional system

Both systems just discussed are highly idealized in that they assume the compensation to be strictly local; that is, they assume that compensation takes place along vertical columns. This presupposes free mobility of the masses to a degree that is obviously unrealistic in this strict form.

For this reason, Vening Meinesz modified the Airy floating theory in 1931, introducing regional instead of local compensation. The principal difference between these two kinds of compensation is illustrated by Fig. 3.11. In Vening

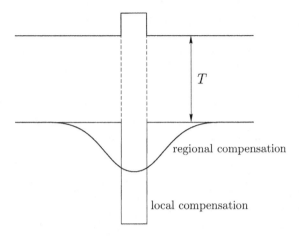

Fig. 3.11. Local and regional compensation

Meinesz' theory, the topography is considered as a load on an unbroken but yielding elastic crust.

In a very sloppy way which is only good for memorizing, we may say that, standing on thin ice, Airy will break through, but under Vening Meinesz the ice is stronger and will bend but not break.

Although Vening Meinesz' refinement of Airy's theory is more realistic, it is more complicated and is, therefore, seldom used by geodesists because, as we will see, any isostatic system, if consistently applied, serves for geodetic purposes as well.

Geophysical and geodetic evidence shows that the earth is about 90% isostatically compensated, but it is difficult to decide, at least from gravimetric evidence alone, which model best accounts for this compensation. Although seismic results indicate an Airy type of compensation, in some places the compensation seems to follow the Pratt model. Nature will never conform to any of these models to the degree of precision which we have assumed above. However, a well-defined and consistent mathematical formulation is certainly a necessary prerequisite for the application of isostasy for geodetic purposes.

For an extensive presentation of several types of isostasy, see Moritz (1990: Chap. 8). The Vening Meinesz model has been treated in detail by Abd-Elmotaal (1995); much information is also available in the internet. A classic on isostasy and its geophysical applications is Heiskanen and Vening Meinesz (1958: Chaps. 5 and 7).

3.6.2 Topographic-isostatic reductions

The objective of the topographic-isostatic reduction of gravity is the *regularization* of the earth's crust according to some model of isostasy. Regularization here means that we are trying to make the earth's crust as homogeneous as possible. The topographic masses are not completely removed as in the Bouguer reduction but are shifted into the interior of the geoid in order to make up the mass deficiencies that exist under the continents. In the topographic-isostatic model of Pratt and Hayford, the topographic masses are distributed between the level of compensation and sea level, in order to bring the crustal density from its original value to the constant standard value ϱ_0. In the Airy–Heiskanen model, the topographic masses are used to fill the roots of the continents, bringing the density from $\varrho_0 = 2.67 \text{ g/cm}^3$ to $\varrho_1 = 3.27 \text{ g/cm}^3$.

In other terms, the topography is removed together with its compensation, and the final result is ideally a homogeneous crust of density ϱ_0 and constant thickness D (Pratt–Hayford) or T (Airy–Heiskanen).

Thus we have three steps:

1. removal of topography,
2. removal of compensation,
3. free-air reduction to the geoid.

Steps 1 and 3 are known from Bouguer reduction, so that the techniques of Sect. 3.4 can be applied to them. Step 2 is new and will be discussed now for the two main topographic-isostatic systems.

Pratt–Hayford system
The method is the same as for the terrain correction, Sect. 3.4, Eq. (3–32). The attraction of the (negative) compensation is again computed by

$$A_C = \sum \Delta A, \tag{3–65}$$

where the attraction of a vertical column representing a compartment is given by (3–22) with

$$b = D, \quad c = D + H_P \tag{3–66}$$

and ϱ replaced by the density defect $\Delta \varrho$. If the preceding Bouguer reduction were done with the original density ϱ of the column expressed by

$$\varrho = \frac{D}{D + H} \varrho_0 \tag{3–67}$$

according to (3–48), then $\Delta \varrho$ would be given by (3–50).

Usually the Bouguer reduction is performed using a constant density ϱ_0; the density defect $\Delta\varrho$ must then be computed by

$$\Delta\varrho = \frac{H}{D}\,\varrho_0\,, \tag{3–68}$$

which differs slightly from (3–50), in order to restore equality of mass according to

$$(\varrho_0 - \Delta\varrho)\,D + \varrho_0 H = \varrho_0 D\,. \tag{3–69}$$

The first term on the left-hand side represents the mass of the layer between the level of compensation and sea level; the second term represents the mass of the topography, now assumed to have a density ϱ_0.

Airy–Heiskanen system

Again we use

$$A_C = \sum \Delta A\,, \tag{3–70}$$

where b and c in (3–22) are, according to Fig. 3.12, given by

$$b = t\,, \quad c = H_P + T + t\,, \tag{3–71}$$

and ϱ is replaced by $\Delta\varrho = \varrho_1 - \varrho_0 = 0.6 \text{ g/cm}^3$.

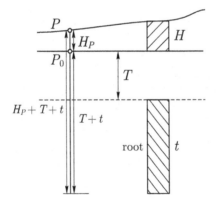

Fig. 3.12. Topography and compensation – Airy–Heiskanen model

Total reduction

In analogy with (3–38), the topographic-isostatically reduced gravity on the geoid becomes

$$g_{\text{TI}} = g - A_T + A_C + F\,, \tag{3–72}$$

where $-A_C$ is the attraction of the compensation which is actually negative, so that its removal is equivalent to the term $+A_C$. The quantity A_T is the attraction of topography, to be computed as the effect of a Bouguer plate combined with terrain correction, Eq. (3–36), or in one step, as described in Sect. 3.4; F is the free-air reduction approximated by (3–26).

Oceanic stations

Here the terms A_T and F of (3–72) are zero, since the station is situated on the geoid, but the term A_C is more complicated.

In the Pratt–Hayford model, the procedure is as follows. The mass surplus (3–53) of a suboceanic column of height $D - H'$ (Fig. 3.9) is removed and used to fill the corresponding oceanic column of height H' to the proper density ϱ_0. In mathematical terms, this is

$$A_C = -A_1 + A_2 \,, \tag{3–73}$$

where both A_1 and A_2 are of the form (3–32), ΔA is given by (3–22). For A_1 we have

$$b = D - H' \,, \quad c = D \,, \tag{3–74}$$

and density $\varrho - \varrho_0$; for A_2 we have

$$b = c = H' \tag{3–75}$$

and density $\varrho_0 - \varrho_w$.

In the Airy–Heiskanen model, the mass surplus of the antiroot, $\varrho_1 - \varrho_0$, is used to fill the oceans to the proper density ϱ_0. The corresponding value is again given by (3–73), where for A_1 we now have

$$b = t' \,, \quad c = T \,, \tag{3–76}$$

and density $\varrho_1 - \varrho_0$; and for A_2 we have, as before,

$$b = c = H' \tag{3–77}$$

and density $\varrho_0 - \varrho_w$.

In both models, Eq. (3–72) reduces for oceanic stations to

$$g_{\text{TI, ocean}} = g + A_C \,. \tag{3–78}$$

Topographic-isostatic anomalies

The topographic-isostatic gravity anomalies are – in analogy to the Bouguer anomalies – defined by

$$\Delta g_{\mathrm{TI}} = g_{\mathrm{TI}} - \gamma \,. \tag{3-79}$$

If any of the topographic-isostatic systems were rigorously true, then the topographic-isostatic reduction would fulfil perfectly its goal of complete regularization of the earth's crust, which would become level and homogeneous. Then, with a properly chosen reference model for γ, the topographic-isostatic gravity anomalies (3–79) would be zero.

The actual topographic-isostatic compensation occurring in nature cannot completely conform to such abstract models. As a consequence, nonzero topo-graphic-isostatic gravity anomalies will be left, but they will be small, smooth, and more or less randomly positive and negative. On account of this smoothness and independence of elevation, they are better suited for interpolation or extrapolation than any other type of anomalies; see Chap. 9, particularly Sect. 9.7.

It may be stressed again that for geodetic purposes the topographic-isostatic model used must be mathematically precise and self-consistent, and the same model must be used throughout. Refinements include the consideration of irregularities of density of the topographic masses and the consideration of the anomalous gradient of gravity.

3.7   The indirect effect

The removal or shifting of masses underlying the gravity reductions change the gravity potential and, hence, the geoid. This change of the geoid is an *indirect effect* of the gravity reductions.

Thus, the surface computed by Stokes' formula from topographic-isostatic gravity anomalies, is not the geoid itself but a slightly different surface, the cogeoid. To every gravity reduction there corresponds a different cogeoid.

Let the undulation of the cogeoid be N^c. Then the undulation N of the actual geoid is obtained from

$$N = N^c + \delta N \tag{3-80}$$

by taking into account the indirect effect on N, which is given by

$$\delta N = \frac{\delta W}{\gamma} \,, \tag{3-81}$$

where δW is the change of potential at the geoid. Equation (3–81) is an application of Bruns' theorem (2–237).

The change of potential, δW, is for the Bouguer reduction expressed by

$$\delta W_B = U_T \tag{3-82}$$

and for the topographic-isostatic reduction by

$$\delta W_{\mathrm{TI}} = U_T - U_C\,, \tag{3-83}$$

the subscripts of the potential U corresponding to those of the attraction A used in the preceding sections.

For the practical determination of U_T and U_C, the template technique, as expressed in (3–32), may again be used (at least, conceptually):

$$U = \sum \Delta U\,, \tag{3-84}$$

where the relevant formulas are the first equation of (3–21), (3–9), (3–12), and (3–15). The point U refers to is always the point P_0 at sea level (Fig. 3.1). For U_T we use U_0, see (3–12), with $b = H$ and density ϱ_0 (see Fig. 3.12). For U_C in the continental case, we use U_e, see (3–9), with the following values: Pratt–Hayford,

$$b = c = d\,, \quad \text{density } \frac{H}{D}\,\varrho_0\,; \tag{3-85}$$

Airy–Heiskanen,

$$b = t\,, \quad c = t + T\,, \quad \text{density } \varrho_1 - \varrho_0\,. \tag{3-86}$$

The corresponding considerations for the oceanic case are left as an exercise for the reader.

The indirect effect with Bouguer anomalies is very large, of the order of ten times the geoidal undulation itself. See the map at the end of Helmert (1884: Tafel I), where the maximum value is 440 m! The reason is that the earth is in general topographic-isostatically compensated. Therefore, the Bouguer anomalies cannot be used for the determination of the geoid.

With topographic-isostatic gravity anomalies, as might be expected, the indirect effect is smaller than N, of the order of 10 m. It is necessary, however, to compute the indirect effect δN_I carefully, using exactly the same topographic-isostatic model as for the gravity reductions.

Furthermore, before applying Stokes' formula, the topographic-isostatic gravity anomalies must be reduced from the geoid to the cogeoid. This is done by a simple free-air reduction, using (3–26), by adding to Δg_I the correction

$$\delta = +0.3086\,\delta N \text{ [mgal]}\,, \tag{3-87}$$

δN in meters. This correction δ is the *indirect effect on gravity*; it is of the order of 3 mgal.

Now the topographic-isostatic gravity anomalies refer strictly to the cogeoid. The application of Stokes' formula gives N^c, which according to (3–80) is to be corrected by the indirect effect δN to give the undulation N of the actual geoid.

Deflections of the vertical

The indirect effect on the deflections of the vertical is, in agreement with Eqs. (2–377), given by

$$\delta \xi = -\frac{1}{R} \frac{\partial \, \delta N}{\partial \varphi} \,,$$

$$\delta \eta = -\frac{1}{R \cos \varphi} \frac{\partial \, \delta N}{\partial \lambda} \,. \tag{3–88}$$

The indirect effect is essentially identical with the so-called topographic-isostatic deflection of the vertical (Heiskanen and Vening Meinesz 1958: pp. 252–255).

The *topographic-isostatic reduction* as such is very much alive, however. *It is practically the only gravity reduction used for geoid determination at the present time* (with the possible exception of free-air reduction, which is a case by itself).

The last purely gravimetric geoid, before the advent of satellites, was the Columbus Geoid (Heiskanen 1957).

3.8   The inversion reduction of Rudzki

It is possible to find a gravity reduction where the indirect effect is zero. This is done by shifting the topographic masses into the interior of the geoid in such a way that

$$U_C = U_T \,. \tag{3–89}$$

Then

$$\delta W = U_T - U_C = 0 \,. \tag{3–90}$$

This procedure was given by M. P. Rudzki in 1905. For the present purpose, we may consider the geoid to be a sphere of radius R (Fig. 3.13). Let the mass element dm at Q be replaced by a mass element dm' at a certain point Q' inside the geoid situated on the same radius vector. The potential due to

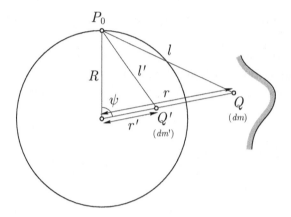

Fig. 3.13. Rudzki reduction as an inversion in a sphere

these mass elements at the geoidal point P_0 is

$$
\begin{aligned}
dU_T &= G\,\frac{dm}{l} = \frac{G\,dm}{\sqrt{r^2 + R^2 - 2R\,r\,\cos\psi}}\,, \\
dU_C &= G\,\frac{dm'}{l'} = \frac{G\,dm'}{\sqrt{r'^2 + R^2 - 2R\,r'\,\cos\psi}}\,.
\end{aligned}
\tag{3-91}
$$

We should have

$$
dU_C = dU_T
\tag{3-92}
$$

if

$$
dm' = \frac{R}{r}\,dm
\tag{3-93}
$$

and

$$
r' = \frac{R^2}{r}\,.
\tag{3-94}
$$

This is readily verified by substitution into the second equation of (3–91). The condition (3–94) means that Q' and Q are related by *inversion in the sphere* of radius R (Kellogg 1929: p. 231). Therefore, this reduction method is called *inversion reduction or Rudzki reduction*.

The condition (3–93) expresses the fact that the compensating mass dm' is not exactly equal to dm but is slightly smaller. Since this relative decrease of mass is of the order of 10^{-8}, it may be safely neglected by setting

$$
dm' = dm\,.
\tag{3-95}
$$

Usually it is even sufficient to replace the sphere by a plane. Then Q' is the ordinary mirror image of Q (Fig. 3.14).

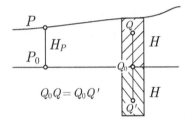

Fig. 3.14. Rudzki reduction as a plane approximation

Rudzki gravity at the geoid becomes, in analogy to (3–72),

$$g_R = g - A_T + A_C + F, \qquad (3\text{–}96)$$

where $A_C = \sum \Delta A$ with $b = H$, $c = H + H_P$, the density being equal to that of topography.

Since the indirect effect is zero, the cogeoid of Rudzki coincides with the actual geoid, but the gravity field outside the earth is changed, which today is in the center of attention. In addition, the Rudzki reduction does not correspond to a geophysically meaningful model. *Nevertheless, it is important conceptually.* Regard it an interesting historic curiosity, but never even consider to use it!

3.9   The condensation reduction of Helmert

Here the topography is condensed so as to form a surface layer (somewhat like a glass sphere made of very thin but very heavy and robust glass) on the geoid so that the total mass remains unchanged. Again, the mass is shifted along the local vertical (Fig. 3.15).

We may consider Helmert's condensation as a limiting case of an isostatic

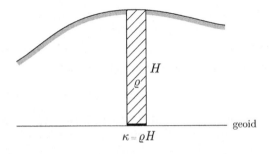

Fig. 3.15. Helmert's method of condensation

reduction according to the Pratt–Hayford system as the depth of compensation D goes to zero. This is sometimes useful.

Again we have

$$g_H = g - A_T + A_C + F, \tag{3-97}$$

where $A_C = \sum \Delta A$ is now to be computed using the second equation of (3–19) with $c = H_P$ and $\kappa = \varrho H$; H_P is the height of the station P and H the height of the compartment.

The indirect effect is

$$\delta W = U_T - U_C. \tag{3-98}$$

The potential $U_C = \sum \Delta U$ is to be computed using the first equation of (3–19) with $\kappa = \varrho H$ as before, but $c = 0$ since it refers to the geoidal point P_0. The corresponding δN is very small, amounting to about $1\,\mathrm{m}$ per $3\,\mathrm{km}$ of average elevation. It may, therefore, usually be neglected so that the cogeoid of the condensation reduction practically coincides with the actual geoid.

Even the "direct effect", $-A_T + A_C$, can usually be neglected, as the attraction of the Helmert layer nearly compensates that of the topography. There remains

$$g_H = g + F, \tag{3-99}$$

that is, the simple free-air reduction. In this sense, *the simple free-air reduction may be considered as giving approximate boundary values at the geoid*, to be used in Stokes' formula. To the same degree of approximation, the "free-air cogeoid" coincides with the actual geoid.

Hence, the free-air anomalies

$$\Delta g_F = g + F - \gamma \tag{3-100}$$

may be considered as approximations of "condensation anomalies"

$$\Delta g_H = g_H - \gamma. \tag{3-101}$$

The many facets of free-air reduction

This is one of the most basic, most difficult, and most fascinating topics of physical geodesy. In fact, the free-air anomaly means several conceptually different but related concepts.

1. The term F above has been seen to be *part of every gravity reduction* rather than a full-fledged gravity reduction itself.

2. Approximately, free-air anomalies may be identified with Helmert's condensation anomalies as we have seen above.

3. *Rigorously*, free-air anomalies can even be considered as resulting from
   a *mass-transporting* gravity reduction, in a similar sense as the isostatic anomaly. Just imagine that you transport the masses above the
   geoid into its interior in such a way that *the external potential remains
   unchanged*! This reminds us of Rudzki's reduction (geoid potential remains constant) but is rather different. The most important advantage
   is that the free-air anomaly in the present sense leaves the external potential unchanged which nowadays is much more important than the
   geoid. The greatest disavantage is that it cannot be computed: we do
   not know how to shift the masses so that the external masses remain
   unchanged. In logical terms, the Rudzki reduction is constructive –
   we are told how to do it –, whereas the present reduction is nonconstructive – we do not know how to do it directly. More about this
   in Sects. 8.2, 8.6, 8.9, and 8.15. We shall, thus, attempt to cut the
   difficult cake into easier pieces.

These are the main methods that have been proposed for the reduction of
gravity. A simple overview is given by Fig. 3.16.

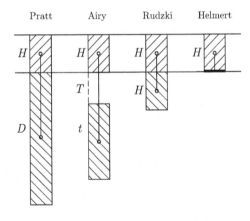

16. Topography and compensation for different gravity reductions

4 Heights

4.1 Spirit leveling

The principle of spirit leveling is well known. To measure the height difference δH_{AB} between two points A and B, vertical rods are set up at each of these two points and a level (leveling instrument) somewhere between them (Fig. 4.1). Since the line $\bar{A}\,\bar{B}$ is horizontal, the difference in the rod readings $l_1 = A\bar{A}$ and $l_2 = B\bar{B}$ is the height difference:

$$\delta H_{AB} = l_1 - l_2 . \tag{4-1}$$

If we measure a circuit, that is, a closed leveling line where we finally return to the initial point, then the algebraic sum of all measured differences in height will not in general be rigorously zero, as one would expect, even if we had been able to observe with perfect precision. This misclosure indicates that leveling is more complicated than it appears at first sight.

Let us look into the matter more closely. Figure 4.2 shows the relevant geometrical principles. Let the points A and B be so far apart that the procedure of Fig. 4.1 must be applied repeatedly. Then the sum of the leveled height differences between A and B will not be equal to the difference in the orthometric heights H_A and H_B. The reason is that the leveling increment δn, as we henceforth denote it, is different from the corresponding increment δH_B of H_B (Fig. 4.2), due to the nonparallelism of the level surfaces. Denoting the corresponding increment of the potential W by δW, we have by (2–21)

$$-\delta W = g\,\delta n = g'\,\delta H_B , \tag{4-2}$$

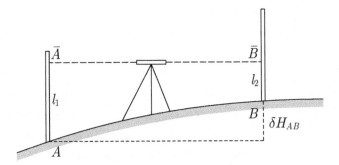

Fig. 4.1. Spirit leveling

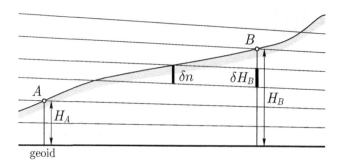

Fig. 4.2. Leveling and orthometric height

where g is the gravity at the leveling station and g' is the gravity on the plumb line of B at δH_B. Hence,

$$\delta H_B = \frac{g}{g'}\,\delta n \neq \delta n\,. \tag{4-3}$$

There is, thus, no direct geometrical relation between the result of leveling and the orthometric height, since (4–3) expresses a physical relation. What, then, if not height, is directly obtained by leveling? If gravity g is also measured, then

$$\delta W = -g\,\delta n \tag{4-4}$$

is determined, so that we obtain

$$W_B - W_A = -\sum_A^B g\,\delta n\,. \tag{4-5}$$

Thus, leveling combined with gravity measurements furnishes *potential differences*, that is, physical quantities.

It is somewhat more rigorous theoretically to replace the sum in (4–5) by an integral, obtaining

$$W_B - W_A = -\int_A^B g\,dn\,. \tag{4-6}$$

Note that this integral is independent of the path of integration; that is, different leveling lines connecting the points A and B (Fig. 4.3) should give the same result. This is evident because W is a function of position only; therefore, to every point there corresponds a unique value W. If the leveling line returns to A, then the total integral must be zero:

$$\oint g\,dn = -W_A + W_A = 0\,. \tag{4-7}$$

Fig. 4.3. Two different leveling lines connecting A and B (taken together, they form a circuit)

The symbol $\oint$ denotes an integral over a circuit.

On the other hand, the measured height difference, that is, the sum of the leveling increments

$$\Delta n_{AB} = \sum_{A}^{B} \delta n = \int_{A}^{B} dn, \qquad (4\text{–}8)$$

depends on the path of integration and is, thus, not in general zero for a circuit:

$$\oint dn = \text{misclosure} \neq 0. \qquad (4\text{–}9)$$

In mathematical terms, dn is not a perfect differential (the differential of a function of position), whereas $dW = -g\,dn$ is perfect, so that dn becomes a perfect differential when it is multiplied by the *integrating factor* $(-g)$.

Thus, potential differences are the result of leveling combined with gravity measurements. They are basic to the whole theory of heights; even orthometric heights must be considered as quantities derived from potential differences. Leveling without gravity measurements, although applied in practice, is meaningless from a rigorous point of view, for the use of leveled heights (4–8) as such leads to contradictions (misclosures); it will not be considered here.

4.2 Geopotential numbers and dynamic heights

Let O be a point at sea level, that is, simplifying speaking, on the geoid; usually a suitable point on the seashore is taken. Let A be another point, connected to O by a leveling line. Then, by formula (4–6), the potential difference between A and O can be determined. The integral

$$\int_{0}^{A} g\,dn = W_0 - W_A = C, \qquad (4\text{–}10)$$

which is the difference between the potential at the geoid and the potential at the point A, has been introduced as the *geopotential number* of A in Sect. 2.4. It is defined so as to be always positive.

As a potential difference, the geopotential number C is independent of the particular leveling line used for relating the point to sea level. It is the same for all points of a level surface; it can, thus, be considered as a *natural measure of height*, even if it does not have the dimension of a length.

The geopotential number C is measured in geopotential units (g.p.u.), where

$$1 \text{ g.p.u.} = 1 \text{ kgal m} = 1000 \text{ gal m.} \tag{4-11}$$

Using $g \doteq 0.98$ kgal in (4-10), we get

$$C \doteq g \, H \doteq 0.98 \, H \,, \tag{4-12}$$

so that the geopotential numbers in g.p.u. are almost equal to the height above sea level in meters.

The geopotential numbers were adopted at a meeting of a Subcommission of the IAG at Florence in 1955. Formerly, the *dynamic heights* were used, defined by

$$H^{\text{dyn}} = \frac{C}{\gamma_0} \,, \tag{4-13}$$

where γ_0 is normal gravity for an arbitrary standard latitude, usually $45°$:

$$\gamma_{45°} = 9.806\,199\,203 \text{ m s}^{-2} = 980.6\,199\,203 \text{ gal} \tag{4-14}$$

for the GRS 1980. Just note and keep in mind that $1 \text{ gal} = 10^{-2} \text{ m s}^{-2}$ and, accordingly, $1 \text{ mgal} = 10^{-5} \text{ m s}^{-2}$.

The dynamic height differs from the geopotential number only in the scale or the unit: The division by the constant γ_0 in (4-13) merely converts a geopotential number into a length. However, the dynamic height has no geometrical meaning whatsoever, so that the division by an arbitrary γ_0 somehow obscures the true physical meaning of a potential difference. Hence, the geopotential numbers are, for reasons of theory and for practically establishing a national or continental height system, preferable to the dynamic heights.

Dynamic correction

It is sometimes convenient to convert the measured height difference Δn_{AB} of (4-8) into a difference of dynamic height by adding a small correction.

Using Eqs. (4-13) and (4-10) gives

$$\Delta H_{AB}^{\text{dyn}} = H_B^{\text{dyn}} - H_A^{\text{dyn}} = \frac{1}{\gamma_0} \left(C_B - C_A \right) = \frac{1}{\gamma_0} \int_A^B g \, dn \,, \tag{4-15}$$

which may be rewritten as

$$\Delta H_{AB}^{\mathrm{dyn}} = \frac{1}{\gamma_0} \int_A^B (g - \gamma_0 + \gamma_0)\, dn = \int_A^B dn + \int_A^B \frac{g - \gamma_0}{\gamma_0}\, dn\,, \qquad (4\text{–}16)$$

so that

$$\Delta H_{AB}^{\mathrm{dyn}} = \Delta n_{AB} + \mathrm{DC}_{AB}\,, \qquad (4\text{–}17)$$

where

$$\mathrm{DC}_{AB} = \int_A^B \frac{g - \gamma_0}{\gamma_0}\, dn \doteq \sum_A^B \frac{g - \gamma_0}{\gamma_0}\, \delta n \qquad (4\text{–}18)$$

is the *dynamic correction*.

As a matter of fact, the dynamic correction may also be used for computing differences of geopotential numbers. We at once obtain

$$C_B - C_A = \gamma_0\, \Delta n_{AB} + \gamma_0\, \mathrm{DC}_{AB}\,. \qquad (4\text{–}19)$$

4.3   Orthometric heights

We denote the intersection of the geoid and the plumb line through point P by P_0 (Fig. 4.4). Let C be the geopotential number of P, that is,

$$C = W_0 - W\,, \qquad (4\text{–}20)$$

and H its orthometric height, that is, the length of the plumb-line segment between P_0 and P. Perform the integration in (4–10) along the plumb line P_0P. This is permitted because the result is independent of the path. We then get

$$C = \int_0^H g\, dH\,. \qquad (4\text{–}21)$$

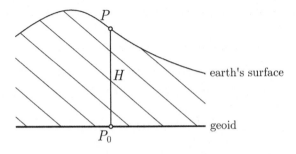

Fig. 4.4. Gravity reduction

This equation contains H in an implicit way. It is also possible to get H explicitly. From

$$dC = -dW = g\,dH\,, \qquad dH = -\frac{dW}{g} = \frac{dC}{g}\,, \tag{4-22}$$

we obtain

$$H = -\int_{W_0}^{W} \frac{dW}{g} = \int_{0}^{C} \frac{dC}{g}\,. \tag{4-23}$$

As before, the integration is extended over the plumb line.

The explicit formula (4–23), however, is of little practical use. It is better to transform (4–21) in a way that at first looks entirely trivial:

$$C = \int_{0}^{H} g\,dH = H \cdot \frac{1}{H} \int_{0}^{H} g\,dH\,, \tag{4-24}$$

so that

$$C = \bar{g}\,H\,, \tag{4-25}$$

where

$$\bar{g} = \frac{1}{H} \int_{0}^{H} g\,dH \tag{4-26}$$

is the mean value of the gravity along the plumb line between the geoid, point P_0, and the surface point P. From (4–25) it follows that

$$H = \frac{C}{\bar{g}}\,, \tag{4-27}$$

which permits H to be computed if the mean gravity $\bar{g}$ is known. Since $\bar{g}$ does not strongly depend on H, Eq. (4–27) is a practically useful formula and not merely a tautology. For determining the mean gravity $\bar{g}$, Eq. (4–26) may be written

$$\bar{g} = \frac{1}{H} \int_{0}^{H} g(z)\,dz\,, \tag{4-28}$$

where $g(z)$ is the actual gravity at the variable point Q which has the height z (Fig. 3.8). The simplest approximation is to use the simplified Prey reduction of (3–45):

$$g(z) = g + 0.0848\,(H - z)\,, \tag{4-29}$$

where g is the gravity measured at the surface point P. The integration (4–28) can now be performed immediately, giving

$$\bar{g} = \frac{1}{H} \int_{0}^{H} \left[g + 0.0848\,(H - z) \right] dz$$

$$= g + \frac{0.0848}{H} \left[H\,z - \frac{z^2}{2} \right]\Big|_{0}^{H} \tag{4-30}$$

or

$$\bar{g} = g + 0.0424\,H \quad (g \text{ in gal, } H \text{ in km}). \tag{4–31}$$

The factor 0.0424 refers to the normal density $\varrho = 2.67$ g/cm^3. The corresponding formula for arbitrary constant density is, by (3–43),

$$\bar{g} = g - \left(\frac{1}{2} \frac{\partial \gamma}{\partial h} + 2\pi\,G\,\varrho \right) H. \tag{4–32}$$

If we use $\bar{g}$ according to (4–31) or (4–32) in the basic formula (4–27), we obtain the so-called Helmert height:

$$H = \frac{C}{g + 0.0424\,H} \tag{4–33}$$

with C in g.p.u., g in gal and H in km.

As we have seen in Sect. 3.5, this approximation replaces the terrain with an infinite Bouguer plate of constant density and of height H. This is often sufficient. Sometimes, in high mountains and for highest precision, it is necessary to apply to g a more rigorous Prey reduction, such as the three steps described in Sect. 3.5. A practical and very accurate method for this purpose has been given by Niethammer in 1932. It takes the topography into account, assuming only that the free-air gradient is normal and the density is constant down to the geoid.

It is also sufficient to calculate $\bar{g}$ as the arithmetic mean of gravity g measured at the surface point P and of gravity g_0 computed at the corresponding geoidal point P_0 by the Prey reduction:

$$\bar{g} = \tfrac{1}{2}\,(g + g_0). \tag{4–34}$$

This presupposes that gravity g varies linearly along the plumb line. This can usually be assumed with sufficient accuracy, even in extreme cases, as shown by Mader (1954) and by Ledersteger (1955).

Orthometric correction

The orthometric correction is added to the measured height difference, in order to convert it into a difference in orthometric height.

We let the leveling line connect two points A and B (Fig. 4.5) and apply a simple trick first:

$$\begin{aligned}
\Delta H_{AB} &= H_B - H_A = H_B - H_A - H_B^{\mathrm{dyn}} + H_A^{\mathrm{dyn}} + (H_B^{\mathrm{dyn}} - H_A^{\mathrm{dyn}}) \\
&= \Delta H_{AB}^{\mathrm{dyn}} + (H_B - H_B^{\mathrm{dyn}}) - (H_A - H_A^{\mathrm{dyn}}).
\end{aligned} \tag{4–35}$$

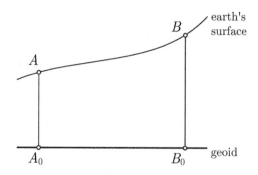

Fig. 4.5. Orthometric and dynamic correction

From (4–17), we have

$$\Delta H_{AB}^{\text{dyn}} = \Delta n_{AB} + DC_{AB}. \tag{4–36}$$

Consider now the differences between the orthometric and dynamic heights, $H_A - H_A^{\text{dyn}}$ and $H_B - H_B^{\text{dyn}}$. Imagine a fictitious leveling line leading from point A_0 at the geoid to the surface point A along the plumb line. Then the measured height difference would be H_A itself: $\Delta n_{A_0 A} = H_A$, so that

$$DC_{A_0 A} = \Delta H_{A_0 A}^{\text{dyn}} - \Delta n_{A_0 A} = H_A^{\text{dyn}} - H_A \tag{4–37}$$

and

$$
\begin{aligned}
H_A - H_A^{\text{dyn}} &= -DC_{A_0 A}, \\
H_B - H_B^{\text{dyn}} &= -DC_{B_0 B}.
\end{aligned}
\tag{4–38}
$$

Substituting (4–36) and (4–38) into (4–35), we finally have

$$\Delta H_{AB} = \Delta n_{AB} + DC_{AB} + DC_{A_0 A} - DC_{B_0 B} \tag{4–39}$$

or

$$\Delta H_{AB} = \Delta n_{AB} + OC_{AB}, \tag{4–40}$$

where

$$OC_{AB} = DC_{AB} + DC_{A_0 A} - DC_{B_0 B} \tag{4–41}$$

is the orthometric correction. This is a remarkable relation between the orthometric and dynamic corrections (Ledersteger 1955). We may write this

$$OC_{AB} = DC_{AB} + DC_{A_0 A} + DC_{B B_0}, \tag{4–42}$$

where we have reversed the sequence of the subscripts of the last term and, consequently, the sign. With $DC_{B_0 A_0} = 0$ (why?), we may write

$$OC_{AB} = DC_{AB} + DC_{B B_0} + DC_{B_0 A_0} + DC_{A_0 A}. \tag{4–43}$$

Accordingly, this may be written

$$OC_{AB} = DC_{ABB_0A_0A}. \tag{4-44}$$

Thus, the orthometric correction from A to B equals the dynamic correction over the loop ABB_0A_0A, a curious, but practically completely useless relation equivalent to (4–41). (*Question:* Why is this independent of γ_0?)

From (4–18), we find

$$DC_{AB} = \int_A^B \frac{g - \gamma_0}{\gamma_0}\, dn = \sum_A^B \frac{g - \gamma_0}{\gamma_0}\, \delta n,$$

$$DC_{A_0A} = \int_{A_0}^A \frac{g - \gamma_0}{\gamma_0}\, dH = \frac{\bar{g}_A - \gamma_0}{\gamma_0}\, H_A, \tag{4-45}$$

$$DC_{B_0B} = \int_{B_0}^B \frac{g - \gamma_0}{\gamma_0}\, dH = \frac{\bar{g}_B - \gamma_0}{\gamma_0}\, H_B,$$

where $\bar{g}_A$ and $\bar{g}_B$ are the mean values of gravity along the plumb lines of A and B. Thus, the orthometric correction (4–41) becomes

$$OC_{AB} = \sum_A^B \frac{g - \gamma_0}{\gamma_0}\, \delta n + \frac{\bar{g}_A - \gamma_0}{\gamma_0}\, H_A - \frac{\bar{g}_B - \gamma_0}{\gamma_0}\, H_B. \tag{4-46}$$

Here again we need the mean values of gravity along the plumb lines, $\bar{g}_A$ and $\bar{g}_B$; γ_0 is an arbitrary constant for which we always take normal gravity at $45°$ latitude.

Accuracy

Let us first evaluate the effect on H of an error in the mean gravity $\bar{g}$. From $H = C/\bar{g}$, we obtain by differentiation

$$\delta H = -\frac{C}{\bar{g}^2}\, \delta\bar{g} = -\frac{H}{\bar{g}}\, \delta\bar{g}. \tag{4-47}$$

Since $\bar{g}$ is about 1000 gal, we have, neglecting the minus sign, the simple formula

$$\delta H_{[mm]} \doteq \delta\bar{g}_{[mgal]}\, H_{[km]}, \tag{4-48}$$

the subscripts denoting the units; δH is the error in H, caused by an error $\delta\bar{g}$ in $\bar{g}$.

For $H = 1$ km,

$$\delta H_{[mm]} \doteq \delta\bar{g}_{[mgal]}, \tag{4-49}$$

which shows that an error $\delta\bar{g}$ in the order of 100 mgal falsifies an elevation of 1000 m by only 10 cm.

Let us now estimate the effect of an error of the density ϱ on $\bar{g}$. Differentiating (4–32) and omitting the minus sign we find

$$\delta\bar{g} = 2\pi\,G\,H\,d\varrho\,. \tag{4–50}$$

If $\delta\varrho = 0.1$ g cm^{-3} and $H = 1$ km, then

$$\delta\bar{g} = 4.2 \text{ mgal}\,, \tag{4–51}$$

which causes an error of 4 mm in H. A density error of 0.6 g/cm^3, which corresponds to the maximum variation of rock density occurring in practice, falsifies $H = 1000$ m by only 25 mm.

Mader (1954) has estimated the difference between the simple computation of mean gravity according to Helmert, Eq. (4–32), and more accurate methods that take the terrain correction into account. He found for Hochtor, in the Alps, $H = 2504$ m:

$$
\begin{array}{lll}
\text{Helmert} & \bar{g} = 980.263 & \text{(Bouguer plate only)},\\[4pt]
\text{Niethammer} & \left.286 \atop \right\} & \\
\bar{g} = \tfrac{1}{2}\,(g + g_0) & 285 & \text{(also terrain correction)}\,.
\end{array}
\tag{4–52}
$$

Mean gravity $\bar{g}$ according to (4–34) differs from Niethammer's value by only 1 mgal, which shows the linearity of g along the plumb line even in an extreme case. This corresponds to a difference in H of 3 mm. The simple Helmert height differs by about 6 cm from these more elaborately computed heights.

Therefore, the differences are very small even in this rather extreme case; we see that orthometric heights can be obtained with very high accuracy. This is of great importance for a discussion of the recent theory of Molodensky from a practical point of view. See Chap. 8, particularly Sect. 8.11.

4.4   Normal heights

Assume for the moment the gravity field of the earth to be normal, that is, $W = U$, $g = \gamma$, $T = 0$. On this assumption compute "orthometric heights"; they will be called *normal heights* and denoted by H^*. Thus, Eqs. (4–21) through (4–26) become

$$W_0 - W = C = \int_0^{H^*} \gamma\,dH^*\,, \tag{4–53}$$

$$H^* = \int_0^C \frac{dC}{\gamma}, \tag{4–54}$$

$$C = \bar{\gamma} H^*, \tag{4–55}$$

where

$$\bar{\gamma} = \frac{1}{H^*} \int_0^{H^*} \gamma \, dH^* \tag{4–56}$$

is the mean normal gravity along the plumb line.

As the normal potential U is a simple analytic function, these formulas can be evaluated straightforwards; but since the potential of the earth is evidently not normal, what does all this mean? Consider a point P on the physical surface of the earth. It has a certain potential W_P and also a certain normal potential U_P, but in general $W_P \neq U_P$. However, there is a certain point Q on the plumb line of P, such that $U_Q = W_P$; that is, the normal potential U at Q is equal to the actual potential W at P. The normal height H^* of P is nothing but the ellipsoidal height of Q above the ellipsoid, just as the orthometric height of P is the height of P above the geoid.

For more details the reader is referred to Sect. 8.3; Fig. 8.2 illustrates the geometric relations.

We now give some practical formulas for the computation of normal heights from geopotential numbers. Writing (4–56) in the form

$$\bar{\gamma} = \frac{1}{H^*} \int_0^{H^*} \gamma(z) \, dz \tag{4–57}$$

corresponding to (4–28), then we can express $\gamma(z)$ by (2–215) as

$$\gamma(z) = \gamma \left[1 - \frac{2}{a} \left(1 + f + m - 2f \sin^2\varphi \right) z + \frac{3}{a^2} z^2 \right], \tag{4–58}$$

where γ is the gravity at the ellipsoid, depending on the latitude φ but not on z. Thus, straightforward integration with respect to z yields

$$\begin{aligned}
\bar{\gamma} &= \frac{1}{H^*} \gamma \left[z - \frac{2}{a} \left(1 + f + m - 2f \sin^2\varphi \right) \frac{z^2}{2} + \frac{3}{a^2} \frac{z^3}{3} \right] \Big|_0^{H^*} \\
&= \frac{1}{H^*} \gamma \left[H^* - \frac{1}{a} \left(1 + f + m - 2f \sin^2\varphi \right) H^{*2} + \frac{1}{a^2} H^{*3} \right]
\end{aligned} \tag{4–59}$$

or

$$\bar{\gamma} = \gamma \left[1 - \left(1 + f + m - 2f \sin^2\varphi \right) \frac{H^{*2}}{a} + \frac{H^{*2}}{a^2} \right]. \tag{4–60}$$

This formula may be used for computing H^* by the formula

$$H^* = \frac{C}{\bar{\gamma}}. \tag{4–61}$$

The mean theoretical gravity itself depends on H^*, by (4–60), but not strongly, so that an iterative solution is very simple.

It is also possible to give a direct expression of H^* in terms of the geopotential number C by substituting (4–60) into (4–61) and expanding into a series of powers of H^*:

$$H^* = \frac{C}{\gamma}\left[1 + \frac{1}{a}\left(1 + f + m - 2f\sin^2\varphi\right)H^* + O(H^{*2})\right]. \qquad (4\text{–}62)$$

Solving this equation for H^* and expanding H^* in powers of C/γ, we obtain

$$H^* = \frac{C}{\gamma}\left[1 + \left(1 + f + m - 2f\sin^2\varphi\right)\frac{C}{a\gamma} + \left(\frac{C}{a\gamma}\right)^2\right], \qquad (4\text{–}63)$$

where γ is normal gravity at the ellipsoid, for the same latitude φ. The accuracy of this formula will be sufficient for almost all practical purposes; still more accurate expressions are given in Hirvonen (1960).

Corresponding to the dynamic and orthometric corrections, there is a *normal correction* NC of the measured height differences. Equation (4–46) immediately yields, on replacing $\bar{g}$ by $\bar{\gamma}$ and H by H^*:

$$\mathrm{NC}_{AB} = \sum_A^B \frac{g - \gamma_0}{\gamma_0}\,\delta n + \frac{\bar{\gamma}_A - \gamma_0}{\gamma_0}H_A^* - \frac{\bar{\gamma}_B - \gamma_0}{\gamma_0}H_B^*, \qquad (4\text{–}64)$$

so that

$$\Delta H_{AB}^* = H_B^* - H_A^* = \Delta n_{AB} + \mathrm{NC}_{AB}. \qquad (4\text{–}65)$$

The normal heights were introduced by Molodensky in connection with his method of determining the physical surface of the earth; see Chap. 8.

4.5   Comparison of different height systems

By means of the geopotential number

$$C = W_0 - W = \int_{\text{geoid}}^{\text{point}} g\, dn, \qquad (4\text{–}66)$$

we can write the different kinds of height in a common form which is very instructive:

$$\text{height} = \frac{C}{G_0}, \qquad (4\text{–}67)$$

where the height systems differ according to how the gravity value G_0 in the denominator is chosen. We have:

dynamic height: $G_0 = \gamma_0 = \text{constant}$,

orthometric height: $G_0 = \bar{g}$,   (4–68)

normal height: $G_0 = \bar{\gamma}$.

It is seen that one can devise an unlimited number of other height systems by selecting G_0 in a different way.

The geopotential number C is, in a way, the most direct result of leveling and is of great scientific importance. However, it is not a height in a geometrical or practical sense. While the dynamic height has at least the dimension of a height, it has no geometrical meaning. One advantage is that points of the same level surface have the same dynamic height; this corresponds to the intuitive feeling that if we move horizontally, we remain at the same height. Note that the orthometric height differs for points of the same level surface because the level surfaces are not parallel. This gives rise to the well-known paradoxes of "water flowing uphill", etc.

The dynamic correction can be very large, because gravity varies from equator to pole by about 5000 mgal. Take, for instance, a leveling line of 1000 m difference of height at the equator, where $g \doteq 978.0$ gal, computed with $\gamma_0 = \gamma_{45°} \doteq 980.6$ gal. Then (4–18) gives a dynamic correction of approximately

$$\text{DC} = \frac{978.0 - 980.6}{980.6} \cdot 1000 \text{ m} = -2.7 \text{ m}.$$ (4–69)

Because of these large corrections, dynamic heights are not suitable as practical heights, and the geopotential numbers are preferable for scientific purposes.

Orthometric heights are the natural "heights above sea level", that is, heights above the geoid. Therefore, they have an unequalled geometrical and physical significance. Their computation is relatively laborious, unless Helmert's simple formula (4–33) is used, which is sufficient in most cases. The orthometric correction is rather small. In the Alpine leveling line of Mader (1954), leading from an elevation of 754 m to 2505 m, the orthometric correction is about 15 cm per 1 km of measured height difference. See also Sect. 8.15.

The physical and geometrical meaning of the *normal heights* is less obvious; they depend on the reference ellipsoid used. Although they are basic in the new theories of physical geodesy, they have a somewhat artificial character as compared to the orthometric heights. They are, however, easy to compute rigorously; the order of magnitude of the normal corrections is about the same as that of the orthometric corrections. In some countries they have replaced the orthometric heights in practice.

For estimates of the difference between orthometric height H and normal height H^*, we refer the reader to Sect. 8.13.

All these height systems based on C are functions of position only. There are, thus, no misclosures, as there are with measured heights. From a purely practical point of view, the desired requirements of a height system are that

1. misclosures be eliminated,
2. corrections to the measured heights be as small as possible.

Empirical height systems have been devised to give smaller corrections than either the orthometric or the normal heights. They have no clear physical significance, however, and are beyond the scope of this book.

Accuracy

Leveling is one of the most accurate geodetic measurements. A standard error of ± 0.1 mm per km distance is possible; it increases with the square root of the distance.

If the error of measurement and interpolation, etc., of gravity is negligible, then the differences in the geopotential number C can be determined with an accuracy of ± 0.1 gal m per km distance; this corresponds to ± 0.1 mm in measured height. Referring to gravity measurements, it is sufficient to measure at distances of some kilometers.

Dynamic heights and normal heights are clearly as accurate as the geopotential numbers, because normal gravity γ is errorless. Orthometric heights, however, are also affected by imperfect knowledge of density, etc., but only slightly; see the end of Sect. 4.3.

Triangulated heights

Historically and for the sake of completeness, the determination of heights by triangulation, that is, by means of zenith angles, should be mentioned.

The main problem is the atmospheric refraction affecting the zenith angles. Thus, the accuracy of triangulated heights is much less than that of leveling. Consequently, triangulated heights are not considered any longer here.

For small distances (e.g., < 1 km), trigonometric height measurements, referred to the local plumb line, have the character of a leveled height difference δn This fact may be used (with care!) to fill small gaps in a leveling network.

Remark on misclosures

All misclosures in any acceptable system of heights denoted for the moment by h (not to be confused with ellipsoidal heights) must be zero:

$$\oint dh = 0 \qquad (4\text{--}70)$$

for any closed path. Height networks consisting of triangles, if computed by least-squares adjustment, thus must satisfy the condition that the sum of height differences must be zero for each triangle. Mathematically, this can be shown to be equivalent to the commutativity of second derivatives:

$$\frac{\partial^2 h}{\partial x\,\partial y} = \frac{\partial^2 h}{\partial y\,\partial x}. \qquad (4\text{--}71)$$

4.6   GPS leveling

Spirit leveling (Fig. 4.6) is a very time-consuming operation. GPS has introduced a revolution also here. The basic equation is

$$H = h - N. \qquad (4\text{--}72)$$

This equation relates the orthometric height H (above the geoid), the ellipsoidal height h (above the ellipsoid), and the geoidal undulation N. If any two of these quantities are measured, then the third quantity can be computed.

If h is measured by GPS, and if there exists a reliable digital geoid map of N, then the orthometric height H can be obtained immediately.

Equation (4–72) can also be used for geoid determination: if h is measured by GPS, and H is available from leveling, then the geoid N can be determined as $N = h - H$. The same principle can be applied even on the oceans as *satellite altimetry*, as we will see later in Chap. 7, e.g., Eq. (7–47).

GPS leveling implies replacing to some extent the classical leveling by GPS. Referring to Fig. 4.6 and applying (4–72) to A and B leads to

$$\begin{aligned} H_A &= h_A - N_A, \\ H_B &= h_B - N_B, \end{aligned} \qquad (4\text{--}73)$$

and the height difference

$$H_B - H_A = h_B - h_A - N_B + N_A. \qquad (4\text{--}74)$$

Introducing the notations $\delta H_{AB} = H_B - H_A$, $\delta h_{AB} = h_B - h_A$, and $\delta N_{AB} = N_B - N_A$, the relation reduces to

$$\delta H_{AB} = \delta h_{AB} - \delta N_{AB}. \qquad (4\text{--}75)$$

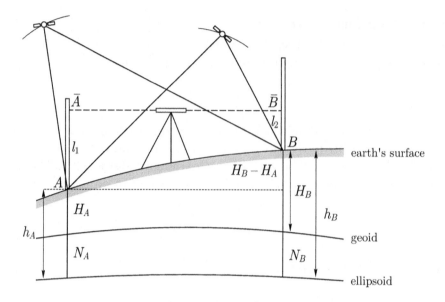

Fig. 4.6. GPS leveling

With GPS leveling, δh_{AB} is obtained, so that with a known geoid, i.e., known δN_{AB}, the orthometric height difference δH_{AB} may be computed according to (4–75). This is a tremendous advantage since otherwise the classical leveling together with gravity measurements is required to determine the orthometric height difference, see Eqs. (4–40) and (4–46).

Note that only the difference of the geoidal undulations impacts the result.

5 The geometry of the earth

5.1 Overview

This chapter consists of three parts.

Part I: Global reference systems after GPS

A fundamental task is to define a global reference system based on a reference ellipsoid which is a good global representation of the earth and which should ideally be characterized by the following properties: Its center coincides with the geocenter (the earth's center of mass), its z-axis represents a suitably defined mean rotation axis of the earth, and the xz-plane is parallel to a mean plane close to the Greenwich meridian. The reference ellipsoid itself is defined to be an ellipsoid of revolution that globally approximates the geoid best in some global sense.

Actually, such a geometric or physical definition cannot be absolutely accurately and unambiguously realized; the final definition will always contain an arbitrary conventional element.

To make things even more complicated, the earth is not a completely rigid body. It can (again approximately!) be regarded as an elastic body with a liquid core. It undergoes small more or less periodic changes. So it must be referred to a mean ellipsoid that does not change with time.

All this will be taken for granted in the present introductory treatment. We shall assume a well-defined geocentric reference ellipsoid with rigid dimensions, a fixed origin, and a time-invariant orientation – close to reality but, in principle, conventionally adopted. For temporal changes in the earth's body and rotation, the reader may be referred to Moritz and Mueller (1987).

Before the advent of satellite geodesy, a geocentric reference system could not be realized. Thus, we had to work with a local geodetic system displaced with respect to the geocenter by an unknown amount on the order of up to a few hundred meters. Therefore, we must take into account a translation (parallel shift) of the local reference ellipsoid with respect to a geocentric system. This implies three translation parameters.

Note that "local" here is used in the sense of "regional", i.e., for a country, territory, or region, in contrast to "global".

Usually, the orientation of a local reference system is accurately known since the direction of the xyz-axes was accessible by astronomical measurements quite accurately at least for the last two centuries. Thus, the orientation of a local geodetic datum is known to the order of $0.1''$ (arc seconds).

Today, we can readily determine the deviation of a local system or *local datum* from a global reference system. We have the deviation of

- size and shape of the reference ellipsoid (a, f),
- translation (x_0, y_0, z_0), and
- orientation (three very small Euler angles $\varepsilon_1, \varepsilon_2, \varepsilon_3$).

Since GPS is very well established (cf. Hofmann-Wellenhof et al. 2001), we assume a general knowledge for granted and recapitulate in this book only some basic facts.

Part II: Three-dimensional geodesy: a transition

This part considers how the concepts of geodesy in the modern sense of Molodensky, Marussi, and Hotine would look shortly before the advent of satellites, but already including electronically measured spatial distances (trilateration). We work with local Cartesian coordinates rotated in a known way by the astronomically measurable quantities Φ, Λ, A (astronomical latitude, longitude, azimuth), considered as Eulerian angles of rotation of the local with respect to the global axes. However, we have no means to determine the geocenter. So the situation is somewhat more complicated but still geometrically well defined and transparent. "Local" here means "strictly local", varying from point to point together with their plumb lines defined by (Φ, Λ).

The main problem with this approach is the impossibility of measuring precise zenith angles because of atmospheric refraction. We may say that the vertical dimension is much worse defined than the horizontal dimension.

Finally, we shall consider how terrestrial and GPS data can be combined.

Part III: Local geodetic datum

The way out of the dilemma of the worse vertical dimension is a complete separation of horizontal and vertical and determining the latter by the differential method of astrogeodetic geoid determination. This was a "2+1-dimensional" rather than a three-dimensional approach, logically more complicated but practically more accurate. In fact, the former (and present) astrogeodetic methods can be understood much better by deriving them from the global situation. Thus, today with GPS we are in a much better position practically as well as theoretically: the classical local datums can be understood best by their relation with the global geometry. "Local geodetic system" or "local geodetic datum" is again meant in the sense of "regional", e.g., the North-American Datum or the European Datum.

GPS permits to separate the geometry from the gravity field, which continues to be a challenge for physical geodesy to be solved by a combination of terrestrial and satellite data.

Part I: Global reference systems after GPS

5.2   Introduction

Geodesy, as the theory of size and shape of the earth, is not a purely geometrical science since the earth's gravity field, a physical entity, is involved in many geodetic measurements, especially terrestrial ones.

The gravimetric methods are usually considered to constitute physical geodesy in the narrower sense. The measurements of triangulation, leveling, and geodetic astronomy, all make essential use of the plumb line, which, being the direction of the gravity vector, is no less physically defined by nature than its magnitude, that is, the gravity g. All determinations of the geoid by various methods and its use as well as the use of deflections of the vertical belong to physical geodesy, quite as well as the gravimetric methods.

Even in the age of GPS, we have many previous geodetic data which continue to be useful and have to be understood in order to be optimally combined with the new satellite data. In precise operations of engineering geodesy such as tunnel surveying, the plumb line and deflections of the vertical must be taken into account.

For an optimal understanding and use of local (or rather regional) geodetic datums, we must know their relation to a global geodetic system as used in GPS. Therefore, it is appropriate to start with global geometry in a rather elementary way.

A few introductory ideas may help in comprehending this subject. To fix the position of a point in space, we need three coordinates. We can use, and have used, a rectangular Cartesian coordinate system. This is the basic geometric coordinate system. It may be easily converted computationally to ellipsoidal coordinates φ, λ, h referred to any given reference ellipsoid.

For many special purposes, however, it is preferable to take what we have called the *natural coordinates*: Φ (astronomical latitude), Λ (astronomical longitude), and H (orthometric height), which directly refer to the gravity field of the earth (Sect. 2.4). The height H may be obtained by geometric leveling, combined with gravity measurements, and Φ and Λ are determined by astronomical measurements.

As long as the geoid can be identified with an ellipsoid, the use of these coordinates for computations is very simple. Since this identification is sufficient only for results of rather low accuracy, the deviations of the geoid from an ellipsoid must be taken into account. As we have seen, the geoid has rather disagreeable mathematical properties. It is a complicated surface with discontinuities of curvature. Thus, it is not suitable as a surface on which to perform mathematical computations directly, as on the ellipsoid.

To repeat, the ellipsoidal coordinates φ, λ, h are defined such as to refer to the ellipsoid exactly as the natural coordinates refer to the geoid, hence their names.

Since the deviations of the geoid from the ellipsoid are small and computable, it is convenient to add small reductions to the original coordinates Φ, Λ, H, so as to get values which refer to an ellipsoid. In this way we shall find in Sect. 5.12:

$$\varphi = \Phi - \xi \,,$$
$$\lambda = \Lambda - \eta \sec \varphi \,, \tag{5–1}$$
$$h = H + N \,;$$

φ and λ are the ellipsoidal coordinates on the ellipsoid, sometimes also called *geodetic latitude* and *geodetic longitude* to distinguish them from the *astronomical latitude* Φ and the *astronomical longitude* Λ. Astronomical and ellipsoidal coordinates differ by the deflection of the vertical (components ξ and η). The quantity h is the *geometric height* above the ellipsoid; it differs from the *orthometric height* H above the geoid by the geoidal undulation N.

Geodetic measurements (angles, distances) are treated similarly. The principle of *triangulation* is well known: historically, distances were obtained indirectly by measuring the angles in a suitable network of triangles; only one baseline was necessary in principle to furnish the scale of the network. Triangulation was indispensable in former times, because angles could be measured much more easily than long distances.

Nowadays, however, long distances can be measured directly just as easily as angles by means of electronic instruments, so that triangulation, using angular measurements, is often replaced or supplemented by *trilateration*, using distance measurements. The computation of triangulations and trilaterations on the ellipsoid is easy. It is, therefore, convenient to reduce the measured angles, baselines, and long distances to the ellipsoid, in much the same way as the astronomical coordinates are treated. Then the ellipsoidal coordinates φ, λ obtained (1) by reducing the astronomical coordinates and (2) by computing triangulations or trilaterations on the ellipsoid can be compared; they should be identical for the same point.

Today, of course, GPS is the best method for determining φ, λ, and h directly.

5.3   The Global Positioning System

The following sections on the Global Positioning System (GPS) are extracted from Hofmann-Wellenhof et al. (2003: Sect. 9.3) which in return is based on

Hofmann-Wellenhof et al. (2001: Chap. 2). For details supplementing the compact description here, the reader is referred to these books.

5.3.1 Basic concept

GPS is the responsibility of the Joint Program Office (JPO), a component of the Space and Missile Center at El Segundo, California. In 1973, the JPO was directed by the U.S. Department of Defense (DOD) to establish, develop, test, acquire, and deploy a spaceborne positioning system. The present navigation system with timing and ranging is the result of this initial directive. GPS was conceived as a ranging system from known positions of satellites in space to unknown positions on land, at sea, in air, and in space. The original objectives of GPS were the instantaneous determination of position and velocity on a continuous basis, and the precise coordination of time (i.e., time transfer).

Based on code or carrier phase measurements, GPS uses pseudoranges derived from the broadcast satellite signal.

Using the code measurements, the pseudorange is derived from measuring the travel time of the coded signal and multiplying it by its velocity. Since the clocks of the receiver and the satellite are never perfectly synchronized, a clock error must be taken into account. Consequently, each equation of this type comprises four unknowns: the three point coordinates contained in the true range and the clock error. Thus, four satellites are necessary to solve for the four unknowns. Indeed, the GPS concept assumes that – without obstruction – four or more satellites are in view at any location on or near the earth 24 hours a day.

Using carrier phase measurements, ambiguities must be taken into account as additional unknowns. For more details see Hofmann-Wellenhof et al. (2001: Sect. 6.1.2).

5.3.2 System architecture

Space segment

Constellation

The GPS satellites have nearly circular orbits with an altitude of about 20200 km above the earth, i.e., they are mean earth orbit (MEO) satellites, yielding a period of nominally 12 sidereal hours. The nominal constellation consists of 24 operational satellites deployed in six evenly spaced planes (A to F) with an inclination of 55° against the equator and with four satellites per plane. Furthermore, active spare satellites for replenishment may be operational. See http://tycho.usno.navy.mil/gpscurr.html for the current status.

With the nominal constellation, the space segment provides global coverage with four to eight simultaneously observable satellites above 15° elevation angle at any time of day. If the elevation mask is reduced to 10°, occasionally up to 10 satellites will be visible; and if the elevation mask is further reduced to 5°, occasionally 12 satellites will be visible.

Satellites categories

Essentially, the GPS satellites provide a platform for radio transceivers, atomic clocks, computers, and various ancillary equipment. The electronic equipment of each satellite allows the user to measure a pseudorange to the satellite, and each satellite broadcasts a message which allows the user to determine the spatial position of the satellite for arbitrary instants. The auxiliary equipment of each satellite, among others, consists of solar panels for power supply and a propulsion system for orbit and stability control.

There are several classes or types of GPS satellites. These are the Block I, Block II, Block IIA, Block IIR, Block IIR-M, and the future Block IIF and Block III satellites. An up-to-date description is difficult because new notations are introduced in a rather arbitrary way; an example is the recently introduced notation Block IIR-M.

Eleven Block I satellites were launched in the period between 1978 to 1985. Today, none of them is in operation anymore.

The essential difference between Block I and Block II satellites is related to U.S. national security. Block I satellite signals were fully available to civilian users. Starting with Block II, satellite signals may be restricted for civilian use. The Block II satellites are equipped with mutual communication capability. Some of them carry retroreflectors and can be tracked by laser ranging.

The Block IIR satellites ("R" denotes replenishment or replacement) have a design life of 10 years. They are equipped with improved facilities for communication and intersatellite tracking. Block IIR-M satellites incorporate two new military signals and a second civil signal. The first Block IIR-M was launched on September 25, 2005.

Currently (April 2006), the first launch of a Block IIF satellite ("F" denotes follow on) is scheduled for 2008 (instead of the previously projected dates mid of 2006 and 2007). These satellites will broadcast a third civil signal on L5 (see Sect 5.3.5).

Presently, the DOD undertakes studies for the next generation of GPS satellites, called Block III satellites. Preliminary dates (likely to change) are 2011/12 for first launches and on-orbit tests (Civil GPS Service Interface

Committee 2002). These satellites will be characterized by an assured and improved level of integrity without the need of augmentation.

Satellite signal

The key to the accuracy of the system is the fact that all signal components are precisely controlled by atomic clocks. These highly accurate frequency standards of GPS satellites produce the fundamental frequency of 10.23 MHz. Coherently derived from this frequency are (presently) two signals in the L-band, the L1 and the L2 carrier waves generated by multiplying the fundamental frequency by 154 and 120, respectively, yielding

$$L1 = 1575.42 \text{ MHz},$$

$$L2 = 1227.60 \text{ MHz}.$$

These dual frequencies are essential for eliminating the major source of error, i.e., the ionospheric refraction.

The pseudoranges that are derived from measured travel times of the signal from each satellite to the receiver use two pseudorandom noise (PRN) codes that are modulated onto the two carriers.

The C/A-code (coarse/acquisition-code) is available for civilian use. Each C/A-code is a unique sequence of 1023 bits, called chips, which is repeated each millisecond. The duration of each C/A-code chip is about $1 \mu s$. Equivalently, the chip length – denoted also as wavelength or chip width (Misra and Enge 2001: Sect. 2.3.1) – is about 300 m. The C/A-code is presently modulated upon L1 only and is purposely omitted from L2. This omission allows the JPO to control the information broadcast by the satellite and, thus, denies full system accuracy to nonmilitary users.

The P-code (precision-code) has been reserved for U.S. military and other authorized users. This is achieved by using the W-code to encrypt the P-code to the Y-code (anti-spoofing). The P-code has an effective chip length of about 30 m. The P-code is modulated on both carriers L1 and L2.

In addition to the PRN codes, a data message is modulated onto the carriers consisting of status information, satellite clock bias, and satellite ephemerides. The orbit data are given as Kepler-like elements and are denoted as broadcast ephemerides. The full set of elements is given in, e.g., Montenbruck and Gill (2001: Sect. A.2.2). It is worth noting that the present signal structure will be improved in the near future (see Sect. 5.3.5).

Control segment

The operational control system (OCS) consists of a master control station, monitor stations, and ground control stations. The main tasks of the OCS are tracking of the satellites for the orbit and clock determination and prediction, time synchronization of the satellites, and upload of the data message to the satellites.

Master control station
The master control station is located at the Consolidated Space Operations Center (CSOC) at Shriver Air Force Base, Colorado Springs, Colorado. CSOC collects the tracking data from the monitor stations and calculates the satellite orbit and clock parameters by a Kalman estimator. These results are then passed to one of the three ground control stations for eventual upload to the satellites. The satellite control and system operation is also the responsibility of the master control station.

Monitor stations
There are five monitor stations located at Hawaii, Colorado Springs, Ascension Island in the South Atlantic Ocean, Diego Garcia in the Indian Ocean, and Kwajalein in the North Pacific Ocean. Each of these stations is equipped with a precise atomic time standard and receivers which continuously measure pseudoranges to all satellites in view. Pseudoranges are measured every 1.5 seconds and, using ionospheric and meteorological data, they are smoothed to produce 15-minute interval data which are transmitted to the master control station.

Ground control stations
These stations collocated with the monitor stations at Ascension, Diego Garcia, and Kwajalein are the communication links to the satellites and mainly consist of the ground antennas. The satellite ephemerides and clock information, calculated at the master control station and received via communication links, are uploaded to each GPS satellite via S-band radio links.

User segment
The diversity of the military and civilian users is matched by the type of receivers available today.

On the basis of the type of observables (i.e., code pseudoranges or phase pseudoranges) and of the availability of codes (i.e., C/A-code, P-code, or Y-code), GPS receivers can be classified. For the majority of navigation applications, C/A-code pseudorange receivers will suffice. With this type of receiver, only code pseudoranges using the C/A-code on L1 are measured. Typical devices output the three-dimensional position either in latitude, longitude, and height or in some map projection systems, e.g., universal transverse Mercator (UTM) coordinates and height.

5.3.3   Satellite signal and observables

Components of the signal

The official description of the GPS signal is given in the GPS Interface Control Document ICD-GPS-200, available at www.navcen.uscg.gov. Details may also be found in Spilker (1996).

The (current) components of the signal are summarized in Table 5.1. Note that the nominal fundamental frequency f_0 is intentionally reduced by about 0.005 Hz to compensate for relativistic effects.

The navigation message essentially contains information about the satellite health status, the satellite clock, the orbit, and various correction data.

The parameters in the block of orbit information are the reference epoch, six parameters to describe a Kepler ellipse at the reference epoch, three secular correction terms and six periodic correction terms.

Observables

In concept, the GPS observables are ranges which are deduced from measured time or phase differences based on a comparison between received signals and receiver-generated signals. As mentioned earlier, the ranges are biased by satellite and receiver clock errors and, consequently, they are denoted as pseudoranges. Essentially, pseudoranges differ from distances by an unknown additive constant.

Apart from the satellite and the receiver clock bias, further error sources can be classified into three groups, i.e., satellite-related errors (e.g., orbital errors), signal propagation medium-related errors (e.g., ionospheric and tropospheric refraction), and receiver-related errors (e.g., antenna phase center variation, multipath), but are omitted in the subsequent simplified models. Extended models are given in Hofmann-Wellenhof et al. (2003: Sect. 10.2.2).

Table 5.1. Components of the satellite signal

Component	Frequency or code chipping rate [MHz]		Wavelength
Fundamental frequency	f_0	$= 10.23$	
Carrier L1	$154\,f_0$	$= 1575.42$	$19.0\,\mathrm{cm}$
Carrier L2	$120\,f_0$	$= 1227.60$	$24.4\,\mathrm{cm}$
P-code	f_0	$= 10.23$	
C/A-code	$f_0/10$	$= 1.023$	
Navigation message	$f_0/204\,600$	$= 50 \cdot 10^{-6}$	

Code pseudoranges
The measured time difference Δt is affected by the satellite clock error δ_S and the receiver clock error δ. The error δ_S of the satellite clock can be modeled by a polynomial with the coefficients being transmitted in the navigation message. Assuming the δ_S correction is applied, the time interval Δt multiplied by the speed of light c yields the code pseudorange R and, hence,

$$R = c\,\Delta t\,.\tag{5-2}$$

Assuming a common time reference for satellite and receiver, e.g., GPS time, the term Δt may be decomposed into the run time $\Delta t(\text{GPS})$ and the receiver clock errors δ leading to

$$R = c\,\Delta t(\text{GPS}) + c\,\delta = \varrho + c\,\delta\,,\tag{5-3}$$

where ϱ is the geometric range between the satellite and the receiver. The receiver module responsible for code pseudorange measurements is denoted as delay lock loop (DLL). Details on the DLL functionality are given in Misra and Enge (2001: Sect. 9.5).

Phase pseudoranges
Assuming again that the satellite clock error correction is applied, the phase pseudorange Φ is modeled by

$$\lambda\,\Phi = \varrho + c\,\delta + \lambda\,N\,,\tag{5-4}$$

where the carrier wavelength λ has been introduced. The range ϱ represents the distance between the satellite at emission epoch t and the receiver at reception epoch $t + \Delta t$. Phase measurements are ambiguous, since the initial integer number N of cycles between satellite and receiver is unknown. As long as the tracking of a satellite is not interrupted, the ambiguity remains constant within the tracking loop of the receiver. The responsible receiver hardware is denoted as phase lock loop (PLL). Compared to (5–3), the phase pseudorange differs from the code pseudorange only by the phase ambiguity term $\lambda\,N$. Dividing the above equation by λ scales the phase to cycles.

As mentioned previously, the majority of navigation applications does not need carrier phase measurements. Only for increased accuracy requirements (e.g., relative positioning; see below), phase measurements become relevant.

Doppler data
Some of the first solution models proposed for GPS were to use the Doppler observable. Considering Eq. (5–4), the equation for the observed Doppler

shift scaled to range rate is given by

$$D = \lambda\,\dot{\Phi} = \dot{\varrho} + c\,\dot{\delta}\,, \qquad\qquad (5\text{--}5)$$

where the derivatives with respect to time are indicated by a dot. The raw Doppler shift is less accurate than integrated Doppler.

The Doppler shift is measured in the carrier tracking loop of a GPS receiver (Misra and Enge 2001: Sect. 9.6). Assuming a known satellite velocity, the Doppler shift can be used to estimate the velocity of the user.

5.3.4 System capabilities and accuracies

Two operational capabilities are distinguished: firstly, the initial operational capability (IOC) and, secondly, the full operational capability (FOC).

IOC was attained in July 1993, when 24 (Block I/II/IIA) GPS satellites were operating and were available for navigation. Officially, IOC was declared by the DOD on December 8, 1993.

FOC was achieved when 24 Block II/IIA satellites were operational in their assigned orbits and the constellation was tested for operational military performance. Even though 24 Block II and Block IIA satellites were available since March 1994, FOC was not declared before July 17, 1995 which indicates an extensive testing phase.

The selection of the GPS observation technique depends upon the particular requirements of the project; especially the desired accuracy plays a dominant role.

Point positioning

When using a single receiver, usually point positioning with code pseudoranges is performed. The concept of point positioning is simple (Fig. 5.2). Without clock errors, trilateration in space (i.e., using three ranges) solves the task to determine the point coordinates. Using pseudoranges, four observations are necessary to account for the three coordinate components and the receiver clock error. For point positioning, GPS provides two levels of service: the standard positioning service (SPS) with access for civilian users and the precise positioning service (PPS) with access for authorized users.

SPS performance standards are based on signal-in-space performance. Contributions of ionosphere, troposphere, receiver, multipath, topography, or interference are not included. Furthermore, SPS is provided on the L1 signal only; the L2 signal is not part of the SPS (Department of Defense 2001). The global average positioning domain accuracy amounts to 13 m horizontal error (95% probability level) and 22 m vertical error (95% probability level).

The PPS has access to both codes and provides accuracies down to the meter level.

Differential GPS

Selective availability (SA), the deliberate degradation of the point positioning accuracy by "dithering" (i.e., distorting on purpose) the satellite clock (called δ-process) and manipulating the ephemerides (called ε-process), has led to the development of differential GPS (DGPS). Only the basic idea is explained here.

DGPS is based on the use of two (or more) receivers, where one (stationary) reference or base receiver is located at a known point and the position of the (mostly moving) remote receiver is to be determined. Using code pseudoranges, at least four common satellites must be tracked simultaneously at both sites. The known position of the reference receiver is used to calculate corrections to the observed pseudoranges. These corrections are then transmitted via telemetry (i.e., controlled radio link) to the roving receiver and allow the computation of the rover position with far more accuracy than for the single-point positioning mode.

Using DGPS based on C/A-code pseudoranges, real-time accuracies at the 1–5 m level can be routinely achieved. Phase-smoothed code ranges yield the submeter level (Lachapelle et al. 1992). Even higher accuracies can be reached by the use of carrier phases (precise DGPS). For ranges up to some 20 km, accuracies at the subdecimeter level can be obtained in real time (De-Loach and Remondi 1991). To achieve this accuracy, the ambiguities must be resolved "on the fly" and, therefore, (generally) dual-frequency receivers are required. Furthermore, five satellites per epoch are required.

After the deactivation of SA in May 2000, DGPS must be seen from a different viewpoint. The increased point positioning accuracy achieved with a single receiver may suffice for some kinds of applications.

Relative positioning

At present, highest accuracies are achieved in the relative-positioning mode with observed carrier phases. Relative positioning is associated with baselines, i.e., the three-dimensional vector between a known reference station and the location to be determined. Processing a baseline requires that the phases are simultaneously observed at both baseline endpoints (Fig. 5.1). Originally, relative positioning was only possible by postprocessing data. Today, (near) real-time data transfer over short baselines is routinely possible, which enables real-time computation of baseline vectors and has led to the real-time kinematic (RTK) technique.

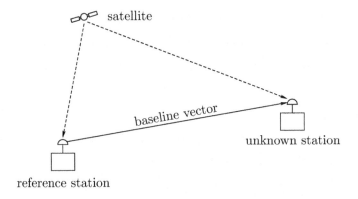

Fig. 5.1. Concept of relative positioning

Static relative positioning

The reference station and the unknown station are static, i.e., no motion occurs between the two points of the baseline. When highest accuracy is an issue, then this is the preferred method. Fully depending on the application and on the length of the baseline, the observation time may amount from several tens of minutes to many hours. Referring to navigation, where usually motion is involved, static relative positioning is of minor importance. The reader is referred to Hofmann-Wellenhof et al. (2001: Sect. 7.1.2) for details.

Kinematic relative positioning

The kinematic method is very productive because the greatest number of points can be determined in the least time.

The drawback is that after initialization a continuous lock on at least four satellites must be maintained.

The semikinematic or stop-and-go technique is characterized by alternatively stopping and moving one receiver to determine the positions of fixed points along the trajectory. The most important feature of this method is the increase in accuracy when several measurement epochs at the stop locations are accumulated and averaged. This technique is often referred to simply as kinematic method. Relative positional accuracies at the centimeter level can be achieved for baselines up to some 20 km.

The kinematic technique requires the resolution of the phase ambiguities by initialization which can be performed by static or kinematic techniques. Currently available commercial software (for dual-frequency receivers) only requires 1–2 minutes of observation for baselines up to 20 km to resolve the ambiguities kinematically ("on the fly").

5.3.5 GPS modernization concept

In January 1999, the USA announced the GPS modernization concept, a
$400 million initiative. The key feature is the implementation of a new signal
structure in future satellites.

Future GPS satellites

The Block IIR satellites increase their presence in the GPS constellation. A
new effort will bring modernized functionality to IIR satellites. These modernized satellites, denoted as IIR-M (replenishment-modernization), will provide new services to military and civilian users. New signals and increased
L-band power will significantly improve the navigation performance (Marquis 2001).

The Block IIF and the Block III satellites are the next generations.

These next generations of satellites will have many improvements over
the present satellites. It is planned to include the capability to transmit data
between satellites to make the system more independent. The autonomous
navigation (auto-nav) capability via intersatellite cross-link ranging will allow the satellites to essentially position themselves without extensive ground
tracking. In summary, the future satellites will have the following mainly military advantages:

- Navigation accuracy will be maintained for six months without ground
  support and control.
- Uplink jamming concerns will be minimized.
- One upload per spacecraft per month instead of one or even more per
  day will be performed.
- Need for overseas stations to support navigation uploads will be re-
  duced.
- Improved navigation accuracy will be achieved.

New signal structure

Referring to codes, presently civil users have unlimited access only to the
C/A-code on the carrier L1. The modernization will provide new signals:
implementing military codes (M-codes) on L1 and on L2 and a civilian code
on L2 (abbreviated as L2c). The M-code will provide the authorized users
with more signal security, improved acquisition options, and more jamming
resistance. The new civilian L2c signal will provide nonauthorized users dual-
frequency operation to perform ionospheric error correction. In addition to
these codes, a new L5 frequency will be provided for civilian users to en-
hance aviation applications. The notation L5 is chosen because, actually,

the satellites transmit additional signals at frequencies referred to as L3 and
L4. These signals are classified and for military purposes only (Misra and
Enge 2001: Sect. 2.3).

According to the modernization initiative released in 1999, the Inter-
agency GPS Executive Board concept will be realized with the following
specifications. Future GPS signals will be transmitted by three carriers where
L1 and L2 remain unchanged, and the new carrier L5 is specified as

$$L5 = 115 f_0 = 1176.45 \text{ MHz},$$

where $f_0 = 10.23$ MHz denotes the basic GPS frequency. The carrier L5,
placed in a protected aeronautical radio navigation service band, was recently allocated by the World Radio Conference organized regularly by the
International Telecommunication Union (Vorhies 2000).

Note that both new civil GPS signals will have two codes. L5 will not
share with military signals and use two equal-length codes in phase quadra-
ture, each clocked at 10.23 MHz. L2 is shared between civil and military
signals. The new L2c signal provides two codes by time multiplexing. The
two codes are of different length (Fontana et al. 2001). The existing military
Y-code will be replaced by new (split) M-codes.

The linear carrier phase combination of L2 with L5 results in a signal
with a wavelength of about 5.9 m. Long wavelengths facilitate ambiguity
resolution. By contrast, the linear combination of L1 with L5 will be used
as ionosphere-free combination because large frequency differences are ad-
vantageous for calculating ionospheric corrections. The common processing
of phase data from all three carriers will be performed in the three-carrier
ambiguity resolution approach (Vollath et al. 1999).

A perspective for the implementation is given in the 2001 Federal Radio-
navigation Plan: IOC (18 satellites in orbit with the new L2c signal and
M-code capability) is planned for 2008 and FOC (24 satellites in orbit) is
planned for 2010. At least one satellite is planned to be operational with the
new L5 capability no later than 2005, with IOC planned for 2012 and FOC
planned for 2014.

5.4 From GPS to coordinates

So far, we have got an introductory GPS overview. Now we are interested in
applying elementary GPS approaches to demonstrate how coordinates are
obtained. Two examples, as simple as possible, are selected: point positioning
and relative positioning.

5.4.1   Point positioning with code pseudoranges

The situation is shown in Fig. 5.2. The coordinates of A are to be determined by using GPS. As we know from Sect. 5.3.4, four pseudoranges to different satellites are necessary to determine the three coordinate components of A and the receiver clock error. Generalizing (5–3), we obtain

$$R_A^j(t) = \varrho_A^j(t) + c\,\delta_A(t)\,. \tag{5–6}$$

This is the code pseudorange at an epoch t, where $R_A^j(t)$ is the measured code pseudorange between the observing site A (as indicated in Fig. 5.2) and the satellite j, and $\varrho_A^j(t)$ is the geometric distance between the satellite and the observing point, and c is the speed of light. The last item is the receiver clock error $\delta_A(t)$. Note that we assume the simplest possible model, thus, we do not consider ionospheric and tropospheric influences, other biases and errors.

Examining Eq. (5–6), the desired point coordinates to be determined are implicitly comprised in the distance $\varrho_A^j(t)$, which can explicitly be written as

$$\varrho_A^j(t) = \sqrt{(X^j(t) - X_A)^2 + (Y^j(t) - Y_A)^2 + (Z^j(t) - Z_A)^2}\,, \tag{5–7}$$

where the WGS 84 (World Geodetic System 1984, see Sect. 2.11) coordinates $X^j(t)$, $Y^j(t)$, $Z^j(t)$ are the components of the geocentric position vector of the satellite at epoch t, and X_A, Y_A, Z_A are the three unknown WGS 84 coordinates of the observing site, which might be denoted $(X_A, Y_A, Z_A)_{\text{WGS 84}}$ or, which means the same, $(X_A, Y_A, Z_A)_{\text{GPS}}$.

How many unknowns are involved? Note that the satellite coordinates $X^j(t)$, $Y^j(t)$, $Z^j(t)$ may always be assumed known (more precisely, are cal-

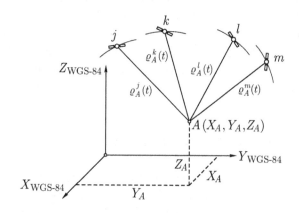

Fig. 5.2. Point positioning

culable) from the information broadcast by the satellite. Therefore, there remain the three unknown station coordinates X_A, Y_A, Z_A and the unknown receiver clock error $\delta_A(t)$. In other terms, at least four satellites are required to set up four equations of type (5–6). Denoting the satellites by j, k, l, m, the corresponding system of equations

$$
\begin{aligned}
R_A^j(t) &= \varrho_A^j(t) + c\,\delta_A(t)\,, \\
R_A^k(t) &= \varrho_A^k(t) + c\,\delta_A(t)\,, \\
R_A^l(t) &= \varrho_A^l(t) + c\,\delta_A(t)\,, \\
R_A^m(t) &= \varrho_A^m(t) + c\,\delta_A(t)
\end{aligned}
\qquad (5\text{–}8)
$$

is obtained or, by substituting (5–7) accordingly,

$$
\begin{aligned}
R_A^j(t) &= \sqrt{(X^j(t) - X_A)^2 + (Y^j(t) - Y_A)^2 + (Z^j(t) - Z_A)^2} + c\,\delta_A(t)\,, \\
R_A^k(t) &= \sqrt{(X^k(t) - X_A)^2 + (Y^k(t) - Y_A)^2 + (Z^k(t) - Z_A)^2} + c\,\delta_A(t)\,, \\
R_A^l(t) &= \sqrt{(X^l(t) - X_A)^2 + (Y^l(t) - Y_A)^2 + (Z^l(t) - Z_A)^2} + c\,\delta_A(t)\,, \\
R_A^m(t) &= \sqrt{(X^m(t) - X_A)^2 + (Y^m(t) - Y_A)^2 + (Z^m(t) - Z_A)^2} + c\,\delta_A(t)
\end{aligned}
$$
$$(5\text{–}9)$$

results. This system of equations comprises only the previously mentioned four unknowns X_A, Y_A, Z_A and the unknown receiver clock error $\delta_A(t)$ and may, thus, be solved. We do not consider linearization, possible redundant measurements, etc. We just intended to demonstrate the principle. The clock error is a by-product, but the desired result obtained from (5–9) are the GPS coordinates X_A, Y_A, Z_A; this means, the resulting coordinates are obatined in the WGS 84.

As described in Sect. 5.3.4, the accuracy of the point positioning method based on code ranges may be expected to amount some 10 m (nominally). A much higher accuracy is achieved by relative positioning treated in the next section.

5.4.2   Relative positioning with phase pseudoranges

The objective of relative positioning is to determine the coordinates of an unknown point with respect to a known point. In other words, relative positioning aims at the determination of the vector between the two points which is often called the baseline vector or simply baseline (Fig. 5.3). Let now A denote the known reference point, B the unknown point, and $\mathbf{b}_{AB}$ the baseline vector. Introducing the corresponding position vectors $\mathbf{X}_A$, $\mathbf{X}_B$,

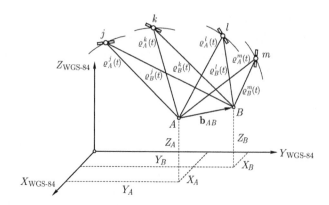

Fig. 5.3. Relative positioning

the relation

$$\mathbf{X}_B = \mathbf{X}_A + \mathbf{b}_{AB} \tag{5-10}$$

may be formulated, and the components of the baseline vector $\mathbf{b}_{AB}$ are

$$\mathbf{b}_{AB} = \begin{bmatrix} X_B - X_A \\ Y_B - Y_A \\ Z_B - Z_A \end{bmatrix} = \begin{bmatrix} \Delta X_{AB} \\ \Delta Y_{AB} \\ \Delta Z_{AB} \end{bmatrix}. \tag{5-11}$$

The coordinates of the reference point must be given in the WGS 84 and are usually approximated by a code pseudorange solution. Relative positioning can be performed with code pseudoranges (cf. Eq. (5–3)) or with phase pseudoranges (cf. Eq. (5–4)). Subsequently, only phase pseudoranges are explicitly considered. We repeat (5–4),

$$\lambda \Phi = \varrho + c \delta + \lambda N, \tag{5-12}$$

where we have already explained the wavelength λ, the phase Φ, the distance ϱ (which is the same as for the code pseudorange model), the speed of light c, the receiver clock error δ, and the ambiguity N in Sect. 5.3.3.

Introducing f, the frequency of the corresponding satellite signal, and taking into account the relation $f = c/\lambda$, we may divide (5–12) by λ obtaining

$$\Phi = \frac{1}{\lambda} \varrho + f \delta + N. \tag{5-13}$$

This may be generalized to

$$\Phi_i^j(t) = \frac{1}{\lambda} \varrho_i^j(t) + f \delta_i(t) + N_i^j, \tag{5-14}$$

where $\Phi_i^j(t)$ is the measured carrier phase expressed in cycles referred to station i and satellite j at epoch t. The time-independent phase ambiguity N_i^j is an integer number and, therefore, often called integer ambiguity or integer unknown or simply ambiguity.

Relative positioning requires simultaneous observations at both the reference and the unknown point. This means that the observation time tags for the two points must be the same. Assuming such observations (5–14) at the two points A and B to satellite j and another satellite k simultaneously at epoch t, the following measurement equations may be set up:

$$
\begin{aligned}
\Phi_A^j(t) &= \frac{1}{\lambda}\varrho_A^j(t) + f\,\delta_A(t) + N_A^j\,, \\[4pt]
\Phi_A^k(t) &= \frac{1}{\lambda}\varrho_A^k(t) + f\,\delta_A(t) + N_A^k\,, \\[4pt]
\Phi_B^j(t) &= \frac{1}{\lambda}\varrho_B^j(t) + f\,\delta_B(t) + N_B^j\,, \\[4pt]
\Phi_B^k(t) &= \frac{1}{\lambda}\varrho_B^k(t) + f\,\delta_B(t) + N_B^k\,.
\end{aligned}
\tag{5–15}
$$

Introducing the short-hand notations

$$
\begin{aligned}
\Phi_{AB}^{jk}(t) &= \Phi_B^k(t) - \Phi_B^j(t) - \Phi_A^k(t) + \Phi_A^j(t)\,, \\[4pt]
\varrho_{AB}^{jk}(t) &= \varrho_B^k(t) - \varrho_B^j(t) - \varrho_A^k(t) + \varrho_A^j(t)\,, \\[4pt]
N_{AB}^{jk} &= N_B^k - N_B^j - N_A^k + N_A^j\,,
\end{aligned}
\tag{5–16}
$$

we form the double-difference model which is defined as

$$
\Phi_{AB}^{jk}(t) = \frac{1}{\lambda}\varrho_{AB}^{jk}(t) + N_{AB}^{jk}\,.
\tag{5–17}
$$

Note that the receiver clock biases have canceled; this is the reason why double-differences are preferably used. This cancellation resulted from the assumptions of simultaneous observations and equal frequencies of the satellite signals (which is justified for GPS).

Assuming A as reference station with known coordinates, the remaining unknowns of the double-difference model are the desired coordinates X_B, Y_B, Z_B – which are comprised in $\varrho_B^j(t)$ and $\varrho_B^k(t)$ – and the ambiguities. To solve for these unknowns, we need more satellites (to set up additional double-differences) and also more epochs.

We do not consider linearization, possible redundant measurements, etc. We just intended to demonstrate the principle. The desired result obtained from (5–17) is the baseline vector $\mathbf{b}_{AB}$ with the components ΔX_{AB}, ΔY_{AB}, ΔZ_{AB} or, finally, the GPS coordinates X_B, Y_B, Z_B derived from (5–10) via the known station A to achieve the high accuracy. Note that the resulting coordinates are obtained in the WGS 84.

This concludes the short introduction how the user of GPS gets WGS 84 coordinates, i.e., geocentric rectangular coordinates X, Y, Z or, computed from them, ellipsoidal coordinates φ, λ, h; see Sect. 5.6.1.

5.5 Projection onto the ellipsoid

Let us establish the position of a point P by means of the natural coordinates Φ, Λ, H. Then we may project it onto the geoid along the (slightly curved) plumb line. The orthometric height is the distance between P and its projection P_0 onto the geoid, measured along the plumb line (Fig. 5.4). Although this mode of projection is entirely natural, the geoid is not suited for performing computations on it directly; the point P_0 is, therefore, projected onto the reference ellipsoid by means of the straight ellipsoidal normal, thus getting a point Q_0 on the ellipsoid. In this way, the earth's surface point P and the corresponding point Q_0 on the ellipsoid are connected by a double projection, that is, by two projections which are performed one after the other and which are quite analogous, the orthometric height $H = PP_0$ corresponding to the geoidal undulation $N = P_0Q_0$. This double projection is called *Pizzetti's projection*.

It is much simpler to project the point P from the physical surface of the earth directly onto the ellipsoid through the straight ellipsoidal normal, thus obtaining a point Q. The distance $PQ = h$ is the ellipsoidal height, i.e., the height above the ellipsoid. The earth's surface point P is then determined by the ellipsoidal height h and the ellipsoidal coordinates φ, λ of Q on the ellipsoid so that the *ellipsoidal coordinates* φ, λ, h take the place of the *natural coordinates* Φ, Λ, H. This is called *Helmert's projection*.

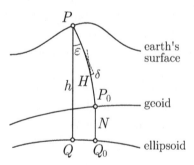

Fig. 5.4. The projection of Helmert and of Pizzetti

The practical difference between Pizzetti's and Helmert's projection is small. The ellipsoidal height h is equal to $H + N$ within a fraction of a millimeter. The ellipsoidal coordinates φ and λ, with respect to the two projections, are related by the equations

$$\varphi_{\text{Helmert}} = \varphi_{\text{Pizzetti}} + \frac{H}{R}\,\xi\,,$$

$$\lambda_{\text{Helmert}} = \lambda_{\text{Pizzetti}} + \frac{H}{R}\,\eta\,\sec\varphi\,, \tag{5-18}$$

which can be read from Fig. 5.4, since $QQ_0 \doteq H\varepsilon$; $R = 6371$ km is the mean radius of the earth. Even if $\varepsilon = 1$ arc minute and $H = 1000\,\text{m}$, the distance QQ_0 is only about 30 cm and the ellipsoidal coordinates differ by less than $0.01''$, which is below the accuracy of astronomical observations. For most purposes, we may, therefore, neglect the difference between the two projections.

Pizzetti's projection is better adapted to the geoid, because there is an exact correspondence between a geoidal point P_0 and an ellipsoidal point Q_0. Helmert's projection has overwhelming practical advantages, notably the straightforward conversion of the ellipsoidal coordinates φ, λ, h into rectangular coordinates x, y, z; it is also simpler in other respects. The decisive advantage of Helmert's projection is its direct relation to GPS. It is, therefore, exclusively used now in practice.

5.6 Coordinate transformations

5.6.1 Ellipsoidal and rectangular coordinates

We now derive the relation between the ellipsoidal coordinates φ, λ, h and the corresponding rectangular coordinates x, y, z.

The equation of the reference ellipsoid in rectangular coordinates is

$$\frac{x^2 + y^2}{a^2} + \frac{z^2}{b^2} = 1\,. \tag{5-19}$$

The representation of this ellipsoid in terms of ellipsoidal coordinates is given by

$$x = N\,\cos\varphi\,\cos\lambda\,,$$

$$y = N\,\cos\varphi\,\sin\lambda\,,$$

$$z = \frac{b^2}{a^2}\,N\,\sin\varphi\,, \tag{5-20}$$

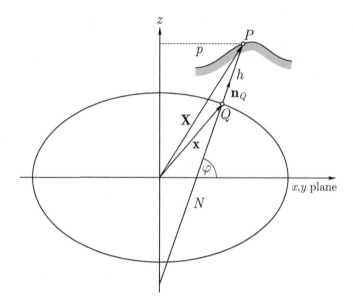

Fig. 5.5. Ellipsoidal and rectangular coordinates

where N is the normal radius of curvature (2–149):

$$N = \frac{a^2}{\sqrt{a^2 \cos^2\varphi + b^2 \sin^2\varphi}} \, . \tag{5–21}$$

These equations are known from ellipsoidal geometry; it may also be verified by direct substitution that a point with xyz-coordinates (5–20) satisfies the equation of the ellipsoid (5–19) and so lies on the ellipsoid. The components of the unit normal vector $\mathbf{n}$ are

$$\mathbf{n} = \left[\cos\varphi \, \cos\lambda, \;\; \cos\varphi \, \sin\lambda, \;\; \sin\varphi \right], \tag{5–22}$$

because φ is the angle between the ellipsoidal normal and the xy-plane, which is the equatorial plane (Fig. 5.5). Now let the coordinates of a point P outside the ellipsoid form the vector

$$\mathbf{X} = [X, Y, Z]; \tag{5–23}$$

similarly we have, for the coordinates of the point Q on the ellipsoid,

$$\mathbf{x} = [x, \; y, \; z]. \tag{5–24}$$

From Fig. 5.5, we read

$$\mathbf{X} = \mathbf{x} + h \, \mathbf{n}, \tag{5–25}$$

that is

$$X = x + h \cos \varphi \cos \lambda,$$

$$Y = y + h \cos \varphi \sin \lambda, \qquad (5\text{--}26)$$

$$Z = z + h \sin \varphi.$$

By (5–20), this becomes

$$X = (N + h) \cos \varphi \cos \lambda,$$

$$Y = (N + h) \cos \varphi \sin \lambda, \qquad (5\text{--}27)$$

$$Z = \left(\frac{b^2}{a^2} N + h \right) \sin \varphi.$$

These equations are the basic transformation formulas between the ellipsoidal coordinates φ, λ, h and the rectangular coordinates X, Y, Z of a point outside the ellipsoid. The origin of the rectangular coordinate system is the center of the ellipsoid, and the z-axis is its axis of rotation; the x-axis has the Greenwich longitude $0°$ and the y-axis has the longitude $90°$ east of Greenwich (i.e., $\lambda = +90°$).

A possible source of confusion is that the normal radius of curvature of the ellipsoid and the geoidal undulation are both denoted by the symbol N; in (5–27), N is, of course, the normal radius of curvature. Generally, let the context decide between quantities of such different magnitude (6000 km and 60 m).

Equations (5–27) permit the computation of rectangular coordinates X, Y, Z from the ellipsoidal coordinates φ, λ, h.

The inverse procedure, the computation of φ, λ, h from given X, Y, Z, is frequently performed iteratively, although a solution in closed form exists. A possible iterative procedure is as follows.

Denoting $\sqrt{X^2 + Y^2}$ by p, we get from the first two equations of (5–27) or from Fig. 5.5

$$p = \sqrt{X^2 + Y^2} = (N + h) \cos \varphi, \qquad (5\text{--}28)$$

so that

$$h = \frac{p}{\cos \varphi} - N. \qquad (5\text{--}29)$$

The third equation of (5–27) may be transformed into

$$Z = \left(N - \frac{a^2 - b^2}{a^2} N + h \right) \sin \varphi = (N + h - e^2 N) \sin \varphi, \qquad (5\text{--}30)$$

where $e^2 = (a^2 - b^2)/a^2$. Dividing this equation by the above expression for p, we find

$$\frac{Z}{p} = \left(1 - e^2 \frac{N}{N+h}\right) \tan \varphi, \tag{5–31}$$

so that

$$\tan \varphi = \frac{Z}{p} \left(1 - e^2 \frac{N}{N+h}\right)^{-1}. \tag{5–32}$$

Given X, Y, Z, and hence p, Eqs. (5–29) and (5–32) may be solved iteratively for h and φ. As a first approximation, we set $h = 0$ in (5–32), obtaining

$$\tan \varphi_{(1)} = \frac{Z}{p} (1 - e^2)^{-1}. \tag{5–33}$$

Using $\varphi_{(1)}$, we compute an approximate value $N_{(1)}$ by means of (5–21). Then (5–29) gives $h_{(1)}$. Now, as a second approximation, we set $h = h_{(1)}$ in (5–32), obtaining

$$\tan \varphi_{(2)} = \frac{Z}{p} \left(1 - e^2 \frac{N_{(1)}}{N_{(1)} + h_{(1)}}\right)^{-1}. \tag{5–34}$$

Using $\varphi_{(2)}$, improved values for N and h are found, etc. This procedure is repeated until φ and h remain practically constant.

The result for λ is immediately obtained from the first two equations of (5–27):

$$\lambda = \arctan \frac{Y}{X}. \tag{5–35}$$

Many other computation methods have been devised. One example for the transformation of X, Y, Z into φ, λ, h without iteration but with an inherent approximation is

$$\varphi = \arctan \frac{Z + e'^2 b \sin^3 \theta}{p - e^2 a \cos^3 \theta},$$

$$\lambda = \arctan \frac{Y}{X}, \tag{5–36}$$

$$h = \frac{p}{\cos \varphi} - N,$$

where

$$\theta = \arctan \frac{Z a}{p b} \tag{5–37}$$

is an auxiliary quantity and

$$e^2 = (a^2 - b^2)/a^2, \quad e'^2 = (a^2 - b^2)/b^2 \tag{5–38}$$

are first and second numerical eccentricity. As introduced in (5–28), $p = \sqrt{X^2 + Y^2}$. Actually, there is no reason why these formulas are less popular than the iterative procedure since there is no significant difference between the two methods. Computation methods with neither iteration nor approximation are, e.g., given by Sünkel (1977) and Zhu (1993).

5.6.2 Ellipsoidal, ellipsoidal-harmonic, and spherical coordinates

Even if we have several times pointed out the different definitions, it is very important to stress once more the need not to confuse the following coordinate triples (see Fig. 5.6):

- ellipsoidal coordinates: φ, λ, h;
- ellipsoidal-harmonic coordinates: β, λ, u, alternatively: $\vartheta_{\text{ellipsoidal-harmonic}}, \lambda, u$;
- spherical coordinates: $\bar{\varphi}, \lambda, r$, alternatively: $\vartheta_{\text{spherical}}, \lambda, r$.

The longitude λ is the same in all triples. The ellipsoidal coordinates latitude φ and longitude λ are sometimes also denoted *geodetic latitude* and *geodetic*

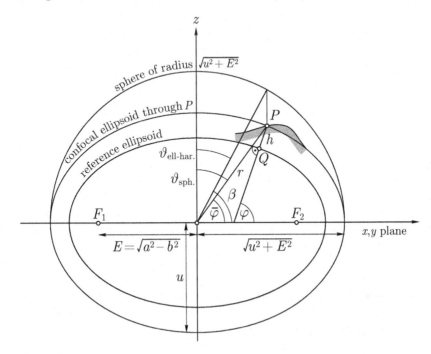

Fig. 5.6. Ellipsoidal, ellipsoidal-harmonic, and spherical coordinates

longitude. The ellipsoidal-harmonic coordinate β is the *reduced* latitude, and the spherical coordinate $\bar{\varphi}$ is the *geocentric* latitude.

The latitude φ refers to the *reference ellipsoid*. The reduced latitude β refers to the *coordinate ellipsoid* $u = $ constant (confocal ellipsoid through P in Fig. 5.6).

So far so clear. Real attention is necessary when using the coordinate ϑ, which has been introduced as complement of the spherical coordinate $\bar{\varphi}$ *and* as the complement of the ellipsoidal harmonic β as well.

Therefore, a correct but clumsy notation would be

$$\vartheta_{\text{ellipsoidal-harmonic}} = 90° - \beta \,,$$

$$\vartheta_{\text{spherical}} \qquad\qquad = 90° - \bar{\varphi} \,. \qquad\qquad (5\text{–}39)$$

Note, however, that we did not use these indications to distinguish between the spherical and the ellipsoidal-harmonic ϑ! Thus, the reader is challenged to attentively distinguish between these quantities. Wherever possible, we tried to avoid conflicts.

Some examples: we used the spherical coordinates r, ϑ, λ in Sects. 1.4, 1.11, 1.12, 1.14, 2.5, 2.6, 2.13, 2.18, etc. We used the ellipsoidal-harmonic coordinates u, ϑ, λ in Sects. 1.15, 1.16; we used the ellipsoidal-harmonic coordinates u, β, λ in Sects. 2.7, 2.8, and we used the spherical coordinates r, ϑ, λ as well as the ellipsoidal-harmonic coordinates u, β, λ in Sect. 2.9.

The following equations express the rectangular coordinates in these three systems:

$$X = (N + h) \cos \varphi \cos \lambda = \sqrt{u^2 + E^2} \cos \beta \cos \lambda = r \cos \bar{\varphi} \cos \lambda \,,$$

$$Y = (N + h) \cos \varphi \sin \lambda = \sqrt{u^2 + E^2} \cos \beta \sin \lambda = r \cos \bar{\varphi} \sin \lambda \,, \qquad (5\text{–}40)$$

$$Z = \left(\frac{b^2}{a^2} N + h\right) \sin \varphi \ = u \sin \beta \qquad\qquad\quad = r \sin \bar{\varphi} \,.$$

These relations, which follow from combining Eqs. (1–26), (1–151), and (5–27), can be used if we wish to compute u and β from h and φ or from r and $\bar{\varphi}$, etc.

5.7 Geodetic datum transformations

5.7.1 Introduction

First we define a *geodetic datum* or a *geodetic reference system*. It is defined by (1) the dimensions of the reference ellipsoid (semimajor axis a and flattening f) and (2) its position with respect to the earth or the geoid. This relative position is most simply defined by the coordinates x_0, y_0, z_0 of the center of the reference ellipsoid with respect to the geocenter. Since the geocenter was not accessible to classical geodetic measurements before the satellite era, a *fundamental* or *initial point* P_1 on the earth surface was chosen, such as Meades Ranch for North America and Potsdam for Central Europe. It turns out that a convenient but conventional choice of the ellipsoidal coordinates $\varphi_1, \lambda_1, h_1$ of the fundamental point P_1 is equivalent to x_0, y_0, z_0 of the geocenter.

Thus, we have 5 defining parameters:

- 2 parameters a (semimajor axis) and f (flattening) as *form parameters*, and
- 3 parameters x_0, y_0, z_0 or $\varphi_1, \lambda_1, h_1$ as *position parameters*.

Later on we shall also admit a scale factor and small rotations around the three coordinate axes.

A (geodetic) datum transformation defines the relationship between a global (geocentric) and a local (in general nongeocentric) three-dimensional Cartesian coordinate system; therefore, a datum transformation transforms one coordinate system of a certain type to another coordinate system of the same type. This is one of the primary tasks when combining GPS data with terrestrial data, i.e., the transformation of geocentric WGS 84 coordinates to local terrestrial coordinates. The terrestrial system is usually based on a locally best-fitting ellipsoid, e.g., the Clarke ellipsoid or the GRS-80 ellipsoid in the U.S. and the Bessel ellipsoid in many parts of Europe. The local ellipsoid is linked to a nongeocentric Cartesian coordinate system, where the origin coincides with the center of the ellipsoid.

5.7.2   Three-dimensional transformation in general form

Consider two arbitrary sets of three-dimensional Cartesian coordinates forming the vectors $\mathbf{X}$ and $\mathbf{X}_T$ (Fig. 5.7). The 7-parameter transformation, also denoted as Helmert transformation or similarity transformation in space, between the two sets can be formulated by the relation

$$\mathbf{X}_T = \mathbf{x}_0 + \mu\,\mathbf{R}\,\mathbf{X}\,, \qquad (5\text{--}41)$$

where $\mathbf{x}_0$ is the translation (or shift) vector, μ is a scale factor, and $\mathbf{R}$ is a rotation matrix.

The components of the shift vector

$$\mathbf{x}_0 = \begin{bmatrix} x_0 \\ y_0 \\ z_0 \end{bmatrix} \qquad (5\text{--}42)$$

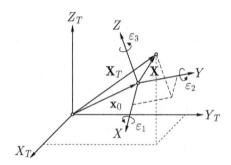

Fig. 5.7. Three-dimensional transformation

account for the coordinates of the origin of the $\mathbf{X}$ system in the $\mathbf{X}_T$ system. Note that a single scale factor is considered. More generally (but with GPS not necessary), three scale factors, one for each axis, could be used. The rotation matrix is an orthogonal matrix which is composed of three successive rotations

$$\mathbf{R} = \mathbf{R}_3\{\varepsilon_3\}\,\mathbf{R}_2\{\varepsilon_2\}\,\mathbf{R}_1\{\varepsilon_1\}\,. \tag{5--43}$$

Explicitly,

$$\mathbf{R} = \begin{bmatrix} \cos\varepsilon_2\cos\varepsilon_3 & \begin{array}{c}\cos\varepsilon_1\sin\varepsilon_3\\ +\sin\varepsilon_1\sin\varepsilon_2\cos\varepsilon_3\end{array} & \begin{array}{c}\sin\varepsilon_1\sin\varepsilon_3\\ -\cos\varepsilon_1\sin\varepsilon_2\cos\varepsilon_3\end{array} \\ -\cos\varepsilon_2\sin\varepsilon_3 & \begin{array}{c}\cos\varepsilon_1\cos\varepsilon_3\\ -\sin\varepsilon_1\sin\varepsilon_2\sin\varepsilon_3\end{array} & \begin{array}{c}\sin\varepsilon_1\cos\varepsilon_3\\ +\cos\varepsilon_1\sin\varepsilon_2\sin\varepsilon_3\end{array} \\ \sin\varepsilon_2 & -\sin\varepsilon_1\cos\varepsilon_2 & \cos\varepsilon_1\cos\varepsilon_2 \end{bmatrix}$$

$$\tag{5--44}$$

is obtained.

In the case of known transformation parameters $\mathbf{x}_0$, μ, $\mathbf{R}$, a point from the $\mathbf{X}$ system can be transformed into the $\mathbf{X}_T$ system by (5--41).

If the transformation parameters are unknown, they can be determined with the aid of common (identical) points, also denoted as control points. This means that the coordinates of the same point are given in both systems. Since each common point (given by $\mathbf{X}_T$ and $\mathbf{X}$) yields three equations, two common points and one additional common component (e.g., height) are sufficient to solve for the seven unknown parameters. In practice, redundant common point information is used and the unknown parameters are calculated by least-squares adjustment.

Since the parameters are mixed nonlinearly in Eq. (5--41), a linearization must be performed, where approximate values $\mathbf{x}_{0\,\mathrm{approx}}$, μ_{approx}, $\mathbf{R}_{\mathrm{approx}}$ are required.

5.7.3   Three-dimensional transformation between WGS 84 and a local system

In the case of a datum transformation between WGS 84 and a local system, some simplifications will arise. Referring to the necessary approximate values, the approximation $\mu_{\text{approx}} = 1$ is appropriate and the relation

$$\mu = \mu_{\text{approx}} + \delta\mu = 1 + \delta\mu \tag{5–45}$$

is obtained. Furthermore, the rotation angles ε_i in (5–44) are small and may be treated as differential quantities. Introducing these quantities into (5–44), setting $\cos\varepsilon_i = 1$ and $\sin\varepsilon_i = \varepsilon_i$, and considering only first-order terms gives

$$\mathbf{R} = \begin{bmatrix} 1 & \varepsilon_3 & -\varepsilon_2 \\ -\varepsilon_3 & 1 & \varepsilon_1 \\ \varepsilon_2 & -\varepsilon_1 & 1 \end{bmatrix} = \mathbf{I} + \delta\mathbf{R}, \tag{5–46}$$

where $\mathbf{I}$ is the unit matrix and $\delta\mathbf{R}$ is a (skewsymmetric) differential rotation matrix. Thus, the approximation $\mathbf{R}_{\text{approx}} = \mathbf{I}$ is appropriate. Finally, the shift vector is split up in the form

$$\mathbf{x}_0 = \mathbf{x}_{0\,\text{approx}} + \delta\mathbf{x}_0, \tag{5–47}$$

where the approximate shift vector

$$\mathbf{x}_{0\,\text{approx}} = \mathbf{X}_T - \mathbf{X} \tag{5–48}$$

follows by substituting the approximations for the scale factor and the rotation matrix into Eq. (5–41).

Introducing Eqs. (5–45), (5–46), (5–47) into (5–41) and skipping details which can be found, for example, in Hofmann-Wellenhof et al. (1994: Sect. 3.3) gives the linearized model for a single point i. This model can be written in the form

$$\mathbf{X}_{T_i} - \mathbf{X}_i - \mathbf{x}_{0\,\text{approx}} = \mathbf{A}_i\,\delta\mathbf{p}, \tag{5–49}$$

where the left side of the equation is known and may formally be considered as an observation. The design matrix $\mathbf{A}_i$ and the vector $\delta\mathbf{p}$, containing the unknown parameters, are given by

$$\mathbf{A}_i = \begin{bmatrix} 1 & 0 & 0 & X_i & 0 & -Z_i & Y_i \\ 0 & 1 & 0 & Y_i & Z_i & 0 & -X_i \\ 0 & 0 & 1 & Z_i & -Y_i & X_i & 0 \end{bmatrix}, \tag{5–50}$$

$$\delta\mathbf{p} = \begin{bmatrix} \delta x_0 & \delta y_0 & \delta z_0 & \delta\mu & \varepsilon_1 & \varepsilon_2 & \varepsilon_3 \end{bmatrix}.$$

Recall that Eq. (5–49) is now a system of linear equations for point i. For n common points, the design matrix A is

$$A = \begin{bmatrix} \mathbf{A}_1 \\ \mathbf{A}_2 \\ \vdots \\ \mathbf{A}_n \end{bmatrix}. \tag{5–51}$$

In detail, for three common points the design matrix is

$$A = \begin{bmatrix}
1 & 0 & 0 & X_1 & 0 & -Z_1 & Y_1 \\
0 & 1 & 0 & Y_1 & Z_1 & 0 & -X_1 \\
0 & 0 & 1 & Z_1 & -Y_1 & X_1 & 0 \\
\\
1 & 0 & 0 & X_2 & 0 & -Z_2 & Y_2 \\
0 & 1 & 0 & Y_2 & Z_2 & 0 & -X_2 \\
0 & 0 & 1 & Z_2 & -Y_2 & X_2 & 0 \\
\\
1 & 0 & 0 & X_3 & 0 & -Z_3 & Y_3 \\
0 & 1 & 0 & Y_3 & Z_3 & 0 & -X_3 \\
0 & 0 & 1 & Z_3 & -Y_3 & X_3 & 0
\end{bmatrix}, \tag{5–52}$$

which leads to a slightly redundant system. Least-squares adjustment yields the parameter vector $\delta\mathbf{p}$ and the adjusted values by (5–45), (5–46), (5–47). Once the seven parameters of the similarity transformation are determined, formula (5–41) can be used to transform other than the common points.

For a specific example, consider the task of transforming GPS coordinates of a network, i.e., global geocentric WGS 84 coordinates, to (three-dimensional) coordinates of a (nongeocentric) local system indicated by the subscript LS. The GPS coordinates are denoted by $(X, Y, Z)_{\text{GPS}}$ and the local system coordinates are the plane coordinates $(y, x)_{\text{LS}}$ and the ellipsoidal height h_{LS}. To obtain the transformation parameters, it is assumed that the coordinates of the common points in both systems are available. The solution of the task is obtained by the following algorithm:

1. Transform the plane coordinates $(y, x)_{\text{LS}}$ of the common points into the ellipsoidal surface coordinates $(\varphi, \lambda)_{\text{LS}}$ by using the appropriate mapping formulas.

2. Transform the ellipsoidal coordinates $(\varphi, \lambda, h)_{\text{LS}}$ of the common points into the Cartesian coordinates $(X, Y, Z)_{\text{LS}}$ by (5–27).

3. Determine the seven parameters of a Helmert transformation by using the coordinates $(X, Y, Z)_{\text{GPS}}$ and $(X, Y, Z)_{\text{LS}}$ of the common points.

4. For network points other than the common points, transform the coordinates $(X, Y, Z)_{\text{GPS}}$ into $(X, Y, Z)_{\text{LS}}$ via Eq. (5–41) using the transformation parameters determined in the previous step.

5. Transform the Cartesian coordinates $(X, Y, Z)_{\text{LS}}$ computed in the previous step into ellipsoidal coordinates $(\varphi, \lambda, h)_{\text{LS}}$, e.g., by the iterative procedure given in (5–28) through (5–34).

6. Map the ellipsoidal surface coordinates $(\varphi, \lambda)_{\text{LS}}$ computed in the previous step into plane coordinates $(y, x)_{\text{LS}}$ by the appropriate mapping formulas.

The advantage of the three-dimensional approach is that no a priori information is required for the seven parameters of the similarity transformation. The disadvantage of the method is that for the common points ellipsoidal heights (and, thus, geoidal heights) are required. However, as reported by Schmitt et al. (1991), incorrect heights of the common points often have a negligible effect on the plane coordinates (y, x). For example, incorrect heights may cause a tilt of a $20\,\text{km} \times 20\,\text{km}$ network by an amount of $5\,\text{m}$ in space; however, the effect on the plane coordinates is only approximately $1\,\text{mm}$.

For large areas, the height problem can be solved by adopting approximate ellipsoidal heights for the common points and performing a three-dimensional affine transformation instead of the similarity transformation.

5.7.4 Differential formulas for other datum transformations

Now we consider simplified cases. Suppose that the geocenter does not coincide with the center of the reference ellipsoid, but that *the geocentric axes and the ellipsoidal axes are parallel*. Such a parallel shift is also called a *translation* (Fig. 5.8). Assume a rectangular coordinate system XYZ whose origin is the geocenter, the axes being directed as usual. Let the coordinates of the center of the ellipsoid with respect to this system be x_0, y_0, z_0, as stated previously. Then Eqs. (5–27) must obviously be modified so that they become

$$X = x_0 + (N + h) \cos \varphi \cos \lambda,$$

$$Y = y_0 + (N + h) \cos \varphi \sin \lambda,$$

$$Z = z_0 + \left(\frac{b^2}{a^2} N + h \right) \sin \varphi. \tag{5–53}$$

These equations form the starting point for various important differential formulas of coordinate transformation.

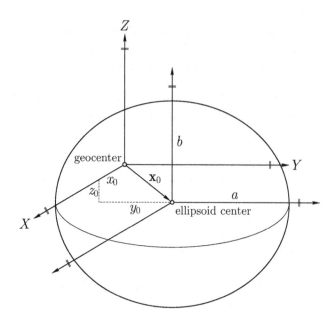

Fig. 5.8. Translation problem

First we ask how the rectangular coordinates X, Y, Z change if we vary the ellipsoidal coordinates φ, λ, h by small amounts $\delta\varphi, \delta\lambda, \delta h$ and if we also alter the geodetic datum, namely, the reference ellipsoid a, f and its position x_0, y_0, z_0, by $\delta a, \delta f$ and $\delta x_0, \delta y_0, \delta z_0$. Note that $\delta x_0, \delta y_0, \delta z_0$ correspond to a small translation (parallel displacement) of the ellipsoid, *its axis remaining parallel to the axis of the earth.*

The solution of this problem is found by differentiating (5–53):

$$\delta X = \delta x_0 + \frac{\partial X}{\partial a}\,\delta a + \frac{\partial X}{\partial f}\,\delta f + \frac{\partial X}{\partial \varphi}\,\delta\varphi + \frac{\partial X}{\partial \lambda}\,\delta\lambda + \frac{\partial X}{\partial h}\,\delta h\,,$$

$$\delta Y = \delta y_0 + \frac{\partial Y}{\partial a}\,\delta a + \frac{\partial Y}{\partial f}\,\delta f + \frac{\partial Y}{\partial \varphi}\,\delta\varphi + \frac{\partial Y}{\partial \lambda}\,\delta\lambda + \frac{\partial Y}{\partial h}\,\delta h\,, \qquad (5\text{–}54)$$

$$\delta Z = \delta z_0 + \frac{\partial Z}{\partial a}\,\delta a + \frac{\partial Z}{\partial f}\,\delta f + \frac{\partial Z}{\partial \varphi}\,\delta\varphi + \frac{\partial Z}{\partial \lambda}\,\delta\lambda + \frac{\partial Z}{\partial h}\,\delta h\,,$$

since, according to Taylor's theorem, small changes can be treated as differentials.

In these differential formulas we shall be satisfied with an approximation.

Since the flattening f is small, we may expand (2–149) as

$$N = \frac{a^2}{b}\,(1 + e'^2 \cos^2\varphi)^{-1/2} = \frac{a^2}{b}\,\left(1 - \tfrac{1}{2}\,e'^2 \cos^2\varphi \,\cdots\right)$$

$$= a\,(1 + f \,\cdots)(1 - f \cos^2\varphi \,\cdots) = a(1 + f - f \cos^2\varphi \,\cdots) \tag{5–55}$$

yielding

$$N \doteq a\,(1 + f \sin^2\varphi) \tag{5–56}$$

and

$$\frac{b^2}{a^2}\,N = (1 - 2f \,\cdots)\,a\,(1 + f \sin^2\varphi \,\cdots) \doteq a\,(1 - 2f + f \sin^2\varphi) \tag{5–57}$$

and

$$b = a\,(1 - f)\,, \quad e'^2 = 2f \,\cdots\,. \tag{5–58}$$

Thus, Eqs. (5–53) are approximated by

$$X = x_0 + (a + af \sin^2\varphi + h)\cos\varphi\,\cos\lambda\,,$$

$$Y = y_0 + (a + af \sin^2\varphi + h)\cos\varphi\,\sin\lambda\,, \tag{5–59}$$

$$Z = z_0 + (a - 2af + af \sin^2\varphi + h)\sin\varphi\,.$$

Now we can form the partial derivatives in (5–54), for instance,

$$\frac{\partial X}{\partial a} = (1 + f \sin^2\varphi)\cos\varphi\,\cos\lambda \doteq \cos\varphi\,\cos\lambda\,, \tag{5–60}$$

since we may neglect the flattening in these coefficients. This amounts to using for the coefficients, and only for them, a spherical approximation analogous to that of Sect. 2.13. Similarly, all coefficients are easily obtained as partial derivatives, and Eqs. (5–54) become

$$\delta X = \delta x_0 - a\,\sin\varphi\,\cos\lambda\,\delta\varphi - a\,\cos\varphi\,\sin\lambda\,\delta\lambda$$
$$+ \cos\varphi\,\cos\lambda\,(\delta h + \delta a + a\,\sin^2\varphi\,\delta f)\,,$$

$$\delta Y = \delta y_0 - a\,\sin\varphi\,\sin\lambda\,\delta\varphi + a\,\cos\varphi\,\cos\lambda\,\delta\lambda \tag{5–61}$$
$$+ \cos\varphi\,\sin\lambda\,(\delta h + \delta a + a\,\sin^2\varphi\,\delta f)\,,$$

$$\delta Z = \delta z_0 + a\,\cos\varphi\,\delta\varphi + \sin\varphi\,(\delta h + \delta a + a\,\sin^2\varphi\,\delta f)$$
$$- 2a\,\sin\varphi\,\delta f\,.$$

These formulas give the changes in the rectangular coordinates X, Y, Z in terms of the variation in the position (x_0, y_0, z_0) and the dimensions (a, f) of the ellipsoid and in the ellipsoidal coordinates φ, λ, h referred to it.

Transformation of the ellipsoidal coordinates

Several important formulas for the transformation of coordinates may be derived from Eqs. (5–61). First, let the position of P in space remain unchanged; that is, let

$$\delta X = \delta Y = \delta Z = 0. \tag{5–62}$$

Determine the change of the ellipsoidal coordinates φ, λ, h if the dimensions of the reference ellipsoid and its position are varied. Geometrically, this is illustrated by Fig. 5.9. The problem is, thus, to solve equations (5–61) for $\delta\varphi, \delta\lambda, \delta h$, the left-hand sides being set equal to zero. To get $\delta\varphi$, multiply the first equation of (5–61) by $-\sin\varphi\cos\lambda$, the second equation of (5–61) by $-\sin\varphi\sin\lambda$, and the third equation of (5–61) by $\cos\varphi$ and add all equations obtained in this way. For $\delta\lambda$, the factors are $-\sin\lambda$, $\cos\lambda$, and 0; for δh, they are $\cos\varphi\cos\lambda$, $\cos\varphi\sin\lambda$, and $\sin\varphi$. The result is

$$a\,\delta\varphi = \sin\varphi\cos\lambda\,\delta x_0 + \sin\varphi\sin\lambda\,\delta y_0 - \cos\varphi\,\delta z_0 + 2a\,\sin\varphi\cos\varphi\,\delta f,$$

$$a\,\cos\varphi\,\delta\lambda = \sin\lambda\,\delta x_0 - \cos\lambda\,\delta y_0,$$

$$\delta h = -\cos\varphi\cos\lambda\,\delta x_0 - \cos\varphi\sin\lambda\,\delta y_0 - \sin\varphi\,\delta z_0 - \delta a + a\,\sin^2\varphi\,\delta f.$$
$$\tag{5–63}$$

These formulas express the variations $\delta\varphi, \delta\lambda, \delta h$ at an arbitrary point in terms of the variations $\delta x_0, \delta y_0, \delta z_0$ at a given point and the changes δa and δf of the parameters of the reference ellipsoid. Thus, they relate two different systems of ellipsoidal coordinates, provided these systems are so close to each other that their differences may be considered as linear. Mathematically, Eqs. (5–63) are infinitesimal coordinate transformations (essentially but not exclusively orthogonal transformations); to the geodesist, they give the effect

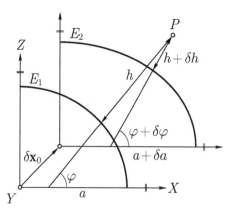

Fig. 5.9. A small change of the reference ellipsoid together with a small parallel shift

of a change in the geodetic datum.

Remark. The differential formulas could also be replaced by a successive application of the original finite formulas. Try!

Part II: Three-dimensional geodesy: a transition

5.8 The three-dimensional geodesy of Bruns and Hotine

The idea of a computation of a triangulation network in space dates back to Bruns (1878). On the basis of his ideas, Hotine (1969), and earlier in 1959, developed extensively the concept of a classical (pre-satellite) geodetic network in a rigorous three-dimensional way. For a comparison, see Levallois (1963).

Consider the polyhedron formed by triangulation benchmarks on the surface of the earth and the straight lines of sight connecting them (Fig. 5.10). Another set of straight lines – one through each corner – represents the plumb line at the stations.

In order to determine this figure, we need five parameters for each station – three coordinates and two parameters defining the direction of the plumb line. The main terrestrial observational data for this purpose are

1. horizontal angles and zenith angles, obtained by theodolite observations;

2. straight spatial distances, obtained by electronic distance measurements; and

3. astronomical observations of latitude and longitude to fix the direction of the plumb line, and of azimuth to determine the orientation of the polyhedron.

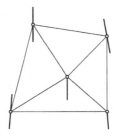

Fig. 5.10. Bruns' polyhedron

We may use a rectangular coordinate system; then the three coordinates
to be determined will be X, Y, Z. The parameters defining the direction of
the plumb line are conveniently taken to be Φ and Λ, astronomical latitude
and longitude. We can express the astronomical azimuth A, the measured
zenith angle z', and the spatial distance s in terms of these five parameters.
This will be the scope of Sect. 5.9.

This information is purely "geometric". We need the terrestrial measurements (especially Φ, Λ, A) in order to link this geometry to the gravity field
as represented by the plumb lines. The Bruns polyhedron is the best way to
show this geometrically.

Today, GPS is the best way to determine global rectangular coordinates
X, Y, Z or ellipsoidal coordinates φ, λ, h directly.

5.9 Global coordinates and local level coordinates

We shall use a Cartesian coordinate system XYZ introduced in Sect. 5.6.1,
global but not necessarily geocentric. The coordinates X, Y, Z form a vector
$\mathbf{X}$. Thus, the vectors $\mathbf{X}_i$ and $\mathbf{X}_j$ represent two terrestrial points P_i and
P_j. We define the vector between these two points in the global coordinate
system by $\mathbf{X}_{ij} = \mathbf{X}_j - \mathbf{X}_i$.

In addition, we introduce a "local level system" referred to the tangential
plane to the level surface at a point P_i and to the local vertical, which is
the tangent at P_i to the natural plumb line defined by the astronomical
coordinates Φ and Λ, see Sect. 2.4. The axes $\mathbf{n}_i$, $\mathbf{e}_i$, $\mathbf{u}_i$ of this local (tangent
plane) coordinate system at P_i corresponding to the north, east, and up

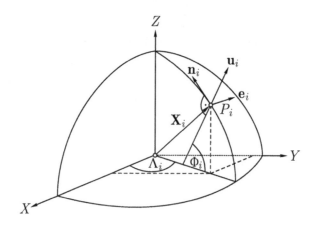

Fig. 5.11. Global and local level coordinates

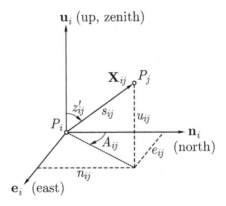

Fig. 5.12. Measurement quantities in the local level system

direction, are thus represented in the global system by

$$\mathbf{n}_i = \begin{bmatrix} -\sin\Phi_i\cos\Lambda_i \\ -\sin\Phi_i\sin\Lambda_i \\ \cos\Phi_i \end{bmatrix}, \quad \mathbf{e}_i = \begin{bmatrix} -\sin\Lambda_i \\ \cos\Lambda_i \\ 0 \end{bmatrix}, \quad \mathbf{u}_i = \begin{bmatrix} \cos\Phi_i\cos\Lambda_i \\ \cos\Phi_i\sin\Lambda_i \\ \sin\Phi_i \end{bmatrix},$$

$$(5\text{--}64)$$

where the vectors $\mathbf{n}_i$ and $\mathbf{e}_i$ span the tangent plane at P_i (Fig. 5.11). The third coordinate axis of the local level system, i.e., the vector $\mathbf{u}_i$, is orthogonal to the tangent plane and has the direction of the plumb line.

Now the components n_{ij}, e_{ij}, u_{ij} of the vector $\mathbf{x}_{ij}$ in the local level system are introduced. These coordinates are sometimes denoted as ENU (east, north, up) coordinates. Considering Fig. 5.12, these components are obtained by a projection of vector $\mathbf{X}_{ij}$ onto the local level axes $\mathbf{n}_i$, $\mathbf{e}_i$, $\mathbf{u}_i$. Analytically, this is achieved by scalar products. Therefore,

$$\mathbf{x}_{ij} = \begin{bmatrix} n_{ij} \\ e_{ij} \\ u_{ij} \end{bmatrix} = \begin{bmatrix} \mathbf{n}_i \cdot \mathbf{X}_{ij} \\ \mathbf{e}_i \cdot \mathbf{X}_{ij} \\ \mathbf{u}_i \cdot \mathbf{X}_{ij} \end{bmatrix} \tag{5--65}$$

is obtained. Assembling the vectors $\mathbf{n}_i$, $\mathbf{e}_i$, $\mathbf{u}_i$ of the local level system as columns in an orthogonal matrix $\mathbf{D}_i$, i.e.,

$$\mathbf{D}_i = \begin{bmatrix} -\sin\Phi_i\cos\Lambda_i & -\sin\Lambda_i & \cos\Phi_i\cos\Lambda_i \\ -\sin\Phi_i\sin\Lambda_i & \cos\Lambda_i & \cos\Phi_i\sin\Lambda_i \\ \cos\Phi_i & 0 & \sin\Phi_i \end{bmatrix}, \tag{5--66}$$

relation (5–65) may be written concisely as

$$\mathbf{x}_{ij} = \mathbf{D}_i^T\, \mathbf{X}_{ij}. \tag{5--67}$$

The components of $\mathbf{x}_{ij}$ may also be expressed by the spatial distance s_{ij}, the azimuth A_{ij}, and the zenith angle z'_{ij}, which is assumed to be corrected for refraction. The appropriate relation is

$$\mathbf{x}_{ij} = \begin{bmatrix} n_{ij} \\ e_{ij} \\ u_{ij} \end{bmatrix} = \begin{bmatrix} s_{ij}\sin z'_{ij}\cos A_{ij} \\ s_{ij}\sin z'_{ij}\sin A_{ij} \\ s_{ij}\cos z'_{ij} \end{bmatrix}, \tag{5-68}$$

where the terrestrial measurement quantities s_{ij}, A_{ij}, z'_{ij} refer to P_i, i.e., the measurements were taken at P_i. Inverting (5–68) gives the measurement quantities explicitly:

$$s_{ij} = \sqrt{n_{ij}^2 + e_{ij}^2 + u_{ij}^2},$$

$$\tan A_{ij} = \frac{e_{ij}}{n_{ij}}, \tag{5-69}$$

$$\cos z'_{ij} = \frac{u_{ij}}{\sqrt{n_{ij}^2 + e_{ij}^2 + u_{ij}^2}}.$$

Substituting (5–65) for n_{ij}, e_{ij} and u_{ij}, the measurement quantities may be expressed by the components of the vector $\mathbf{X}_{ij}$ in the global system.

A note on azimuth and zenith distance
Since the local level coordinates refer to the local plumb line defined by the astronomical coordinates Φ, Λ (Sect. 2.4), A and z' are called astronomical azimuth and astronomical zenith distance (zenith angle). They will also play a basic role in Part III.

A final word on the zenith distance. The measured ("astronomical") azimuth is denoted by A, and the corresponding ellipsoidal azimuth is denoted by α. Since the ellipsoidal zenith distance is conventionally denoted by z, it would be consistent to indicate the measured ("astronomical") zenith distance by Z. This symbol, however, is firmly reserved for the third axis of the XYZ system, so we exceptionally, but consistently with the rest of the book, use the symbol z'. (Both A and z' will return in the following sections.)

5.10   Combining terrestrial data and GPS

5.10.1   Common coordinate system

So far, GPS and terrestrial networks have been considered separately with respect to the adjustment. The combination, for example, by a datum transformation, was supposed to be performed after individual adjustments. Now the common adjustment of GPS observations and terrestrial data is investigated. The problem encountered here is that GPS data refer to the three-dimensional geocentric Cartesian system WGS 84, whereas terrestrial data refer to the individual local level (tangent plane) systems at each measurement point referenced to plumb lines. Furthermore, terrestrial data are traditionally separated into position and height, where the position refers to an ellipsoid and the (orthometric) height to the geoid.

For a joint adjustment, a common coordinate system is required to which all observations are transformed. In principle, any arbitrary system may be introduced as common reference. One possibility is to use two-dimensional (plane) coordinates in the local system as proposed by Daxinger and Stirling (1995). Here, a three-dimensional coordinate system is chosen. The origin of the coordinate system is the center of the ellipsoid adopted for the local system, the Z-axis coincides with the semiminor axis of the ellipsoid, the X-axis is obtained by the intersection of the ellipsoidal Greenwich meridian plane and the ellipsoidal equatorial plane, and the Y-axis completes the right-handed system. Position vectors referred to this system are denoted by $\mathbf{X}_{\mathrm{LS}}$, where LS indicates the reference to the local system.

After the decision on the common coordinate system, the terrestrial measurements referring to the individual local level systems at the observing sites must be represented in this common coordinate system. Similarly, GPS baseline vectors regarded as measurement quantities are to be transformed to this system.

5.10.2   Representation of measurement quantities

Distances

The spatial distance s_{ij} as function of the local level coordinates is given in (5–69). If n_{ij}, e_{ij}, u_{ij}, the components of $\mathbf{x}_{ij}$, are substituted by (5–65), the relation

$$
\begin{aligned}
s_{ij} &= \sqrt{n_{ij}^2 + e_{ij}^2 + u_{ij}^2} \\
&= \sqrt{(X_j - X_i)^2 + (Y_j - Y_i)^2 + (Z_j - Z_i)^2}
\end{aligned}
\tag{5–70}
$$

is obtained, where (5–64) has also been taken into account, namely, the fact that $\mathbf{n}_i$, $\mathbf{e}_i$, $\mathbf{u}_i$ are unit vectors. Obviously, the second expression arises immediately from the Pythagorean theorem. Differentiation of (5–70) yields

$$
ds_{ij} = \frac{X_{ij}}{s_{ij}}\left(dX_j - dX_i\right) + \frac{Y_{ij}}{s_{ij}}\left(dY_j - dY_i\right) + \frac{Z_{ij}}{s_{ij}}\left(dZ_j - dZ_i\right),
$$

$$
\tag{5–71}
$$

where

$$X_{ij} = X_j - X_i \,,$$
$$Y_{ij} = Y_j - Y_i \,, \tag{5–72}$$
$$Z_{ij} = Z_j - Z_i$$

have been introduced accordingly. The relation (5–71) may also be expressed as

$$\delta s_{ij} = \frac{X_{ij}}{s_{ij}} (\delta X_j - \delta X_i) + \frac{Y_{ij}}{s_{ij}} (\delta Y_j - \delta Y_i) + \frac{Z_{ij}}{s_{ij}} (\delta Z_j - \delta Z_i)$$

$$\tag{5–73}$$

if the differentials are replaced by differences.

Azimuths

Again the same principle applies: the measured azimuth A_{ij} as a function of the local level coordinates is given in (5–69). If n_{ij}, e_{ij}, u_{ij}, the components of $\mathbf{x}_{ij}$, are substituted by (5–65), the relation

$$\tan A_{ij} = e_{ij}/n_{ij}$$
$$= \frac{-X_{ij} \sin \Lambda_i + Y_{ij} \cos \Lambda_i}{-X_{ij} \sin \Phi_i \cos \Lambda_i - Y_{ij} \sin \Phi_i \sin \Lambda_i + Z_{ij} \cos \Phi_i} \tag{5–74}$$

is obtained. After a lengthy derivation, the relation

$$\delta A_{ij} = \frac{\sin \varphi_i \cos \lambda_i \sin \alpha_{ij} - \sin \lambda_i \cos \alpha_{ij}}{s_{ij} \sin z_{ij}} (\delta X_j - \delta X_i)$$
$$+ \frac{\sin \varphi_i \sin \lambda_i \sin \alpha_{ij} + \cos \lambda_i \cos \alpha_{ij}}{s_{ij} \sin z_{ij}} (\delta Y_j - \delta Y_i)$$
$$- \frac{\cos \varphi_i \sin \alpha_{ij}}{s_{ij} \sin z_{ij}} (\delta Z_j - \delta Z_i) \tag{5–75}$$
$$+ \cot z_{ij} \sin \alpha_{ij} \, \delta \Phi_i$$
$$+ (\sin \varphi_i - \cos \alpha_{ij} \cos \varphi_i \cot z_{ij}) \, \delta \Lambda_i$$

is obtained. Approximate values are sufficient *in the coefficients*, denoted by $\varphi, \lambda, \alpha, z$ instead of Φ, Λ, A, z'.

Directions

Measured directions R_{ij} are related to azimuths A_{ij} by the orientation unknown o_i. The relation reads

$$R_{ij} = A_{ij} - o_i \,, \tag{5–76}$$

and the expression

$$\delta R_{ij} = \delta A_{ij} - \delta o_i \tag{5–77}$$

is immediately obtained.

Zenith angles

The zenith angle z'_{ij} as function of the local level coordinates is given in (5–69). If n_{ij}, e_{ij}, u_{ij}, the components of $\mathbf{x}_{ij}$, are substituted by (5–65), the relation

$$\cos z'_{ij} = u_{ij}/s_{ij}$$

$$= \frac{X_{ij} \cos \Phi_i \cos \Lambda_i + Y_{ij} \cos \Phi_i \sin \Lambda_i + Z_{ij} \sin \Phi_i}{\sqrt{X_{ij}^2 + Y_{ij}^2 + Z_{ij}^2}} \tag{5–78}$$

is obtained, where (5–70) and (5–72) have been used. After a lengthy derivation, the relation

$$\begin{aligned}
\delta z'_{ij} = \; & \frac{X_{ij} \cos z_{ij} - s_{ij} \cos \varphi_i \cos \lambda_i}{s_{ij}^2 \sin z_{ij}} (\delta X_j - \delta X_i) \\[2mm]
& + \frac{Y_{ij} \cos z_{ij} - s_{ij} \cos \varphi_i \sin \lambda_i}{s_{ij}^2 \sin z_{ij}} (\delta Y_j - \delta Y_i) \\[2mm]
& + \frac{Z_{ij} \cos z_{ij} - s_{ij} \sin \varphi_i}{s_{ij}^2 \sin z_{ij}} (\delta Z_j - \delta Z_i) \\[2mm]
& - \cos \alpha_{ij} \, \delta \Phi_i - \cos \varphi_i \sin \alpha_{ij} \, \delta \Lambda_i
\end{aligned} \tag{5–79}$$

is obtained.

It is presupposed that the zenith angles are reduced to the chord of the light path. This reduction may be modeled by

$$z'_{ij} = z'_{ij\,\mathrm{meas}} + \frac{s_{ij}}{2R} \, k \,, \tag{5–80}$$

where $z_{ij\,\mathrm{meas}}$ is the measured zenith angle, R is the mean radius of the earth, and k is the coefficient of refraction. For k either a standard value may be substituted or the coefficient of refraction is estimated as additional unknown. In the case of estimation, there are several choices, e.g., one value for k for all zenith angles or one value for a group of zenith angles or one value per day. (It is known that measured zenith angles are "weaker" than other observations, which can be taken into account by giving them lower weights.)

Ellipsoidal height differences

The "measured" ellipsoidal height difference is represented by

$$h_{ij} = h_j - h_i \, . \tag{5-81}$$

The heights involved are obtained by transforming the Cartesian coordinates into ellipsoidal coordinates according to (5–36) or by using the iterative procedure given in Sect. 5.6.1. The height difference is approximately (neglecting the curvature of the earth) given by the third component of $\mathbf{x}_{ij}$ in the local level system. Hence,

$$h_{ij} = \mathbf{u}_i \cdot \mathbf{X}_{ij} \tag{5-82}$$

or, by substituting $\mathbf{u}_i$ according to (5–64), the relation

$$h_{ij} = \cos \Phi_i \cos \Lambda_i \, X_{ij} + \cos \Phi_i \sin \Lambda_i \, Y_{ij} + \sin \Phi_i \, Z_{ij} \tag{5-83}$$

is obtained. This equation may be differentiated with respect to the Cartesian coordinates. If the differentials are replaced by the corresponding differences,

$$\begin{aligned}
\delta h_{ij} = {} & \cos \Phi_j \cos \Lambda_j \, \delta X_j + \cos \Phi_j \sin \Lambda_j \, \delta Y_j + \sin \Phi_j \, \delta Z_j \\
& - \cos \Phi_i \cos \Lambda_i \, \delta X_i - \cos \Phi_i \sin \Lambda_i \, \delta Y_i - \sin \Phi_i \, \delta Z_i
\end{aligned} \tag{5-84}$$

is obtained, where the coordinate differences were decomposed into their individual coordinates.

Baselines

From relative GPS measurements, baselines $\mathbf{X}_{ij_{(\text{GPS})}} = \mathbf{X}_{j_{(\text{GPS})}} - \mathbf{X}_{i_{(\text{GPS})}}$ in the WGS 84 are obtained. The position vectors $\mathbf{X}_{i_{(\text{GPS})}}$ and $\mathbf{X}_{j_{(\text{GPS})}}$ may be transformed by a three-dimensional (7-parameter) similarity transformation to a local system indicated by LS. According to Eq. (5–41), the transformation formula reads

$$\mathbf{X}_{\text{LS}} = \mathbf{x}_0 + \mu \, \mathbf{R} \, \mathbf{X}_{\text{GPS}} \, , \tag{5-85}$$

where the meaning of the individual quantities is the following:

$$\begin{aligned}
\mathbf{X}_{\text{LS}} \quad &\dots \quad \text{position vector in the local system} \, , \\
\mathbf{X}_{\text{GPS}} \quad &\dots \quad \text{position vector in the WGS 84} \, , \\
\mathbf{x}_0 \quad &\dots \quad \text{shift vector} \, , \\
\mathbf{R} \quad &\dots \quad \text{rotation matrix} \, , \\
\mu \quad &\dots \quad \text{scale factor} \, .
\end{aligned}$$

Forming the difference of two position vectors, i.e., the baseline $\mathbf{X}_{ij}$, the shift vector $\mathbf{x}_0$ is eliminated. Using (5–85), there results

$$\mathbf{X}_{ij_{(\text{LS})}} = \mu \, \mathbf{R} \, \mathbf{X}_{ij_{(\text{GPS})}} \tag{5-86}$$

for the baseline. Similar to (5–49), the linearized form is

$$\mathbf{X}_{ij(\text{LS})} = \mathbf{X}_{ij(\text{GPS})} + \mathbf{A}_{ij}\,\delta\mathbf{p}\,, \qquad (5\text{–}87)$$

where now the vector $\delta\mathbf{p}$ and the design matrix $\mathbf{A}_{ij}$ are given by

$$\delta\mathbf{p} = [\delta\mu \quad \varepsilon_1 \quad \varepsilon_2 \quad \varepsilon_3]^T,$$

$$\mathbf{A}_{ij} = \begin{bmatrix} X_{ij} & 0 & -Z_{ij} & Y_{ij} \\ Y_{ij} & Z_{ij} & 0 & -X_{ij} \\ Z_{ij} & -Y_{ij} & X_{ij} & 0 \end{bmatrix}_{(\text{GPS})}. \qquad (5\text{–}88)$$

Note that the rotations ε_i refer to the axes of the system used in GPS. If they should refer to the local system, then the signs of the rotations must be changed, i.e., the signs of the elements of the last three columns of matrix $\mathbf{A}_{ij}$ must be reversed.

The vector $\mathbf{X}_{ij(\text{LS})}$ on the left side of (5–87) contains the points $\mathbf{X}_{i(\text{LS})}$ and $\mathbf{X}_{j(\text{LS})}$ in the local system. If these points are unknown, then they are replaced by known approximate values and unknown increments

$$\mathbf{X}_{i(\text{LS})} = \mathbf{X}_{i0(\text{LS})} + \delta\mathbf{X}_{i(\text{LS})}\,,$$
$$\mathbf{X}_{j(\text{LS})} = \mathbf{X}_{j0(\text{LS})} + \delta\mathbf{X}_{j(\text{LS})}\,, \qquad (5\text{–}89)$$

where the coefficients of these unknown increments ($+1$ or -1) together with matrix $\mathbf{A}_{ij}$ form the design matrix.

The vector $\mathbf{X}_{ij(\text{GPS})}$ in (5–87) is regarded as measurement quantity. Thus, finally,

$$\mathbf{X}_{ij(\text{GPS})} = \delta\mathbf{X}_{j(\text{LS})} - \delta\mathbf{X}_{i(\text{LS})} - \mathbf{A}_{ij}\,\delta\mathbf{p} + \mathbf{X}_{j0(\text{LS})} - \mathbf{X}_{i0(\text{LS})} \qquad (5\text{–}90)$$

is the linearized observation equation.

In principle, any type of geodetic measurement can be employed if the integrated geodesy adjustment model is used. The basic concept is that any geodetic measurement can be expressed as a function of one or more position vectors $\mathbf{X}$ and of the gravity field W of the earth. The usually nonlinear function must be linearized where the gravity field W is split into the normal potential U of an ellipsoid and the disturbing potential T, thus, $W = U + T$. Applying a minimum principle leads to the collocation formulas (Moritz 1980 a: Chap. 11).

Many examples integrating GPS and other data can be found in technical publications. For example, there are attempts to detect earth deformations from GPS and terrestrial data.

Part III: Local geodetic datums

5.11   Formulation of the problem

As we have remarked several times, the weak point of the Bruns–Hotine method is the insufficient accuracy of the zenith angle measurement precluding the practical use of this method for larger triangulations. The trigonometric heights obtained in this way are significantly less accurate than the horizontal positions.

A practical solution of this problem was to separate positions and heights. The horizontal position was calculated on the reference ellipsoid in the way we shall see later. Accurate heights were obtained by leveling referred to the "actual" level surfaces, in particular to the geoid.

Thus, this theoretically and practically unsatisfactory procedure used two different reference surfaces: the ellipsoid for horizontal position and the geoid for heights. The mutual position of these two surfaces was not even known because of lack of knowledge of the geoidal height N. It has been rightfully ridiculed as "2+1-dimensional geodesy".

There is a way out of this dilemma even for local (or rather regional) geodetic systems. The trigonometric height h is not determined by zenith-angle measurements but by using the simple formula

$$h = H + N \tag{5–91}$$

from leveled orthometric heights H by adding the geoid height N!

But how do we get the geoid? Even before the satellite era, there existed two methods:

1. the *astrogeodetic method*, determining N from deflections of the vertical ξ and η;

2. the *gravimetric method*, using for this purpose gravity anomalies Δg.

The theories of both methods were known as early as 1850, but what was lacking were data, especially gravimetric ones. Serious practical applications started not much before 1950, a hundred years later, just before the advent of satellites. This will be discussed in detail later in this book.

A reasonable measuring accuracy was achievable, but another difficulty appeared. Both methods require the evaluation of integrals of the data (ξ and η, or Δg) as continuous functions. The data, however, are always measured at discrete points only. Interpolation is necessary and introduces additional errors. If the data are distributed uniformly and densely, resulting errors may be kept small. The fundamental problem exists, however.

Summarizing, we may say: (1) The method of zenith angles is theoretically rigorous but not in general sufficiently accurate; (2) the astrogeodetic method using integration of vertical deflections is not theoretically rigorous in this sense but still may be accurate enough.

Method 1 has been treated in Part II of this chapter, so method 2 warrants detailed considerations in the present Part III.

5.12   Reduction of the astronomical measurements to the ellipsoid

Now we establish the relation between the natural coordinates Φ, Λ, H and the ellipsoidal coordinates φ, λ, h referring to an ellipsoid according to Helmert's projection.

The ellipsoidal height h and the orthometric height H have been considered, e.g., in Sect. 4.6 (see also Fig. 5.4 and Eq. (5–91)). They are related by $h = H + N$.

Thus, there remains the *reduction of the astronomical coordinates* Φ *and* Λ *to the ellipsoid* and, if we also include the astronomical observation of the azimuth, the astronomical azimuth A to the ellipsoid in order to obtain the ellipsoidal coordinates φ and λ and the ellipsoidal azimuth α.

We introduce the auxiliary quantities

$$\Delta\varphi = \Phi - \varphi \,,$$
$$\Delta\lambda = \Lambda - \lambda \,, \tag{5–92}$$
$$\Delta\alpha = A - \alpha \,.$$

The reduction of Φ and Λ to the corresponding ellipsoidal coordinates φ and λ is implicitly contained in Eq. (2–230):

$$\xi = \Phi - \varphi = \Delta\varphi \,,$$
$$\eta = (\Lambda - \lambda)\cos\varphi = \Delta\lambda \cos\varphi \,, \tag{5–93}$$

where we have substituted the respective auxiliary quantities. Thus, the conversion formulas from natural coordinates Φ, Λ, H to ellipsoidal coordinates φ, λ, h are

$$\varphi = \Phi - \xi \,,$$
$$\lambda = \Lambda - \eta/\cos\varphi \,, \tag{5–94}$$
$$h = H + N \,.$$

Now we turn to the reduction of the azimuth. Thus, the question is which $\Delta\alpha$ arises from $\Delta\varphi$ and $\Delta\lambda$. The answer is found in Eq. (5–75), where we only consider the last two terms on the right-hand side (i.e., we do not take into account changes of the point coordinates). Omitting all subscripts and introducing the auxiliary quantities of (5–92), we immediately get

$$\Delta\alpha = \cot z \sin\alpha\,\Delta\varphi + (\sin\varphi - \cos\alpha\,\cos\varphi\,\cot z)\,\Delta\lambda \qquad (5\text{--}95)$$

or, using $\Delta\varphi = \xi$ and $\Delta\lambda\,\cos\varphi = \eta$, yields

$$\Delta\alpha = \xi\sin\alpha\,\cot z + \sin\varphi\,\Delta\lambda - \eta\cos\alpha\,\cot z\,. \qquad (5\text{--}96)$$

This equation may be rearranged to

$$\Delta\alpha = \sin\varphi\,\Delta\lambda + (\xi\sin\alpha - \eta\cos\alpha)\cot z\,. \qquad (5\text{--}97)$$

Alternatively, by using $\Delta\lambda = \eta/\cos\varphi$, we get

$$\Delta\alpha = \eta\tan\varphi + (\xi\sin\alpha - \eta\cos\alpha)\cot z\,. \qquad (5\text{--}98)$$

In first-order triangulation, the lines of sight are usually almost horizontal so that $z \doteq 90°$, $\cot z \doteq 0$. Therefore, the corresponding term can in general be neglected and we get

$$\Delta\alpha = \eta\tan\varphi = \Delta\lambda\sin\varphi\,. \qquad (5\text{--}99)$$

This is *Laplace's equation* in its usual simplified form. It is remarkable that the differences $\Delta\alpha = A - \alpha$ and $\Delta\lambda = \Lambda - \lambda$ should be related in such a simple way. Laplace's equation is fundamental for the classical astrogeodetic computation of triangulations (Sect. 5.14).

For later reference we note that the total deflection of the vertical – that is, the angle ϑ between the actual plumb line and the ellipsoidal normal – is given by

$$\vartheta = \sqrt{\xi^2 + \eta^2} \qquad (5\text{--}100)$$

and that the deflection component ε in the direction of the azimuth α is

$$\varepsilon = \xi\cos\alpha + \eta\sin\alpha\,. \qquad (5\text{--}101)$$

It is clear that ϑ in (5–100) has nothing to do with the two different ϑ used for spherical and ellipsoidal-harmonic coordinates (polar distances).

Returning to the reduction of astronomical to the corresponding ellipsoidal quantities, we have (5–94) for the reduction of Φ, Λ, H to φ, λ, h and, finally, the formula

$$\alpha = A - \eta\tan\varphi \qquad (5\text{--}102)$$

reduces the astronomical azimuth A to the ellipsoidal azimuth α.

For the application of these formulas, we need the geoidal undulation N and the deflection components ξ and η with respect to the reference ellipsoid used. Two points should be noted:

1. The vertical axis of the reference ellipsoid is parallel to the earth's axis of rotation, but it need not be in an absolute position, its center coinciding with the earth's center of gravity. To repeat the reason: the earth axis is accessible to (astronomical) observation, whereas the geocenter is physically defined and inaccessible to direct geometrical observation.

2. The geocenter is accessible in two physically defined ways: (1) gravimetrically through Stokes' formula and (2) physically by the first Kepler law applied to satellite motion and responsible for the geocentricity of GPS orbits.

Note that unless otherwise stated, we always assume that our observations are made at sea level. This is not so unnatural for an inhabitant of a large plain region but causes headache to a geodesist working in the Alps or in the Rocky Mountains. We have already been confronted with this situation before, in gravity reduction, and will meet it repeatedly later, most prominently under the heading of Molodensky's problem.

It should also be mentioned that the ellipsoidal azimuth α in (5–102) refers to the actual target, which does not in general lie on the ellipsoid. For the conventional method of computation on the ellipsoid, one wishes the azimuth to refer to a target on the ellipsoid, which is the point at the foot of the normal through the actual target. Furthermore, α refers to what is called a normal section of the ellipsoid, rather than to a geodesic line, which is used in computation. In either case very small azimuth reductions are necessary; since these reductions are purely problems in ellipsoidal geometry, the reader is referred to any appropriate textbook.

Effect of polar motion

The direction of the earth's axis of rotation is not rigorously fixed, neither in space nor with respect to the earth, but undergoes very small, more or less periodic variations. Astronomers know it by the name of *nutation* (with respect to inertial space), geodesists know it by the name of *polar motion* (with respect to the earth's body). This phenomenon arises from a minute difference between the axes of rotation and of maximum inertia, the angle between these axes being about $0.3''$, and is somewhat similar to the precession of a spinning top. This motion of the pole has a main period of about 430 days, the Chandler period, but is rather irregular, presumably because of the movement of masses, atmospheric variations, etc. (Fig. 5.13).

The International Earth Rotation Service (IERS), initially International Latitude Service and then Polar Motion Service, which is maintained by the International Astronomical Union and by the International Union of

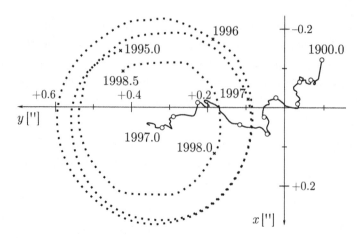

Fig. 5.13. Polar motion: mean pole displacement 1900–1997 (solid line), detailed polar motion 1995–1998 (dotted line)

Geodesy and Geophysics, continuously observes the variation of a number of parameters at a considerable number of stations distributed over the whole earth. Thus, it monitors variations of the earth's axis (polar motion) and of its angular speed of rotation.

The results are published as the rectangular coordinates of the instantaneous pole P_N with respect to a mean pole P_N^0. The astronomically observed values of Φ, Λ, and A naturally refer to the instantaneous pole P_N and must, therefore, be reduced to the mean pole, using the published values of x and y.

This is accomplished by means of the equations

$$\Phi = \Phi_{\text{obs}} - x \cos \lambda + y \sin \lambda \,,$$

$$\Lambda = \Lambda_{\text{obs}} - (x \sin \lambda + y \cos \lambda) \tan \varphi + y \tan \varphi_{\text{Gr}} \,, \qquad (5\text{–}103)$$

$$A = A_{\text{obs}} - (x \sin \lambda + y \cos \lambda) \sec \varphi \,.$$

Now Φ, Λ, A are referred to the mean pole; these values are used in geodesy because they do not vary with time. Longitude, throughout this book, is reckoned positive to the east, as is usual in geodesy; it should be mentioned that in the past literature these formulas are often written for west longitude, according to the former practice of astronomers. Since the correction terms containing x and y are extremely small (of the order of $0.1''$), we may use either the ellipsoidal values φ and λ or the astronomical values Φ and Λ in these terms. The term containing φ_{Gr} (the latitude of Greenwich) in the formula for Λ is usually omitted, so that the mean meridian of Greenwich remains fixed as the conventional *zero meridian*, rather than the astronomical longitude of Greenwich itself.

These formulas (5–103) are Eqs. (7-13), (7-14), and (7-15) of Moritz and Mueller (1987: pp. 419–420). It is interesting to note the close similarity between the azimuth reduction (5–98) because of the "zenith variation" – that is, the deflection of the vertical – and the longitude reduction of (5–103) because of the polar variation. Actually, the geometry for both cases is the same. The quantities $\xi, \eta, 90° - z, \varphi$ correspond to $x, y, \varphi, \varphi_{\mathrm{Gr}}$; the difference in sign of $\sin \alpha$ and $\sin \lambda$ is due to the fact that, when viewed from the zenith, azimuth is reckoned clockwise and, when viewed from the pole, east longitude is reckoned counterclockwise.

5.13   Reduction of horizontal and vertical angles and of distances

Horizontal angles

To reduce an observed horizontal angle ω to the ellipsoid, we note that every angle may be considered as the difference between two azimuths:

$$\omega = \alpha_2 - \alpha_1 . \tag{5–104}$$

Hence, we can apply formula (5–98). In the difference $\alpha_2 - \alpha_1$, the main term $\eta \tan \varphi$ drops out, so that for nearly horizontal lines of sight the whole reduction may be neglected.

Vertical angles

The relation between the measured zenith angle z' and the corresponding ellipsoidal zenith angle z may be given as

$$z = z' + \varepsilon = z' + \xi \cos \alpha + \eta \sin \alpha , \tag{5–105}$$

where α is the azimuth of the target.

Spatial distances

Electronic measurement of distance yields straight spatial distances l between two points A and B (Fig. 5.14). These distances may either be used directly for computations in the ellipsoidal coordinate system φ, λ, h, as in "three-dimensional geodesy" (see Sect. 5.9), or they may be reduced to the surface of the ellipsoid to obtain chord distances l_0 or geodesic distances s_0.

We again approximate the ellipsoidal arc $A_0 B_0$ by a circular arc of radius R that is the mean ellipsoidal radius of curvature along $A_0 B_0$. By applying the law of cosines to the triangle OAB, we find

$$l^2 = (R + h_1)^2 + (R + h_2)^2 - 2(R + h_1)(R + h_2) \cos \psi . \tag{5–106}$$

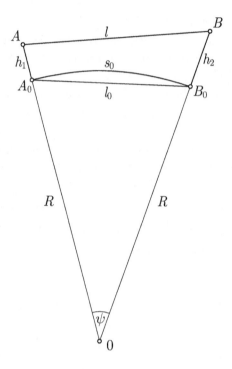

Fig. 5.14. Reduction of spatial distances

With

$$\cos \psi = 1 - 2 \sin^2 \frac{\psi}{2} , \tag{5–107}$$

this is transformed into

$$l^2 = (h_2 - h_1)^2 + 4R^2 \left(1 + \frac{h_1}{R}\right) \left(1 + \frac{h_2}{R}\right) \sin^2 \frac{\psi}{2} ; \tag{5–108}$$

and with

$$l_0 = 2R \sin \frac{\psi}{2} \tag{5–109}$$

and the abbreviation $\Delta h = h_2 - h_1$, we obtain

$$l^2 = \Delta h^2 + \left(1 + \frac{h_1}{R}\right) \left(1 + \frac{h_2}{R}\right) l_0^2 . \tag{5–110}$$

Hence, the chord l_0 and the arc s_0 are expressed by

$$l_0 = \sqrt{\frac{l^2 - \Delta h^2}{\left(1 + \frac{h_1}{R}\right) \left(1 + \frac{h_2}{R}\right)}} ; \tag{5–111}$$

$$s_0 = R\psi = 2R\,\sin^{-1}\frac{l_0}{2R}\,. \tag{5–112}$$

Ellipsoidal refinements of these formulas may be found in Rinner (1956).

As a matter of fact, spatial distances are independent of the vertical. Therefore, the reduction formula (5–111) does not contain the deflection of the vertical ε.

5.14   The astrogeodetic determination of the geoid

Helmert's formula

The shape of the geoid can be determined if the deflections of the vertical are given. *Helmert's formula*

$$dN = -\varepsilon\,ds \tag{5–113}$$

as given in (2–372) is the basic equation (Fig. 5.15). Integrating this relation, we get

$$N_B = N_A - \int_A^B \varepsilon\,ds\,, \tag{5–114}$$

where

$$\varepsilon = \xi\,\cos\alpha + \eta\,\sin\alpha \tag{5–115}$$

is the component of the deflection of the vertical along the profile AB, whose azimuth is α (see Eq. (5–101)).

Formula (5–114) expresses the geoidal undulation as an integral of the vertical deflections along a profile. Since N is a function of position, this integral is independent of the form of the line that connects the points A and B. This line need not necessarily be a geodesic on the ellipsoid, and α may in the general case be variable. In practice, north-south profiles ($\varepsilon = \xi$) or east-west profiles ($\varepsilon = \eta$) are often used. The integral (5–114) is to be evaluated

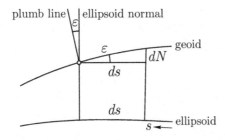

Fig. 5.15. Relation between geoidal undulation and deflection of the vertical

by a numerical or graphical integration. The deflection component ε must
be given at enough stations along the profile such that the interpolation
between these stations can be done reliably. Sometimes a map of ξ and η
is available for a certain area. Such a map is constructed by interpolation
between well-distributed stations at which ξ and η have been determined
(Grafarend and Offermanns 1975). Then the profiles of integration may be
suitably selected; loops may be formed to obtain redundancies which must
be adjusted.

If the deflection components ξ and η are obtained directly from the equations

$$\xi = \Phi - \varphi, \quad \eta = (\Lambda - \lambda) \cos \varphi, \tag{5-116}$$

that is, by comparing the astronomical and ellipsoidal (or geodetic) coordinates of the same point, then this method is called the *astrogeodetic determination of the geoid*.

The astronomical coordinates are directly observed; the ellipsoidal coordinates are obtained in the following way.

Determination of a local astrogeodetic datum

This is of historic interest only, but indispensible for an understanding of
the present classical triangulation system. In agreement with Part I, but in
contrast to Part II, "local" again means "regional", referring to a country
(e.g., France) or even a continent (e.g., European Datum or North-American
Datum). In a larger triangulation system, a certain "initial point" P_1 is
chosen for which the undulation N_1 and the components ξ_1 and η_1 of the
deflection of the vertical are prescribed. Here ξ_1, η_1, and N_1 may be assumed
arbitrarily in principle; the position of the reference ellipsoid with respect
to the earth is thereby fixed. For the sake of definiteness let us consider
the case that has been of greatest practical importance, that is, the case
in which $\xi_1 = \eta_1 = N_1 = 0$. In this case, because $\xi_1 = \eta_1 = 0$, the geoid
and the ellipsoid have the same surface normal so that, because $N_1 = 0$, the
ellipsoid is tangent to the geoid below P_1 (Fig. 5.16). The condition that
the axis of the reference ellipsoid be parallel to the earth's axis of rotation
finally determines the orientation of the triangulation net because Laplace's
equation (5–99) then gives $\Delta\alpha_1 = \eta_1 \tan \varphi_1 = 0$, so that $\alpha_1 = A_1$; that is, at
the initial point the ellipsoidal azimuth is equal to the astronomical azimuth.

Now we can reduce the measured distances and angles to the ellipsoid
and compute on it the position of the points of the triangulation net (their
ellipsoidal coordinates φ and λ) in the usual way. After measuring the coordinates Φ and Λ astronomically at the same points, we can then compute the
deflection components ξ and η by (5–116). Starting from the assumed value

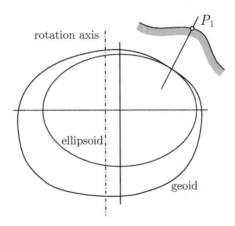

Fig. 5.16. The reference ellipsoid is tangent to the geoid at P_1

N_1 at the initial point P_1 (in our case, $N_1 = 0$), we can finally compute the geoidal heights N of any point of the triangulation net by repeated application of (5–114). These geoidal heights refer to the ellipsoid that was fixed by prescribing ξ_1, η_1, N_1, and, of course, its semimajor axis a and its flattening f. To employ a frequently used term, they refer to the given *astrogeodetic datum* $(a, f; \xi_1, \eta_1, N_1)$.

By means of N and the orthometric height H, the height h above the ellipsoid is obtained via $h = H + N$, so that the rectangular spatial coordinates X, Y, Z can be computed by (5–27). But unless ξ and η are absolute (geocentric) deflections, the origin of the coordinate system will not be at the center of the earth (see Sect. 5.7).

A flaw in the procedure described above apparently is that N, ξ, η are already needed for the reduction of the measured angles and distances to the ellipsoid. However, for this purpose approximate values of N, ξ, η are sufficient. These are obtained by performing the process just explained with unreduced angles and distances. We can also get suitable values for N, ξ, η in other ways, for instance, by Stokes' formula.

Use and misuse of Laplace's equation

It should be mentioned that in practice the component η has been often obtained from azimuth measurements using (5–102) in rearranged form, that is,

$$\eta = (A - \alpha) \cot \varphi, \qquad (5\text{–}117)$$

because astronomical measurements of azimuth are simpler than those of longitude. This is a misuse which may lead to a systematic distortion of the net. Longitude and azimuth are often measured at the same point. Then Laplace's condition

$$\Delta\alpha = \Delta\lambda \sin\varphi \qquad (5\text{--}118)$$

furnishes an important check on the correct orientation of the net and forces the axis of the ellipsoid to be parallel with the earth's axis of rotation. Thus it may be used for adjustment purposes. Astronomical stations with longitude and azimuth observations are, therefore, called *Laplace stations*. For these purposes, the measuring accuracy of astronomical field observations is sufficient, in contrast to the use for directly determining horizontal positions by $\varphi = \Phi - \xi$, etc. in Sect. 2.21.

The astrogeodetic determination of the geoid, also called *astronomical leveling*, was known to Helmert (1880) and even before.

Comparison with the Stokes method

It is quite instructive to compare Helmert's formula

$$N = N_A - \int_A^B \varepsilon \, ds \qquad (5\text{--}119)$$

for the astrogeodetic method with Stokes' formula

$$N = \frac{R}{4\pi\,\gamma_0} \iint_\sigma \Delta g \; S(\psi) \, d\sigma \qquad (5\text{--}120)$$

for the gravimetric method. Both methods use the gravity vector $\mathbf{g}$. It is compared with a normal gravity vector $\boldsymbol{\gamma}$. The components $\xi = \Delta\varphi$ and $\eta = \Delta\lambda \cos\varphi$ of the deflection of the vertical represent the differences in *direction*, and the gravity anomaly Δg represents the difference in *magnitude* of the two vectors. Helmert's formula determines the geoidal undulation N from ξ and η, that is, by means of the direction of $\mathbf{g}$, and Stokes' formula determines N from Δg, that is, by means of the magnitude of $\mathbf{g}$. Both formulas are somewhat similar: they are integrals which contain ε, or ξ and η, and Δg in linear form.

Otherwise, the two formulas show marked differences which are characteristic for the respective method. In Helmert's formula, the integration is extended over part of a profile; thus, it is sufficient to know the deflection of the vertical in a limited area. The position of the reference ellipsoid with respect to the earth's center of gravity is unknown, however, and can be determined only by means of the gravimetric method or, more practically, the analysis of satellite orbits (Sect. 7.2). Furthermore, the astrogeodetic method can be used only on land, because the necessary measurements are impossible at sea.

In Stokes' formula, however, the integration should be extended over the whole earth. The gravity anomaly Δg must be known all over the earth; however, accurate gravity measurements at sea are possible. The gravimetric method yields, for the whole earth, absolute geoidal undulations: the center of the reference ellipsoid coincides with the center of the earth. Nowadays, this is only a theoretical possibility because the required complete coverage of the whole earth is not available; again, GPS helps. Nevertheless, the gravimetric method is still basic: it furnishes, not the geocenter, but details of the geoid, together with the astrogeodetic method!

The astrogeodetic method has often been applied to the determination of geoidal sections. We mention, because of its pioneering character and its romantic title, "Das Geoid im Harz" by Galle (1914). In the years following 1970 it is becoming rare to use Helmert's integral formula in its original form, and deflections of the vertical are more and more combined with other data (gravity, GPS, and other satellite data) for a uniform determination of geoid and gravity field (see Chaps. 10 and 11).

Adjustment of nets of astrogeodetic geoidal heights

With a sufficiently dense net of astrogeodetic stations (preferably Laplace points) with an average station distance of 10–20 km, the Helmert integral (5–119) can be approximated by

$$\Delta N_{AB} \equiv N_B - N_A = - \int_A^B \varepsilon \, ds = - \frac{\varepsilon_A + \varepsilon_B}{2} \int_A^B ds \qquad (5\text{--}121)$$

or

$$\Delta N_{AB} = - \frac{\varepsilon_A + \varepsilon_B}{2} \, s_{AB} . \qquad (5\text{--}122)$$

Thus, the undulation difference can be computed for the line AB, and similarly for other lines BC and CA in the triangle ABC (Fig. 5.17). The closure condition

$$\Delta N_{AB} + \Delta N_{BC} + \Delta N_{CA} = 0 \qquad (5\text{--}123)$$

must be satisfied and imposed as a condition in the least-squares adjustment

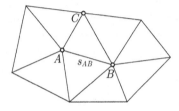

Fig. 5.17. Triangular net for an astrogeodetic geoid

of the net. Accordingly, the other triangles can be computed as in any other
height network (e.g., leveling net).

It is curious that it may be shown that such closures are mathematically
equivalent to the well-known relation

$$\frac{\partial^2 N}{\partial x\, \partial y} = \frac{\partial^2 N}{\partial y\, \partial x}.$$
(5–124)

See also Sect. 4.5.

5.15 Reduction for the curvature of the plumb line

Motivation
The astronomical coordinates Φ and Λ, as observed on the surface of the
earth, are not rigorously equal to their corresponding values at the geoid
because the plumb line, the line of force, is not straight, or in other words,
because the level surfaces are not parallel. Thus, if we wish our astronomical
coordinates to refer to the geoid, we must reduce our observations accordingly.

Examples of such cases are the following:

1. The gravimetric deflections have usually been computed by Vening
   Meinesz' formula for the geoid, so that either the gravimetric deflections must be reduced upward to the ground point or the astronomical
   observations must be reduced downward to the geoid, in order to make
   the two quantities comparable.

2. If astronomical observations are used for the determination of the
   geoid, the same reduction, in principle, must be applied.

Important remark
The principle of reduction of the plumb line is of fundamental theoretical
importance for understanding the geometry of the earth's gravity field. In
practice, it is usually disregarded if the topography is sufficiently flat, or
replaced by more sophisticated methods in mountainous areas, as we shall
see later (Sects. 8.12 and 8.13). The present section may be skimmed at first
reading, except for the *normal curvature of the plumb line* at its very end.

Principles
Consider the projection of the plumb line onto the meridian plane. According
to the well-known definition of the curvature of a plane curve, the angle

between two neighboring tangents of this projection of the plumb line is

$$d\varphi = -\kappa_1 \, dh \,, \qquad (5\text{--}125)$$

where the minus sign is conventional and the curvature κ_1 is given by (2–50):

$$\kappa_1 = \frac{1}{g} \frac{\partial g}{\partial x} \,. \qquad (5\text{--}126)$$

The x-axis is horizontal and points northward. Hence, the total change of latitude along the plumb line between a point on the ground, P, and its projection onto the geoid, P_0, is given by

$$\delta\varphi = \int_{P_0}^{P} d\varphi = - \int_{P_0}^{P} \kappa_1 \, dh \qquad (5\text{--}127)$$

or

$$\delta\varphi = - \int_{P_0}^{P} \frac{1}{g} \frac{\partial g}{\partial x} \, dh \,. \qquad (5\text{--}128)$$

Using κ_2 of (2–51), we similarly find for the change of longitude

$$\delta\lambda \, \cos\varphi = - \int_{P_0}^{P} \frac{1}{g} \frac{\partial g}{\partial y} \, dh \,, \qquad (5\text{--}129)$$

where the y-axis is horizontal and points eastward.

Alternative formulas

There is a close relationship between the curvature reduction of astronomical coordinates and the orthometric reduction of leveling, considered in Sect. 4.3.

The orthometric correction $d(\mathrm{OC})$ has been defined as the quantity that must be added to the leveling increment dn in order to convert it into the orthometric height difference dH:

$$d(\mathrm{OC}) = dH - dn \,. \qquad (5\text{--}130)$$

From Fig. 5.18, we see that, for a north-south profile, the curvature reduction and the orthometric correction are related by the simple formula

$$\delta\varphi = \frac{\partial(\mathrm{OC})}{\partial x} \,. \qquad (5\text{--}131)$$

Similarly, we find

$$\delta\lambda \, \cos\varphi = \frac{\partial(\mathrm{OC})}{\partial y} \,. \qquad (5\text{--}132)$$

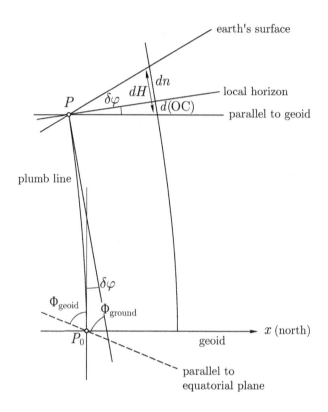

Fig. 5.18. Plumb-line curvature and orthometric correction

According to Sect. 4.3, we have

$$dC = g\, dn = -dW\,, \qquad H = \frac{C}{\bar{g}}\,. \tag{5-133}$$

Hence, (5–130) becomes

$$d(\text{OC}) = dH - \frac{1}{g}\, dC = dH + \frac{1}{g}\, dW\,, \tag{5-134}$$

so that

$$\delta\varphi = \frac{\partial H}{\partial x} + \frac{1}{g}\frac{\partial W}{\partial x}\,,$$

$$\delta\lambda\,\cos\varphi = \frac{\partial H}{\partial y} + \frac{1}{g}\frac{\partial W}{\partial y}\,. \tag{5-135}$$

These equations relate the reduction for the curvature of the plumb line to the orthometric height H and the potential W. In view of the irregular shape of the plumb lines, it is remarkable that such simple general relations as (5–131), (5–132), and (5–135) exist.

These relations may be used to find computational formulas for the curvature reductions $\delta\varphi$ and $\delta\lambda$. We have

$$
d(\text{OC}) = dH - \frac{dC}{g} = d\left(\frac{C}{\bar{g}}\right) - \frac{dC}{g}
$$
$$
= \frac{dC}{\bar{g}} - \frac{C}{\bar{g}^2}\, d\bar{g} - \frac{dC}{g} = -\frac{C}{\bar{g}^2}\, d\bar{g} + \frac{g - \bar{g}}{\bar{g}}\,\frac{dC}{g}
$$
(5–136)

or

$$
d(\text{OC}) = -\frac{H}{\bar{g}}\, d\bar{g} + \frac{g - \bar{g}}{\bar{g}}\, dn .
$$
(5–137)

By substituting this into (5–131) and (5–132), we obtain

$$
\delta\varphi = -\frac{H}{\bar{g}}\frac{\partial \bar{g}}{\partial x} + \frac{g - \bar{g}}{\bar{g}} \tan\beta_1 ,
$$
$$
\delta\lambda \cos\varphi = -\frac{H}{\bar{g}}\frac{\partial \bar{g}}{\partial y} + \frac{g - \bar{g}}{\bar{g}} \tan\beta_2 ,
$$
(5–138)

where we have set

$$
\tan\beta_1 = \frac{\partial n}{\partial x} , \qquad \tan\beta_2 = \frac{\partial n}{\partial y} ,
$$
(5–139)

so that β_1 and β_2 are the angles of inclination of the north-south and east-west profiles with respect to the local horizon; $\bar{g}$ is the mean value of gravity between the geoid and the ground. In these formulas, we need only this mean value $\bar{g}$, together with its horizontal derivatives, and the ground value g, whereas in (5–128) and (5–129), we must know the horizontal derivatives of gravity all along the plumb line. The detailed shape of the plumb lines does not directly enter into (5–138) as it does into (5–128) and (5–129).

The mean value $\bar{g}$ is found by a Prey reduction of the measured gravity g. In order that the numerical differentiations $\partial g/\partial x$ and $\partial g/\partial y$ give reliable results, a dense gravity net around the station is necessary, and the Prey reduction must be performed carefully. The inclination angles β_1 and β_2 are taken from a topographical map.

The sign of these corrections may be found in the following way. If g decreases in the x-direction, then formulas (5–128) and (5–138) give $\delta\varphi > 0$ and Fig. 5.18 shows that Φ at P_0 is then greater than at P. The same holds for Λ, so that we have

$$
\Phi_{\text{geoid}} = \Phi_{\text{ground}} + \delta\varphi ,
$$
$$
\Lambda_{\text{geoid}} = \Lambda_{\text{ground}} + \delta\lambda .
$$
(5–140)

Integrated form

In formula (5–114), the deflection components ξ and η refer to the geoid. This means that the astronomical observations of Φ and Λ must be reduced to the geoid.

It is also possible and often more convenient to apply this correction for plumb-line curvature not to the astronomical coordinates Φ and Λ but to the geoidal height differences computed from the unreduced deflection components.

These N values, denoted by N', are obtained by using in (5–116) the directly observed Φ and Λ, which define the direction of the plumb line at the station P (Fig. 5.19). The notation N will be reserved for the correct geoidal heights. Then we read from Fig. 5.19:

$$dh = dN + dH = dN' + dn\,, \tag{5–141}$$

where h is the geometric height above the ellipsoid. Thus, we see that the difference between the unreduced and the correct element of geoidal height,

$$dN' - dN = dH - dn = d(\text{OC})\,, \tag{5–142}$$

is equal to the difference between the element dH of orthometric height and

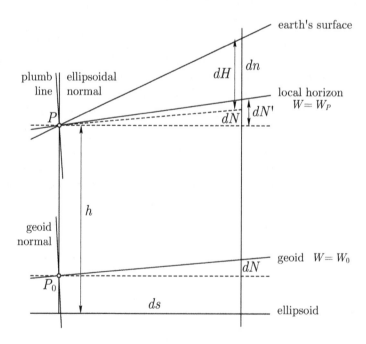

Fig. 5.19. Reduction of astronomical leveling

the leveling increment dn, which is the orthometric reduction $d(\text{OC})$. Thus,

$$N_B - N_A = N_B' - N_A' - \text{OC}_{AB}, \qquad (5\text{-}143)$$

so that we can immediately apply Eq. (4–46):

$$N_B - N_A = - \int_A^B \varepsilon\, ds - \int_A^B \frac{g - \gamma_0}{\gamma_0}\, dn + \frac{\bar{g}_B - \gamma_0}{\gamma_0} H_B - \frac{\bar{g}_A - \gamma_0}{\gamma_0} H_A, \quad (5\text{-}144)$$

where γ_0 is our usual constant $\gamma_{45°}$; the deflection components ε are computed from the observed ground values Φ and Λ by (5–116) and (5–115). These ideas go back to Helmert, but they are hardly used anymore.

Curvature of the normal plumb line

If, instead of the actual gravity g, the normal gravity γ is applied for the computation of the plumb-line curvature, we find, using

$$\gamma = \gamma_a \left(1 + f^* \sin^2 \varphi - \frac{2}{a} h \cdots \right), \qquad (5\text{-}145)$$

that

$$\frac{\partial \gamma}{\partial x} \doteq \frac{1}{R} \frac{\partial \gamma}{\partial \varphi} \doteq \frac{2\gamma_a}{R} f^* \sin \varphi \cos \varphi \doteq \frac{2\gamma}{R} f^* \sin \varphi \cos \varphi,$$

$$\frac{\partial \gamma}{\partial y} \doteq \frac{1}{R \cos \varphi} \frac{\partial \gamma}{\partial \lambda} = 0. \qquad (5\text{-}146)$$

Hence, the integrand $(1/\gamma)(\partial \gamma / \partial x)$ in (5–128) does not depend on h, so that the integration can be performed immediately. We find

$$\delta \varphi_{\text{normal}} = - \frac{f^*}{R} h \sin 2\varphi = -0.17'' \, h_{\text{[km]}} \sin 2\varphi,$$

$$\delta \lambda_{\text{normal}} = 0. \qquad (5\text{-}147)$$

The curvature of the normal plumb line in the east-west direction is zero, owing to the rotational symmetry of the ellipsoid of revolution. The *normal reduction* (5–147) *is very simple and practically important*, see especially Sect. 8.13.

5.16   Best-fitting ellipsoids and the mean earth ellipsoid

We define the mean earth ellipsoid physically as that ellipsoid of revolution which shares with the earth the mass M, the potential W_0, the difference between the principal moments of inertia $G(C - \bar{A})$, where $\bar{A} = (A + B)/2$, and the angular velocity ω.

It is also possible to define the mean earth ellipsoid geometrically as that ellipsoid which approximates the geoid most closely. This definition is perhaps more appealing to the geodesist; it may, for instance, be formulated by the condition that the sum of the squares of the deviations N of the geoid from the ellipsoid be a minimum:

$$\iint_\sigma N^2 \, d\sigma = \text{minimum} \qquad (5\text{--}148)$$

(this integral is to be considered the limit of a sum). The condition of closest approximation may also be expressed in terms of the deflections of the vertical:

$$\iint_\sigma (\xi^2 + \eta^2) \, d\sigma = \text{minimum} , \qquad (5\text{--}149)$$

minimizing the sum of the squares of the total deflection of the vertical

$$\vartheta = \sqrt{\xi^2 + \eta^2} . \qquad (5\text{--}150)$$

Many other similar definitions of closest approximation are possible.

The first definition, based on the condition (5–148), is the most plausible and the most appropriate intuitively, as has been already noted by Helmert; in principle, however, all definitions are more or less conventional and are equivalent theoretically as we shall see below.

The second definition, based on the condition (5–149), uses deflections of the vertical and is, thus, particularly well adapted to the astrogeodetic method. However, since this method can be applied only over limited areas, at most spanning the continents, the integral (5–149) must be replaced by a sum covering the astronomical stations of a restricted region:

$$\sum (\xi^2 + \eta^2) \, d\sigma = \text{minimum} . \qquad (5\text{--}151)$$

In this way, we can get only the best-fitting ellipsoid for the region considered, rather than a general earth ellipsoid. As Fig. 5.20 indicates, a *locally best-fitting ellipsoid* may be quite different from the mean earth ellipsoid, which can be considered a best-fitting ellipsoid for the whole earth.

If a reasonably good approximation of the earth ellipsoid by a local best-fitting ellipsoid is desired, it is advisable to subtract the effect of the topography and of its isostatic compensation from the astrogeodetic deflections of the vertical before the minimum condition (5–151) is applied. The purpose of this procedure is to smooth the irregularities of the geoid. In this way,

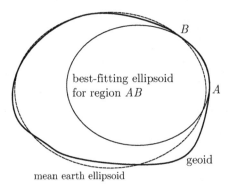

Fig. 5.20. A locally best-fitting ellipsoid and the mean earth ellipsoid

Hayford computed the international ellipsoid as ellipsoid that best fits the isostatically reduced vertical deflections in the United States. Rapp (1963) made an interesting recomputation.

Please note: Don't use formula (5–151) in spite of its historical importance: the determination of local best-fitting ellipsoids is hopelessly obsolete now!

The previously described method is impaired by unknown density anomalies and by the lack of complete isostatic compensation. Therefore, it is better to go still one step further and subtract the gravimetrically computed values ξ^g, η^g from the astrogeodetic deflections ξ^a, η^a. Then the minimum condition

$$\sum \left[(\xi^a - \xi^g)^2 + (\eta^a - \eta^g)^2 \right] = \text{minimum} \qquad (5\text{–}152)$$

results. Thus, we may say that Hayford's method is equivalent to the use of (5–152), the gravimetric values ξ^g, η^g being approximated by deflections that represent the effect of topography and of its isostatic compensation only. If the isostatic compensation were complete, and if we had perfect knowledge of the density above the geoid, both methods would give exactly the same result if applied properly.

Equivalence of different definitions of the earth ellipsoid

It is quite remarkable that the minimum definitions (5–148) or (5–149) and a similar definition due to Rudzki, using the condition

$$\iint_\sigma (\Delta g)^2 \, d\sigma = \text{minimum} \,, \qquad (5\text{–}153)$$

yield results which, to the usual spherical approximation, are identical with each other and with the physical definition in terms of M, W_0, $C - \bar{A}$, and

ω. This can be seen as follows. We write the spherical-harmonic expansion of the disturbing potential in the form

$$T = \frac{G\,\delta M}{R} + \sum_{n=1}^{\infty} \sum_{m=0}^{n} \left[a_{nm} R_{nm}(\vartheta, \lambda) + b_{nm} S_{nm}(\vartheta, \lambda) \right]. \qquad (5\text{-}154)$$

Then, according to Sect. 2.17, Eqs. (2–351) and (2–359) or (2–363), we have

$$N = \frac{G\,\delta M}{R\,\gamma_0} - \frac{\delta W}{\gamma_0} + \frac{1}{\gamma_0} \sum_{n=1}^{\infty} \sum_{m=0}^{n} \left[a_{nm} R_{nm}(\vartheta, \lambda) + b_{nm} S_{nm}(\vartheta, \lambda) \right] \quad (5\text{-}155)$$

and

$$\Delta g = -\frac{G\,\delta M}{R^2} + \frac{2\delta W}{R}$$

$$+ \frac{1}{R} \sum_{n=1}^{\infty} \sum_{m=0}^{n} \left[(n-1)\, a_{nm} R_{nm}(\vartheta, \lambda) + (n-1)\, b_{nm} S_{nm}(\vartheta, \lambda) \right];$$

$$(5\text{-}156)$$

remember that γ_0 denotes a global mean value of gravity. The condition of equal masses, $\delta M = 0$, is very natural and will be assumed. If we square the formulas for N and Δg and integrate over the whole earth, then all the integrals of products of different harmonics R_{nm} and S_{nm} will be zero, according to the orthogonality property (1–83), and the remaining integrals will be given by (1–84). Thus, we find

$$\iint_\sigma N^2 \, d\sigma = \frac{4\pi}{\gamma_0^2}\, \delta W^2$$

$$+ \frac{4\pi}{\gamma_0^2} \sum_{n=1}^{\infty} \frac{1}{2n+1} \left[a_{n0}^2 + \sum_{m=1}^{n} \frac{(n+m)!}{2(n-m)!} \left(a_{nm}^2 + b_{nm}^2 \right) \right],$$

$$(5\text{-}157)$$

$$\iint_\sigma (\Delta g)^2 \, d\sigma = \frac{16\pi}{R^2}\, \delta W^2$$

$$+ \frac{4\pi}{R^2} \sum_{n=1}^{\infty} \frac{(n-1)^2}{2n+1} \left[a_{n0}^2 + \sum_{m=1}^{n} \frac{(n+m)!}{2(n-m)!} \left(a_{nm}^2 + b_{nm}^2 \right) \right].$$

$$(5\text{-}158)$$

By a more complicated derivation, which we omit here but which can be found in Molodenskii et al. (1962: p. 87), one gets the similar formula

$$\iint_\sigma (\xi^2 + \eta^2) \, d\sigma = \frac{4\pi}{R^2 \gamma_0^2} \sum_{n=1}^{\infty} \frac{n\,(n+1)}{2n+1} \left[a_{n0}^2 + \sum_{m=1}^{n} \frac{(n+m)!}{2(n-m)!} \left(a_{nm}^2 + b_{nm}^2 \right) \right].$$

$$(5\text{-}159)$$

Varying the size and shape of the reference ellipsoid and its position with respect to the earth changes only the coefficients δW, a_{10}, a_{11}, b_{11}, and a_{20}, leaving the other coefficients practically invariant. Thus, the minimum of any of the integrals (5–157), (5–158), (5–159) is obtained if all these coefficients are equal to zero. Now, $\delta W = 0$ means equal potential $U_0 = W_0$; $a_{10} = a_{11} = b_{11} = 0$ means absolute position (coincident centers of gravity); and $a_{20} = 0$ means equality of J_2 or of $C - (A + B)/2$.

Therefore, the equivalence of the physical definition by means of M, W_0, $C - \bar{A}$, ω and of the condition of closest approximation in any of the forms (5–148), (5–149), or (5–153) has been established. (It may be noted that (5–158) contains no first-degree term, because of the factor $(n - 1)^2$, and that (5–159) contains no term of degree zero, so that these equations do not determine the missing terms.)

Best-fitting ellipsoid and World Geodetic System

It should be remembered, however, that the mean earth ellipsoid, defined in this manner, is not necessarily the best reference surface for practical geodetic purposes. It is essentially defined empirically by means of empirical determinations of GM, W_0, etc. Its parameters will change with every improvement in the quality or the number of the relevant measurements (gravity, distances, etc.). Since an enormous amount of numerical data is based on an assumed reference ellipsoid, it would be highly impractical to change it very often, for this would involve repeated transformations of all the data. It is much better to use a fixed reference ellipsoid with rigidly assumed parameters, which can be more or less arbitrary if only they give a reasonably good approximation. In this respect, the Geodetic Reference System 1980 is still (2005) perfectly acceptable.

A certain amount of conflict exists between the interests of geodesists and astronomers regarding the earth ellipsoid. The geodesist needs a permanent reference surface, whereas the astronomer wants the best approximation of the earth by an ellipsoid. A good compromise is to use a fixed geodetic reference ellipsoid, but from time to time to compute the "best" corrections to the assumed parameters for astronomical and other purposes. This has been the practice of the IAG since 1974.

6 Gravity field outside the earth

6.1 Introduction

The gravity field outside the earth is particularly important at satellite altitude; this will be treated mainly in Chap. 7. The considerations of the present chapter are applicable to gravitational forces also at satellites (see Sect. 7.2), but their main practical purpose is to compute test values for the gravity vector, gravity disturbances, and gravity anomalies at flight elevations for comparison with airborne gravimetry for reference and calibration purposes. Airborne gravimetry is much faster than both terrestrial and shipborne gravimetry, so it is of interest also for geophysical prospecting.

For computational reasons, it is again convenient to split the gravity potential W and the gravity vector

$$\mathbf{g} = \operatorname{grad} W \qquad (6\text{–}1)$$

into a normal potential U and a normal gravity vector

$$\boldsymbol{\gamma} = \operatorname{grad} U \,, \qquad (6\text{–}2)$$

and the disturbing potential $T = W - U$ and the gravity disturbance vector

$$\delta\mathbf{g} = \operatorname{grad} T = \mathbf{g} - \boldsymbol{\gamma} \,. \qquad (6\text{–}3)$$

The normal gravity field is usually taken to be the gravity field of a suitable equipotential ellipsoid. This permits closed formulas and offers other advantages of mathematical simplicity (see Sect. 2.12).

Thus, U and $\boldsymbol{\gamma}$ are computed first, and W and $\mathbf{g}$ are then obtained by

$$W = U + T \,,$$
$$\mathbf{g} = \boldsymbol{\gamma} + \delta\mathbf{g} \,. \qquad (6\text{–}4)$$

For some purposes, we need the vector of gravitation, grad V (pure attraction without centrifugal force), rather than the vector of gravity. The gravitational vector is computed from the gravity vector by subtracting the vector of centrifugal force:

$$\operatorname{grad} V = \mathbf{g} - \operatorname{grad} \Phi = \mathbf{g} - \begin{bmatrix} \omega^2 x \\ \omega^2 y \\ 0 \end{bmatrix} \,, \qquad (6\text{–}5)$$

where the notations of Sect. 2.1 are used. The rectangular coordinate system x, y, z will be applied in this chapter in the usual sense: it is geocentric, the x- and y-axes lying in the equatorial plane with Greenwich longitudes $0°$ and $90°$ East, respectively, and the z-axis being the rotation axis of the earth.

The sign of the components of $\mathbf{g}$, $\boldsymbol{\gamma}$, $\delta\mathbf{g}$, etc., will always be chosen so that they are positive in the direction of increasing coordinates.

6.2 Normal gravity vector

The gravity field of an equipotential ellipsoid is best expressed in terms of ellipsoidal-harmonic coordinates u, β, λ, introduced in Sects. 1.15 and 2.7. They are related to rectangular coordinates x, y, z by

$$x = \sqrt{u^2 + E^2} \, \cos\beta \cos\lambda \,,$$

$$y = \sqrt{u^2 + E^2} \, \cos\beta \sin\lambda \,, \tag{6-6}$$

$$z = u \sin\beta \,.$$

If x, y, z are given, then u, β, λ can be computed by closed formulas. First we find

$$x^2 + y^2 = (u^2 + E^2)\cos^2\beta \,, \quad z^2 = u^2 \sin^2\beta \,. \tag{6-7}$$

Eliminating β between these two equations, we obtain a quadratic equation for u^2, whose solution is

$$u^2 = (x^2 + y^2 + z^2 - E^2)\left[\frac{1}{2} + \frac{1}{2}\sqrt{1 + \frac{4E^2 z^2}{(x^2 + y^2 + z^2 - E^2)^2}}\,\right]. \tag{6-8}$$

Then β is given by

$$\tan\beta = \frac{z\sqrt{u^2 + E^2}}{u\sqrt{x^2 + y^2}} \,, \tag{6-9}$$

and for λ we simply have

$$\tan\lambda = \frac{y}{x} \,. \tag{6-10}$$

With known ellipsoidal-harmonic coordinates, the normal potential U is given by (2–126):

$$U(u,\beta) = \frac{GM}{E}\tan^{-1}\frac{E}{u} + \tfrac{1}{2}\omega^2 a^2 \frac{q}{q_0}\left(\sin^2\beta - \tfrac{1}{3}\right) + \tfrac{1}{2}\omega^2(u^2 + E^2)\cos^2\beta \,. \tag{6-11}$$

The components of $\boldsymbol{\gamma}$ along the coordinate lines are, by (2–131) and (2–132),

$$\gamma_u = \frac{1}{w}\frac{\partial U}{\partial u} = -\frac{1}{w}\left[\frac{GM}{u^2 + E^2} + \frac{\omega^2 a^2 E}{u^2 + E^2}\frac{q'}{q_0}\left(\tfrac{1}{2}\sin^2\beta - \tfrac{1}{6}\right) - \omega^2 u \cos^2\beta\right],$$

$$\gamma_\beta = \frac{1}{w\sqrt{u^2 + E^2}}\frac{\partial U}{\partial\beta} = -\frac{1}{w}\left[-\frac{\omega^2 a^2}{\sqrt{u^2 + E^2}}\frac{q}{q_0} + \omega^2\sqrt{u^2 + E^2}\right]\sin\beta\cos\beta,$$

$$\gamma_\lambda = \frac{1}{\sqrt{u^2 + E^2}\cos\beta}\frac{\partial U}{\partial\lambda} = 0.$$

$$(6\text{--}12)$$

To get the components of $\boldsymbol{\gamma}$ in the xyz-system, we compute

$$\frac{\partial U}{\partial u} = \frac{\partial U}{\partial x}\frac{\partial x}{\partial u} + \frac{\partial U}{\partial y}\frac{\partial y}{\partial u} + \frac{\partial U}{\partial z}\frac{\partial z}{\partial u}, \quad \text{etc.} \qquad (6\text{--}13)$$

The partial derivatives of x, y, z with respect to u, β, λ are obtained by differentiating equations (6–6); we find

$$\frac{\partial U}{\partial u} = \frac{u}{\sqrt{u^2 + E^2}}\cos\beta\cos\lambda\frac{\partial U}{\partial x} + \frac{u}{\sqrt{u^2 + E^2}}\cos\beta\sin\lambda\frac{\partial U}{\partial y} + \sin\beta\frac{\partial U}{\partial z},$$

$$\frac{\partial U}{\partial\beta} = -\sqrt{u^2 + E^2}\sin\beta\cos\lambda\frac{\partial U}{\partial x} - \sqrt{u^2 + E^2}\sin\beta\sin\lambda\frac{\partial U}{\partial y} + u\cos\beta\frac{\partial U}{\partial z},$$

$$\frac{\partial U}{\partial\lambda} = -\sqrt{u^2 + E^2}\cos\beta\sin\lambda\frac{\partial U}{\partial x} + \sqrt{u^2 + E^2}\cos\beta\cos\lambda\frac{\partial U}{\partial y}.$$

$$(6\text{--}14)$$

Introducing the components

$$\gamma_x = \frac{\partial U}{\partial x}, \cdots; \quad \gamma_u = \frac{1}{w}\frac{\partial U}{\partial u}, \cdots, \qquad (6\text{--}15)$$

we obtain

$$\gamma_u = \frac{u}{w\sqrt{u^2 + E^2}}\cos\beta\cos\lambda\,\gamma_x + \frac{u}{w\sqrt{u^2 + E^2}}\cos\beta\sin\lambda\,\gamma_y + \frac{1}{w}\sin\beta\,\gamma_z,$$

$$\gamma_\beta = -\frac{1}{w}\sin\beta\cos\lambda\,\gamma_x - \frac{1}{w}\sin\beta\sin\lambda\,\gamma_y + \frac{u}{w\sqrt{u^2 + E^2}}\cos\beta\,\gamma_z,$$

$$\gamma_\lambda = -\sin\lambda\,\gamma_x + \cos\lambda\,\gamma_y.$$

$$(6\text{--}16)$$

These are the formulas of an orthogonal rectangular coordinate transformation. The inverse transformation is obtained by interchanging the rows and columns in the matrix of this equation system. Thus, we obtain

$$\gamma_x = \frac{u}{w\sqrt{u^2+E^2}}\cos\beta\cos\lambda\,\gamma_u - \frac{1}{w}\sin\beta\cos\lambda\,\gamma_\beta - \sin\lambda\,\gamma_\lambda ,$$

$$\gamma_y = \frac{u}{w\sqrt{u^2+E^2}}\cos\beta\sin\lambda\,\gamma_u - \frac{1}{w}\sin\beta\sin\lambda\,\gamma_\beta + \cos\lambda\,\gamma_\lambda , \qquad (6\text{--}17)$$

$$\gamma_z = \frac{1}{w}\sin\beta\,\gamma_u + \frac{u}{w\sqrt{u^2+E^2}}\cos\beta\,\gamma_\beta .$$

This follows from the definition of these coefficients as direction cosines. Equations (6–17) may also be found by solving the linear Eqs. (6–16) with respect to γ_x, γ_y, γ_z in some other way.

The formulas of the present section are completely rigorous. They can easily be programmed. Here it would not be appropriate to use the spherical approximation because they are relatively large quantities of the normal ellipsoidal field.

6.3   Gravity disturbance vector from gravity anomalies

In Sect. 1.4, we have introduced spherical coordinates: r (radius vector), ϑ (polar distance), λ (geocentric longitude) (see Fig. 1.3). Now we use these coordinates again but replace the polar distance ϑ by its complement, the geocentric latitude $\bar\varphi$ (Fig. 6.3). In analogy to (1–26), these spherical coor-

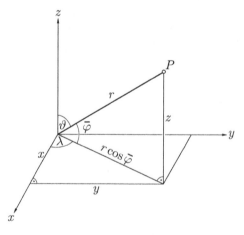

Fig. 6.1. Spherical coordinates $r, \bar\varphi$ (or ϑ, respectively), λ and rectangular coordinates x, y, z

dinates are related to rectangular coordinates x, y, z by the equations

$$x = r \cos \bar{\varphi} \cos \lambda \,,$$
$$y = r \cos \bar{\varphi} \sin \lambda \,, \tag{6–18}$$
$$z = r \sin \bar{\varphi}$$

or inversely by

$$r = \sqrt{x^2 + y^2 + z^2} \,,$$
$$\bar{\varphi} = \tan^{-1} \frac{z}{\sqrt{x^2 + y^2}} \,, \tag{6–19}$$
$$\lambda = \tan^{-1} \frac{y}{x} \,.$$

Now it is convenient to start with the components δg_r, $\delta g_{\bar{\varphi}}$, δg_λ of the gravity disturbance vector $\delta \mathbf{g}$, Eq. (6–3), in the spherical coordinates r, $\bar{\varphi}$, λ. In analogy to (2–377), we have

$$\delta g_r = \frac{\partial T}{\partial r} \,, \qquad \delta g_{\bar{\varphi}} = \frac{1}{r} \frac{\partial T}{\partial \bar{\varphi}} \,, \qquad \delta g_\lambda = \frac{1}{r \cos \bar{\varphi}} \frac{\partial T}{\partial \lambda} \,. \tag{6–20}$$

Since we are dealing with the relatively small quantities of the disturbing field, a spherical approximation may be sufficient (Sect. 2.13), as it was in the case of Stokes' formula.

The disturbing potential T may be expressed in terms of the free-air anomalies at the earth's surface by the formula of Pizzetti, Eqs. (2–302) and (2–303),

$$T_P = T(r, \bar{\varphi}, \lambda) = \frac{R}{4\pi} \iint_\sigma \Delta g \, S(r, \psi) \, d\sigma \,, \tag{6–21}$$

where $S(r, \psi)$ is the extended Stokes function,

$$S(r, \psi) = \frac{2R}{l} + \frac{R}{r} - 3 \frac{Rl}{r^2} - \frac{R^2}{r^2} \cos \psi \left(5 + 3 \ln \frac{r - R \cos \psi + l}{2r} \right), \tag{6–22}$$

and

$$l = \sqrt{r^2 + R^2 - 2Rr \cos \psi} \,. \tag{6–23}$$

According to (6–20), we must differentiate (6–21) with respect to r, $\bar{\varphi}$, and λ. Here we note that the integral on the right-hand side of (6–21) depends on r, $\bar{\varphi}$, λ only through the function $S(r, \psi)$. Thus, Δg being constant with respect to the differentiation, we have

$$\delta g_r = \frac{R}{4\pi} \iint\limits_{\sigma} \Delta g \, \frac{\partial S(r, \psi)}{\partial r} \, d\sigma \,,$$

$$\delta g_{\bar{\varphi}} = \frac{R}{4\pi \, r} \iint\limits_{\sigma} \Delta g \, \frac{\partial S(r, \psi)}{\partial \bar{\varphi}} \, d\sigma \,, \qquad (6\text{--}24)$$

$$\delta g_\lambda = \frac{R}{4\pi \, r \, \cos \bar{\varphi}} \iint\limits_{\sigma} \Delta g \, \frac{\partial S(r, \psi)}{\partial \lambda} \, d\sigma \,.$$

The point P at which $\delta\mathbf{g}$ is to be computed has the coordinates $\bar{\varphi}$, λ; let the corresponding coordinates of the variable point P', to which Δg and $d\sigma$ refer, be denoted by $\bar{\varphi}'$, λ'. Then $d\sigma$ will be expressed by

$$d\sigma = \cos \bar{\varphi}' \, d\bar{\varphi}' \, d\lambda' \qquad (6\text{--}25)$$

and ψ, the angular distance between P and P', is represented via

$$\cos \psi = \sin \bar{\varphi} \sin \bar{\varphi}' + \cos \bar{\varphi} \cos \bar{\varphi}' \cos(\lambda' - \lambda) \,. \qquad (6\text{--}26)$$

We have

$$\frac{\partial S(r, \psi)}{\partial \bar{\varphi}} = \frac{\partial S(r, \psi)}{\partial \psi} \frac{\partial \psi}{\partial \bar{\varphi}} \,, \qquad \frac{\partial S(r, \psi)}{\partial \lambda} = \frac{\partial S(r, \psi)}{\partial \psi} \frac{\partial \psi}{\partial \lambda} \,. \qquad (6\text{--}27)$$

Now we recall the corresponding derivations in Sect. 2.19, leading to Vening Meinesz' formula. As a spherical approximation which is sufficient for T, $\delta\mathbf{g}$, etc., we may identify the geocentric latitude $\bar{\varphi}$ with the ellipsoidal latitude φ. Thus, Eqs. (6–27) and (2–380) are completely analogous, and (2–383) may be borrowed from Sect. 2.19:

$$\frac{\partial \psi}{\partial \bar{\varphi}} = -\cos \alpha \,, \qquad \frac{\partial \psi}{\partial \lambda} = -\cos \bar{\varphi} \sin \alpha \,. \qquad (6\text{--}28)$$

The azimuth α is given by formula (2–388):

$$\tan \alpha = \frac{\cos \bar{\varphi}' \sin(\lambda' - \lambda)}{\cos \bar{\varphi} \sin \bar{\varphi}' - \sin \bar{\varphi} \cos \bar{\varphi}' \cos(\lambda' - \lambda)} \,. \qquad (6\text{--}29)$$

By means of (6–27) and (6–28), Eqs. (6–24) become

$$\delta g_r = \frac{R}{4\pi} \iint\limits_{\sigma} \Delta g \, \frac{\partial S(r, \psi)}{\partial r} \, d\sigma \,,$$

$$\delta g_{\bar{\varphi}} = -\frac{R}{4\pi \, r} \iint\limits_{\sigma} \Delta g \, \frac{\partial S(r, \psi)}{\partial \psi} \cos \alpha \, d\sigma \,, \qquad (6\text{--}30)$$

$$\delta g_\lambda = -\frac{R}{4\pi \, r} \iint\limits_{\sigma} \Delta g \, \frac{\partial S(r, \psi)}{\partial \psi} \sin \alpha \, d\sigma \,.$$

Now we form the derivatives of the extended Stokes function (6–22) with respect to r and ψ. By differentiating (6–23), we get

$$\frac{\partial l}{\partial r} = \frac{r - R \cos \psi}{l}, \qquad \frac{\partial l}{\partial \psi} = \frac{R r}{l} \sin \psi. \qquad (6\text{--}31)$$

By means of these auxiliary relations, we find

$$\frac{\partial S}{\partial r} = -\frac{R(r^2 - R^2)}{r\,l^3} - \frac{4R}{r\,l} - \frac{R}{r^2} + \frac{6R\,l}{r^3}$$
$$+ \frac{R^2}{r^3} \cos \psi \left(13 + 6 \ln \frac{r - R \cos \psi + l}{2r} \right), \qquad (6\text{--}32)$$

$$\frac{\partial S}{\partial \psi} = \sin \psi \left[-\frac{2R^2 r}{l^3} - \frac{6R^2}{r\,l} + \frac{8R^2}{r^2} \right.$$
$$\left. + \frac{3R^2}{r^2} \left(\frac{r - R \cos \psi - l}{l \sin^2 \psi} + \ln \frac{r - R \cos \psi + l}{2r} \right) \right].$$

Somewhat more convenient expressions are obtained by substituting

$$t = \frac{R}{r}, \qquad (6\text{--}33)$$

$$D = \frac{l}{r} = \sqrt{1 - 2t \cos \psi + t^2}. \qquad (6\text{--}34)$$

Then the extended Stokes function (6–22) and its derivatives (6–32) become

$$S(r, \psi) = t \left[\frac{2}{D} + 1 - 3D - t \cos \psi \left(5 + 3 \ln \frac{1 - t \cos \psi + D}{2} \right) \right], \qquad (6\text{--}35)$$

$$\frac{\partial S(r, \psi)}{\partial r} = -\frac{t^2}{R} \left[\frac{1 - t^2}{D^3} + \frac{4}{D} + 1 - 6D \right.$$
$$\left. - t \cos \psi \left(13 + 6 \ln \frac{1 - t \cos \psi + D}{2} \right) \right], \qquad (6\text{--}36)$$

$$\frac{\partial S(r, \psi)}{\partial \psi} = -t^2 \sin \psi \left[\frac{2}{D^3} + \frac{6}{D} - 8 \right.$$
$$\left. - 3 \frac{1 - t \cos \psi - D}{D \sin^2 \psi} - 3 \ln \frac{1 - t \cos \psi + D}{2} \right].$$

These expressions are used in (6–21) and (6–30) to compute T and $\delta \mathbf{g}$.

The separation N_P of the geopotential surface through P, $W = W_P$, and the corresponding spheropotential surface $U = W_P$ is according to Bruns' theorem given by

$$N_P = \frac{T_P}{\gamma_Q}; \qquad (6\text{--}37)$$

see also Sect. 2.14 and Fig. 2.15.

The deflection of the vertical, which is the deviation of the actual plumb line from the normal plumb line at P, is represented by its north-south and east-west components,

$$\xi_P = -\frac{1}{r}\frac{\partial N_P}{\partial \bar\varphi}, \qquad \eta_P = -\frac{1}{r\cos\bar\varphi}\frac{\partial N_P}{\partial \lambda};\qquad (6\text{-}38)$$

these equations correspond to (2–377). Since γ varies very little with latitude and is independent of longitude, we have

$$\frac{\partial N_P}{\partial \bar\varphi} = \frac{\partial}{\partial \bar\varphi}\left(\frac{T_P}{\gamma_Q}\right) = \frac{1}{\gamma_Q}\frac{\partial T_P}{\partial \bar\varphi} - \frac{T_P}{\gamma_Q^2}\frac{\partial \gamma_Q}{\partial \bar\varphi} \doteq \frac{1}{\gamma_Q}\frac{\partial T_P}{\partial \bar\varphi} \qquad (6\text{-}39)$$

and

$$\frac{\partial N_P}{\partial \lambda} = \frac{1}{\gamma_Q}\frac{\partial T_P}{\partial \lambda}. \qquad (6\text{-}40)$$

Substituting the results of (6–39) and (6–40) into (6–38) and comparing then with (6–20) shows that

$$\xi_P = -\frac{1}{\gamma_Q}\delta g_{\bar\varphi}, \qquad \eta_P = -\frac{1}{\gamma_Q}\delta g_\lambda. \qquad (6\text{-}41)$$

We see that N_P, ξ_P, η_P are given by Eqs. (6–21) and (6–30), apart from the factor $\pm 1/\gamma_Q$. Hence, these equations are the extensions of Stokes' and Vening Meinesz' formulas for points outside the earth and reduce to these formulas for $r = R$, $t = 1$.

Writing Eqs. (6–41) in the form

$$\delta g_{\bar\varphi} = -\gamma\,\xi, \qquad \delta g_\lambda = -\gamma\,\eta, \qquad (6\text{-}42)$$

we see that the horizontal components of $\delta\mathbf{g}$ are directly related to the deflection of the vertical, which is the difference *in direction* of the vectors $\mathbf{g}$ and $\boldsymbol{\gamma}$. The radial component δg_r, however, represents the difference *in magnitude* of these vectors, since as a spherical approximation

$$-\delta g_r = \delta g = g_P - \gamma_P, \qquad (6\text{-}43)$$

which is the scalar gravity disturbance (see Sect. 2.12).

Note that here the gravity disturbance δg is the basic quantity to be computed, rather than the gravity anomaly Δg, because both g and γ refer to the computation point P.

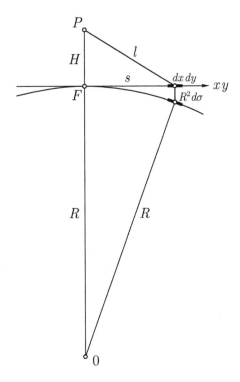

Fig. 6.2. Plane approximation

6.4   Gravity disturbances by upward continuation

We apply Poisson's integral formula (1–123) to the harmonic function T:

$$T_P = \frac{R\,(r^2 - R^2)}{4\pi} \iint_\sigma \frac{T}{l^3}\, d\sigma\,. \tag{6-44}$$

In the neighborhood of P (Fig. 6.2), the sphere practically coincides with its tangent plane at F. Since the value of the integrand is very small at greater distances from P, we may extend the integration over the tangent plane instead of over the sphere. Then, according to Fig. 6.2,

$$l = \sqrt{s^2 + H^2}\,. \tag{6-45}$$

We introduce a rectangular coordinate system x, y, z, the x-axis pointing north and the y-axis pointing east in the tangent plane. Then we may also write

$$l = \sqrt{x^2 + y^2 + H^2}\,, \tag{6-46}$$

the surface element becomes

$$R^2 \, d\sigma \doteq dx \, dy \,, \tag{6–47}$$

and we further have

$$r = R + H \,,$$
$$r^2 - R^2 = (r + R)(r - R) \doteq 2RH \,. \tag{6–48}$$

Thus, (6–44) becomes the plane formula

$$T_P = \frac{H}{2\pi} \int_{-\infty}^{\infty} \int_{-\infty}^{\infty} \frac{T}{l^3} \, dx \, dy = \frac{H}{2\pi} \int_{-\infty}^{\infty} \int_{-\infty}^{\infty} \frac{T}{(x^2 + y^2 + H^2)^{3/2}} \, dx \, dy \,. \tag{6–49}$$

This important formula is called the *"upward continuation integral"*. It performs the computation of the value of the harmonic function T at a point above the xy-plane from the values of T given on the plane, that is, the upward continuation of a harmonic function. Both T and its partial derivatives, $\partial T/\partial x$, $\partial T/\partial y$, $\partial T/\partial z$, are harmonic, because if

$$\frac{\partial^2 T}{\partial x^2} + \frac{\partial^2 T}{\partial y^2} + \frac{\partial^2 T}{\partial z^2} = 0 \,, \tag{6–50}$$

then we also have

$$\frac{\partial^2}{\partial x^2}\left(\frac{\partial T}{\partial x}\right) + \frac{\partial^2}{\partial y^2}\left(\frac{\partial T}{\partial x}\right) + \frac{\partial^2}{\partial z^2}\left(\frac{\partial T}{\partial x}\right) = \frac{\partial}{\partial x}\left(\frac{\partial^2 T}{\partial x^2} + \frac{\partial^2 T}{\partial y^2} + \frac{\partial^2 T}{\partial z^2}\right) = 0 \,. \tag{6–51}$$

Thus, the upward continuation integral (6–49), which applies for any harmonic function, may also be applied to $\partial T/\partial x$, $\partial T/\partial y$, and $\partial T/\partial z$.

As T is the disturbing potential, its partial derivatives are the components of the gravity disturbance:

$$\frac{\partial T}{\partial x} = \delta g_{\bar\varphi} \,, \qquad \frac{\partial T}{\partial y} = \delta g_\lambda \,, \qquad \frac{\partial T}{\partial z} = \delta g_r \,. \tag{6–52}$$

We are not writing δg_x, δg_y, δg_z because we wish to reserve this notation for the components in the geocentric global coordinate system, which should not be confused with the local system introduced in this section. As usual, $r, \bar\varphi, \lambda$ denote geocentric spherical coordinates (see Sect. 6.3) corresponding to the spherical approximation.

Thus, we have in addition to (6–49)

$$\delta g_r = \frac{H}{2\pi} \int\!\!\int_{-\infty}^{\infty} \frac{\delta g_r}{l^3} \, dx \, dy \,, \tag{6–53}$$

$$\delta g_{\bar{\varphi}} = \frac{H}{2\pi} \iint\limits_{-\infty}^{\infty} \frac{\delta g_{\bar{\varphi}}}{l^3} \, dx \, dy \,,$$

$$\delta g_{\lambda} = \frac{H}{2\pi} \iint\limits_{-\infty}^{\infty} \frac{\delta g_{\lambda}}{l^3} \, dx \, dy \,. \tag{6-54}$$

On the left-hand side of these equations, the components of $\delta\mathbf{g}$ refer to the elevated point P; in the integral on the right-hand side, they are taken at sea level and are to be computed from the expressions

$$\delta g_r = -\delta g = - \left(\Delta g + \frac{2\gamma_0}{R} N \right) , \tag{6-55}$$

$$\delta g_{\bar{\varphi}} = -\gamma_0 \, \xi \,,$$

$$\delta g_{\lambda} = -\gamma_0 \, \eta \,, \tag{6-56}$$

which follow from (2–264) together with (6–42) and (6–43) applied to sea level. The symbols R and γ_0 denote, as usual, a mean earth radius and a mean value of gravity on the earth's surface.

Hence, we may compute T and $\delta\mathbf{g}$ by means of the upward continuation integral if the geoidal undulations N and the deflection components ξ and η at the earth's surface are given.

The plane approximation is sufficient except for very high altitudes (e.g., $> 250\,\mathrm{km}$). Otherwise, we must use the spherical formula (6–44) for T. For the radial component δg_r, formula (6–44) may also be applied with T replaced by $r\,\delta g$, since $r\,\delta g$ and $r\,\Delta g$ are harmonic as we know from Sect. 2.14. The corresponding spherical formulas for the upward continuation of the horizontal components $\delta g_{\bar{\varphi}}$ and δg_{λ} are not known. The reason why the same formula, the upward continuation integral, applies for T and the components of $\delta\mathbf{g}$ in the planar case only is that the derivatives of T are harmonic only when referred to a Cartesian coordinate system.

6.5   Additional considerations

Reference surface

The preceding formulas for the disturbing potential T and the gravity disturbance vector $\delta\mathbf{g}$ are rigorously valid if the reference surface is a sphere. In practice, the gravity anomalies are referred to an ellipsoid. The above formulas for T and $\delta\mathbf{g}$ are also valid for an ellipsoidal reference surface if a relative error of the order of the flattening $f \doteq 0.3\%$ is neglected, that is, as a spherical approximation. The reader is reminded that *this does not mean that the ellipsoid is replaced by a sphere in any geometrical sense*; rather it means that in the originally elliptical formulas the first and higher powers of the flattening are neglected, whereby they formally become spherical formulas.

Since the gravity anomalies, etc., are referred to an ellipsoid, we must be very careful in computing t, which enters into the formulas of Sect. 6.3. If an exact sphere of radius R were used as a reference surface, then we should have $r = R + H$, where H is the elevation of the computation point above the sphere. Actually, we use a reference ellipsoid; then we again have

$$r = R + H\,, \qquad t = \frac{R}{R + H}\,, \tag{6–57}$$

but H is now the elevation *above the ellipsoid* (or, to a sufficient accuracy, *above sea level*), the constant $R = 6371$ km being the earth's mean radius. Thus, r as computed by (6–57) differs from the geocentric radius vector $r = \sqrt{x^2 + y^2 + z^2}$. We have already mentioned that we may replace the geocentric latitude $\bar{\varphi}$ by the ellipsoidal latitude φ, as far as T and $\delta\mathbf{g}$ are concerned – for instance, by putting $\bar{\varphi} = \varphi$ in (6–26) or (6–29).

Data

For all computations dealing with the external gravity field of the earth, *free-air gravity anomalies* must be used for Δg, since all other types of gravity anomalies correspond to some removal or transport of masses whereby the external field is changed. If, in addition to Δg, deflections of the vertical ξ, η (in the upward continuation) are used, then these quantities should be computed from free-air anomalies. If, as usually done, the normal free-air gradient $\partial y / \partial h \doteq 0.3086$ mgal/m is used for the free-air reduction, then the free-air anomalies refer, strictly speaking, to the earth's physical surface (to ground level) rather than to the geoid (to sea level). The N values computed from them by Stokes' formula are height anomalies ζ, referring to the ground, rather than heights of the actual geoid. However, this distinction is insignificant and can be ignored in most cases, so that we may consider Δg as sea-level anomalies (see Sect. 8.6).

If we cannot neglect this distinction, aiming at highest accuracy in high and steep mountains for low altitudes H, then we may proceed as follows. We reduce the free-air anomaly Δg from the ground point A to the corresponding point A_0 at sea level (Fig. 6.3):

$$\Delta g^{\text{harmonic}} = \Delta g - \frac{\partial \Delta g}{\partial h}\, h\,, \tag{6–58}$$

and use the sea level anomaly $\Delta g^{\text{harmonic}}$ so obtained. The vertical gradient $\partial \Delta g / \partial h$ may be computed by applying formula (2–394) using the ground-

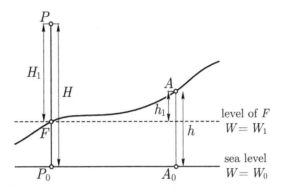

Fig. 6.3. Reduction to sea level and to the level of F

level anomalies Δg. Or we may reduce to any other level surface $W = W_1$, for instance, to that passing through F (Fig. 6.3), using h_1 instead of h in (6–58). Then we should also use H_1, rather than H, in (6–57). For large-scale purposes, reduction to sea level appears to be preferable. Probably such a reduction will attain a considerable amount only in exceptional cases so that it can usually be neglected and H in the formulas of Sects. 6.3 and 6.4 may be taken as the height of P above sea level or above ground. See also Sect. 8.6.

Computation of the gravity vector

After computing the components δg_r, $\delta g_{\bar{\varphi}}$, δg_λ by numerical integration, we may transform them into Cartesian coordinates δg_x, δg_y, δg_z with respect to the global coordinate system.

We may go via ellipsoidal-harmonic coordinates according to Sect. 6.2. For the small quantities δg_u, δg_β, δg_λ, we may apply the spherical approximation, neglecting a relative error of the order of the flattening. If the flattening is neglected, then the ellipsoidal-harmonic coordinates u, β, λ reduce to the spherical coordinates r, $\bar{\varphi}$, λ so that as a spherical approximation

$$\delta g_u = \delta g_r , \qquad \delta g_\beta = \delta g_{\bar{\varphi}} , \qquad (6\text{–}59)$$

δg_λ being rigorously the same in both systems. Thus, δg_r, $\delta g_{\bar{\varphi}}$, δg_λ may also be considered as the components of $\delta\mathbf{g}$ in ellipsoidal-harmonic coordinates.

Then we have

$$g_u = \gamma_u + \delta g_r , \qquad g_\beta = \gamma_\beta + \delta g_{\bar{\varphi}} , \qquad g_\lambda = \delta g_\lambda ; \qquad (6\text{–}60)$$

and g_x, g_y, g_z are obtained by (6–17), the components of $\mathbf{g}$ replacing the corresponding components of $\boldsymbol{\gamma}$. It is evident that the spherical approximation can only be used for $\delta \mathbf{g}$ so that γ_u and γ_β *must be computed by the rigorous formulas* (6–12).

The gravity potential W may be computed by the first equation of (6–4); the gravitational potential V is obtained by subtracting the centrifugal potential $\omega^2(x^2 + y^2)/2$; and the vector of gravitation is given by (6–5).

6.6 Gravity anomalies and disturbances compared

Suppose gravity g is to be computed at some point P outside the earth (Fig. 6.4); we consider here only the *magnitude* of the gravity vector. This is conveniently done by adding a correction to the normal gravity γ. From Sect. 2.12 and later, we recall the two different kinds of such a correction, $g - \gamma$:

1. the *gravity disturbance* δg, in which g and γ both refer to the same point P;
2. the *gravity anomaly* Δg, in which g refers to P, but γ refers to the corresponding point Q, which is situated on the plumb line of P and whose normal potential U is the same as the actual potential W of P, that is, $U_Q = W_P$.

These two quantities are connected by

$$\Delta g = \delta g - \frac{2\gamma_0}{R} N_P \, ; \tag{6–61}$$

this simple relation is sufficient for moderate altitudes.

The gravity disturbance is used when the spatial position of P is given, that is, its geocentric rectangular coordinates x, y, z are measured. With GPS measurements of the position of the aircraft, the use of gravity disturbances is natural.

The use of gravity anomalies Δg had been traditional. This is the case, for instance, in airborne gravity measurements, where the height of the aircraft

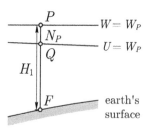

Fig. 6.4. Gravity anomalies and disturbancies

above ground is measured. This case seems rather to belong to the past. If the case should arise, gravity anomalies Δg can be upward continued just as δg as described in Sect. 6.4.

Again, free-air anomalies referred to ground level or, more accurately, to some level surface, are to be used. If the ground is elevated above sea level but reasonably flat, it is somewhat better to regard H as elevation above ground rather than above sea level, because the ground may then be considered locally part of a level surface.

The inverse problem, the downward continuation of gravity anomalies or rather gravity disturbances, occurs in the reduction of gravity measured on board an aircraft. There is, of course, a relation to harmonic downward continuation in the solution of Molodensky's problem as described in Sect. 8.6.

Upward and downward continuation are also tools of geophysical exploration, but here the objective is quite different. Several methods have been developed in this connection, some of which are also applicable for geodetic purposes; see, e.g., Dobrin and Savit (1988) or Telfort et al. (1990).

Upward and downward continuation are related as direct and inverse problems in the theory of inverse problems, see Anger et al. (1993) and also www.inas.tugraz.at under forschung/InverseProblems/AngerMoritz.html, where additional references can be found.

7 Space methods

7.1 Introduction

The subject of this chapter is the use of satellite observations for determining features of the gravity field and the figure of the earth. Only the barest essentials can be presented within the scope of a chapter. The reader will find more information in special textbooks such as Hofmann-Wellenhof et al. (2001), Montenbruck and Gill (2001), and Seeber (2003).

Historical remarks

Immediately after the first launch of artifical satellites (Sputnik 1957, Explorer 1958), their use for geodetic purposes was initiated, and by now the Global Positioning System (GPS) has become the most important method for a fast and precise determination of geodetic positions (see Sect. 5.3). Historically, the first observational methods were intended to determine the spatial direction and the distance to the satellite. Most of these methods are now obsolete, but some principles may be still useful.

Directions

They may be measured by photographing the satellite against the background of stars, or by means of radio waves transmitted from the satellite, using the principle of interference. Photography can only achieve an accuracy of about 0.2 arc seconds and is not used any more in its original sense. The principle of the photographic method was as follows. On the photographic plate, the image of the satellite is surrounded by images of stars. The directions to the surrounding stars are defined by their right ascensions α and declinations δ, which are known from astronomy. Therefore, by interpolation we find the right ascension and declination of the satellite representing the desired direction. This technique is now obsolete.

Ranges

They are measured by radar or by laser. Radar is used for measuring ranges to space probes orbiting in the solar system, which is important to space sciences rather than to geodesy. Lunar Laser Ranging (LLR) and Satellite Laser Ranging (SLR) are useful for determining the earth rotation parameters because of their high (subcentimeter) accuracy; however, their use is restricted to a limited number of fundamental stations.

Range rates

This measurement quantity is found by observing the Doppler effect with radio waves transmitted from a satellite. It is still used within GPS and in satellite-to-satellite tracking (SST).

Satellite altimetry

Here a short-wave electronic ray is sent, from a satellite flying over the oceans, vertically down to the ocean surface, reflected there and received by the satellite again. The measured travel time immediately gives the height H of the satellite above the ocean surface. Knowing the orbital position of the satellite with respect to the global reference system, we can compute the satellite height h above the ellipsoid. Then the difference $h - H$ is the geoidal height N. This is the case if the sea surface is assumed to coincide with the geoid. In reality, because of ocean currents, etc., both surfaces are separated by the "sea surface topography", which may reach the order of $1\,\mathrm{m}$ and is interesting to oceanography. It can be determined if an accurate ocean geoid is known from the gravitational field.

The principles of these methods are illustrated in Fig. 7.1, where $\mathbf{e}$ indicates the direction observation, s between tracking station and satellite refers to the range measurement, and, accordingly, ds/dt corresponds to Doppler observation, whereas ds/dt between the two satellites is obtained by SST; finally, H is measured by satellite altimetry.

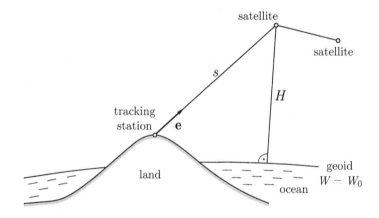

Fig. 7.1. Principles of satellite techniques

7.2   Satellite orbits

The first spectacular result from satellite observations, well advertised by NASA around 1960, was the "discovery" that "the earth is not an ellipsoid but rather shaped like a pear". This pear shape is caused by the spherical harmonic J_3. Its effect, at the North and South Poles, is on the order of 30 m, by three orders of magnitude less than the ellipticity coming from J_2, whose linear effect $a - b$ is about 20 km (!).

The first real result, also found around 1960, was a dramatic improvement in the accuracy of the flattening f itself, which lead to a change from $1/297.3$, generally believed before, to $1/298.25$, corresponding to a linear improvement of the earth size of about 70 m!

The earth's flattening causes the largest but not the only deviation of the earth gravitational field from that of a homogeneous sphere. Generally, the gravitational potential can be expanded into a series of spherical harmonics according to Sect. 2.5, Eq. (2–78):

$$
V = \frac{GM}{r} \left\{ 1 - \sum_{n=2}^{\infty} \left(\frac{a}{r} \right)^n J_n \, P_n(\cos \vartheta) \right.
$$
$$
\left. + \sum_{n=2}^{\infty} \sum_{m=1}^{n} \left(\frac{a}{r} \right)^n [C_{nm} \, \cos m\lambda + S_{nm} \, \sin m\lambda] \, P_{nm}(\cos \vartheta) \right\}.
$$
$$
(7\text{–}1)
$$

Here the terms containing J_n are the zonal harmonics, and those containing S_{nm} and C_{nm} are the tesseral harmonics.

The former notations $J_{nm} = -C_{nm}$ and $K_{nm} = -S_{nm}$ are not used any more for the *tesseral* harmonic coefficients; for the *zonal* harmonics, the use of J_n has prevailed so far, but also $C_{n0} = -J_n$ is being used.

Considering the moon, the only term of appreciable influence is J_2, which represents the flattening. Artificial satellites are, compared to the moon, much closer to the earth; typical heights above ground of a geodetically used satellite range from some 300 km up to 20 000 km. Hence, they are also influenced by harmonics other than J_2 and can, therefore, be used to determine harmonics of low degree. For this purpose, we must study the effect of gravitational disturbances on the orbits of close satellites.

Before we can do this, we must briefly review the theory of an undisturbed orbit, which means that the gravitational potential has the form

$$
V = \frac{GM}{r} ,
$$
$$
(7\text{–}2)
$$

all C's and S's being zero. This represents the gravitational field of a point mass or a homogeneous sphere. Then the motion of a satellite is described

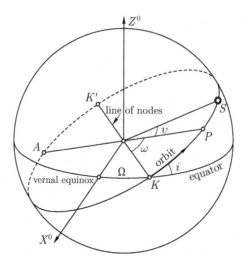

Fig. 7.2. Satellite orbit as projected onto a unit sphere

by *Kepler's three laws for planetary motion.* Satellites with parabolic or hyperbolic orbits are of no interest in this context.

According to *Kepler's first law,* the orbit is an ellipse of which the center of the earth occupies one focus. The position of the orbit in space is defined by the six *orbital elements:*

$$
\begin{array}{ll}
a & \text{semimajor axis,} \\
e & \text{eccentricity,} \\
i & \text{inclination,} \\
\Omega & \text{right ascension of the node,} \\
\omega & \text{argument of perigee,} \\
T & \text{time of perigee passage}.
\end{array}
\qquad (7\text{--}3)
$$

If a and b are the semiaxes of the orbital ellipse (there is no danger of confusion with those of the terrestrial ellipsoid!), then the eccentricity is defined by

$$
e = \frac{\sqrt{a^2 - b^2}}{a} . \qquad (7\text{--}4)
$$

Figure 7.2 shows the projection of the orbit onto a geocentric unit sphere, where P is the perigee, A the apogee, K is the ascending node, K' the descending node, S is the instantaneous position of satellite. The line of nodes is the intersection of the orbital plane with the plane of the equator; it connects the ascending node K and the descending node K'. The right ascension of the node, Ω, is the angle between the line of nodes and the direction to the vernal equinox. The symbol Ω is also called longitude of the node, but in conformity with astronomical terminology it is the right ascension of the (ascending) node. The major axis of the orbit intersects the orbital ellipse at the perigee P, the position where the satellite is closest to the earth, and at the apogee A, where the satellite is farthest away. The angle ω between the nodes and the major axis is the argument of perigee.

The angular distance of the satellite S from perigee is called *true anomaly* and denoted by v; it is a function of time. Note that this strange name comes from the history of astronomy; there is nothing anomalous with it!

The equation of the orbital ellipse may be written

$$r = \frac{p}{1 + e \, \cos v} \, , \tag{7-5}$$

where r is the distance of the satellite from the earth's center of mass and

$$p = \frac{b^2}{a} = a \, (1 - e^2) \tag{7-6}$$

is the length of the radius vector r for $v = 90°$. The radius vector r and the true anomaly v form a pair of polar coordinates in the orbital plane, and (7–5) is the well-known polar equation of an ellipse. See Fig. 7.3 for an illustration of these quantities, where F, the focal point, is the earth's center of mass.

According to *Kepler's second law*, the area of the elliptical sector swept by the radius vector r between any two positions of the satellite is proportional to the time it takes the satellite to pass from one position to the other. In other words, the time rate of change of the area swept by the radius vector is constant. Since the element of area of a sector in polar coordinates r and

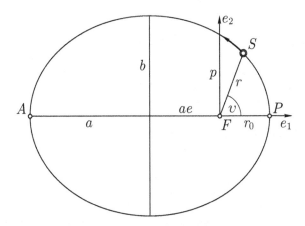

Fig. 7.3. Orbital ellipse

v is $\frac{1}{2}r^2 dv$, this law may be formulated mathematically as

$$r^2 \frac{dv}{dt} = \sqrt{GM\,a\,(1 - e^2)}\,, \tag{7–7}$$

where the constant has already been given its proper value.

Kepler's third law reads

$$n^2 a^3 = GM\,, \tag{7–8}$$

where the satellite mass has been neglected and where

$$n = \frac{2\pi}{P} \tag{7–9}$$

is the "mean motion" (mean angular velocity) of the satellite, P being its period.

So far we have assumed that all J_n, C_{nm}, and S_{nm} in (7–1) are zero. This is not true because of the irregularities of the earth's gravitational field, even though these coefficients are small. Therefore, the satellite is subject to small perturbing forces. We may still consider the satellite orbit as an ellipse, but then the parameters of this ellipse, the orbital elements, will no longer be constant but will change slowly. At each instant, this *osculating ellipse* will be slightly different. It is defined as follows. Imagine that at the instant under consideration all perturbing forces suddenly vanish. Then the satellite will continue its motion along an exact ellipse; this is the osculating ellipse.

If we resolve the total perturbing force into rectangular components S, T, and W, where S is directed along the radius vector, W is normal to the orbital plane, and T is normal to S and W – note that this notation follows astronomical usage; there is no relation to the geodetic use of T and W for potentials! –, then the time rate of change of the orbital parameters can be expressed in terms of these components:

$$\dot{a} = \frac{2a^2}{b} \sqrt{\frac{a}{GM}} \left(e\,S\sin v + \frac{p}{r}\,T \right),$$

$$\dot{e} = \frac{b}{a} \sqrt{\frac{a}{GM}} \left[S\sin v + \left(\frac{r+p}{r}\cos v + \frac{e\,r}{p} \right) T \right],$$

$$\dot{i} = \frac{r}{b} \sqrt{\frac{a}{GM}}\, W\cos(\omega + v)\,,$$

$$\dot{\Omega} = \frac{r}{b} \sqrt{\frac{a}{GM}}\, W\,\frac{\sin(\omega + v)}{\sin i}\,,$$

$$\dot{\omega} = \frac{b}{a} \sqrt{\frac{a}{GM}} \left[-\frac{1}{e}\,S\cos v + \frac{r+p}{e\,p}\,T\sin v - \frac{r}{p}\,W\sin(\omega + v)\cot i \right]. \tag{7–10}$$

As usual, $\dot{a}$ denotes da/dt, etc. The derivation of these equations may be found in any textbook on celestial mechanics, e.g., Plummer (1918: p. 151), Brouwer and Clemence (1961: p. 301), and Seeber (2003: Sect. 3.2.1.3), who uses the symbols K_1, K_2, K_3 instead of W, S, R.

7.3 Determination of zonal harmonics

The effect of the zonal harmonics on satellite orbits is much greater than that of the tesseral harmonics. Only zonal harmonics (J_2, J_3, J_4, ...) will give observable variations of the orbital elements themselves. The tesseral harmonics cause oscillatory disturbances that rapidly change their sign, whereas the effect of the zonal harmonics is cumulative. For this reason, we consider first the effect of zonal harmonics, that is, the effect of those independent of longitude λ. Hence we set

$$V = \frac{GM}{r} + R, \qquad (7\text{–}11)$$

where the *perturbing potential*

$$R = -\frac{GM}{a_e} \sum_{n=2}^{\infty} \left(\frac{a_e}{r}\right)^{n+1} J_n P_n(\cos\vartheta) \qquad (7\text{–}12)$$

is a function of r and ϑ only. Note that the main difference between the perturbing potential R of celestial mechanics and the disturbing potential T of physical geodesy is that R, but not T, also incorporates the effect of the flattening through J_2. There are also other perturbing forces acting on a satellite, such as the resistance of the atmosphere (atmospheric drag), radiation pressure exerted by the sunlight, etc. These nongravitational perturbances must be taken into account separately and will not be considered here.

Note that the equatorial radius of the earth (the semimajor axis of the terrestrial ellipsoid) has been denoted by a_e, in order to distinguish it from a, which now denotes the semimajor axis of the orbital ellipse. This notation will be used in what follows.

Since S is the component of the perturbing force along the radius vector, we have

$$S = \frac{\partial R}{\partial r}. \qquad (7\text{–}13)$$

The components of the perturbing force along the meridian and the prime vertical are

$$-\frac{1}{r}\frac{\partial R}{\partial \vartheta} \quad \text{and} \quad \frac{1}{r\sin\vartheta}\frac{\partial R}{\partial \lambda}. \qquad (7\text{–}14)$$

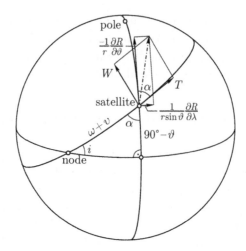

Fig. 7.4. Components of the perturbing force

The components T and W are obtained from them by a plane rotation (Fig. 7.4):

$$T = -\frac{1}{r}\frac{\partial R}{\partial \vartheta}\cos\alpha + \frac{1}{r\sin\vartheta}\frac{\partial R}{\partial \lambda}\sin\alpha,$$

$$W = -\frac{1}{r}\frac{\partial R}{\partial \vartheta}\sin\alpha - \frac{1}{r\sin\vartheta}\frac{\partial R}{\partial \lambda}\cos\alpha.$$

$$(7\text{--}15)$$

From the rectangular spherical triangle in Fig. 7.4 it follows that

$$\cos\alpha = \frac{\cos(\omega + v)\sin i}{\sin\vartheta}, \qquad \sin\alpha = \frac{\cos i}{\sin\vartheta}, \qquad (7\text{--}16)$$

so that finally

$$T = -\frac{\cos(\omega + v)\sin i}{r\sin\vartheta}\frac{\partial R}{\partial \vartheta} + \frac{\cos i}{r\sin^2\vartheta}\frac{\partial R}{\partial \lambda},$$

$$W = -\frac{\cos i}{r\sin\vartheta}\frac{\partial R}{\partial \vartheta} - \frac{\cos(\omega + v)\sin i}{r\sin^2\vartheta}\frac{\partial R}{\partial \lambda}.$$

$$(7\text{--}17)$$

We have included $\partial R/\partial \lambda$ because of the presence of longitude-dependent tesseral harmonics in the general case (see Sect. 7.5). In our present case, where R is given by (7–12), $\partial R/\partial \lambda$ is zero.

Now we must differentiate (7–12) with respect to r and ϑ, compute the components S, T, W from Eqs. (7–13) and (7–17), and substitute them into the system (7–10). In this way, we can express the rates of change $\dot{a}$, $\dot{e}$, ... of the orbital elements in terms of the coefficients J_2, J_3, J_4, ... We cannot, however, observe these rates of change directly. Rather, we observe the changes of the orbital elements after several revolutions. The changes after *one* revolution, with period P, are

$$\Delta a = \int_{t_0}^{t_0+P} \dot{a}\, dt\,, \quad \Delta e = \int_{t_0}^{t_0+P} \dot{e}\, dt\,, \quad \Delta i = \int_{t_0}^{t_0+P} \dot{i}\, dt\,, \quad \text{etc.} \quad (7\text{--}18)$$

The t_0 is an arbitrary "epoch" (instant of time). In order to perform these integrations, we must express $\dot{a}$, $\dot{e}$, ... in terms of one independent variable. For this independent variable, we may take the time t or the true anomaly v. The second possibility will be adopted here.

The polar distance ϑ is expressed as a function of v through the relation

$$\cos\vartheta = \sin(\omega + v)\,\sin i\,, \qquad (7\text{--}19)$$

which follows from the rectangular spherical triangle in Fig. 7.4. The radius vector r is also a function of v according to (7–5). Finally, Kepler's second law (7–7) furnishes the relation between v and the time t:

$$\frac{dt}{dv} = \frac{r^2}{\sqrt{G\,M\,a\,(1-e^2)}}\,. \qquad (7\text{--}20)$$

Hence, we may change the integration variable from t to v, obtaining, for instance,

$$\Delta a = \int_{t_0}^{t_0+P} \dot{a}\, dt = \int_{v=0}^{2\pi} \frac{da}{dv}\, dv\,, \qquad (7\text{--}21)$$

where

$$\frac{da}{dv} = \frac{da}{dt}\frac{dt}{dv} = \frac{r^2}{\sqrt{G\,M\,a\,(1-e^2)}}\,\dot{a}\,. \qquad (7\text{--}22)$$

Analogous formulas result for the other orbital elements.

After performing all these operations, which are lengthy but not too difficult, we find

$$\Delta a = 0\,,$$

$$\Delta e = -\frac{1-e^2}{e}\,\tan i\ \Delta i\,,$$

$$\Delta i = 3\pi e \left(\frac{a_e}{p}\right)^3 \left(1 - \frac{5}{4}\sin^2 i\right)\cos i\,\cos\omega\ J_3$$

$$+ \frac{45}{16}\pi e \left(\frac{a_e}{p}\right)^4 \left(1 - \frac{7}{6}\sin^2 i\right)\sin 2i\,\sin 2\omega\ e\,J_4\ \cdots\,,$$

$$\Delta\Omega = -3\pi \left(\frac{a_e}{p}\right)^2 \cos i\ J_2$$

$$+ 3\pi \left(\frac{a_e}{p}\right)^3 \left(1 - \frac{15}{4}\sin^2 i\right)\cot i\,\sin\omega\ e\,J_3 \tag{7-23}$$

$$+ \frac{15}{2}\pi \left(\frac{a_e}{p}\right)^4 \left(1 - \frac{7}{4}\sin^2 i\right)\cos i\ J_4\ \cdots\,,$$

$$\Delta\omega = 6\pi \left(\frac{a_e}{p}\right)^2 \left(1 - \frac{5}{4}\sin^2 i\right)\ J_2$$

$$+ 3\pi \left(\frac{a_e}{p}\right)^3 \left(1 - \frac{5}{4}\sin^2 i\right)\sin i\,\sin\omega\ e\,J_3$$

$$- 15\pi \left(\frac{a_e}{p}\right)^4 \left[\left(1 - \frac{31}{8}\sin^2 i + \frac{49}{19}\sin^4 i\right)\right.$$

$$\left. + \left(\frac{3}{8} - \frac{7}{16}\sin^2 i\right)\sin^2 i\,\cos 2\omega\right]J_4\ \cdots\,.$$

Terms of the order of $e^2 J_3$ and $e^2 J_4$, which are very small, have been neglected in these equations. The proportionality of Δe and Δi is more or less accidental: it applies only with respect to long-periodic disturbances; $\dot{e}$ and di/dt themselves are not proportional. The quantity p is defined by (7–6); it is hardly necessary to repeat that a, p, e, etc., refer to the orbital ellipse and not to the terrestrial ellipsoid, of which a_e is the equatorial radius.

By integrating over one revolution, we have removed the *short-periodic* terms of periods P, $2P$, $3P$, ..., such as $\cos v$, $\cos 2v$, etc. What remains are *secular* terms, which are constant for one revolution and increase steadily with the number of revolutions, and the *long-periodic* terms, which change very slowly with time in a periodic manner. The argument of perigee ω increases slowly but steadily, so that the perigee of a satellite orbit also rotates around the earth, but much slower than the satellite itself; a typical period of ω is two months. Therefore, terms containing $\cos \omega$, $\sin \omega$, or $\sin 2\omega$ are called long-periodic.

The first equation of (7–23) shows that the semimajor axis of the orbit does not change secularly or long-periodically. The eccentricity and the inclination undergo long-period, but not secular, variations, whereas Ω and ω change both secularly and long-periodically.

Equations (7–23) are linear in J_2, J_3, J_4, ... For practical applications, nonlinear terms containing J_2^2, $J_2 J_3$, $J_2 J_4$, etc., must also be taken into account, since J_2^2 is of the order of J_4. The derivation of these nonlinear terms is much more difficult, and their expressions are different in the various orbital theories that have been proposed. For these reasons, such expressions will not be given here.

Equations (7–23), supplemented by certain nonlinear terms, can be used to determine coefficients J_2, J_3, J_4, etc. Since the secular or long-periodic variations $\Delta\Omega$, $\Delta\omega$, Δe, Δi are known from observation for a sufficient number of satellites, we obtain equations of the form

$$a_2 J_2 + a_3 J_3 + a_4 J_4 + \cdots + a_{22} J_2^2 + a_{23} J_2 J_3 + \cdots = A \,,$$

$$b_2 J_2 + b_3 J_3 + b_4 J_4 + \cdots + b_{22} J_2^2 + b_{23} J_2 J_3 + \cdots = B \,, \qquad (7\text{–}24)$$

$$\vdots \qquad \vdots \qquad \vdots \qquad \qquad \vdots \qquad \vdots \qquad \qquad \vdots$$

which can be solved for J_2, J_3, J_4, ... Since there can be only a finite number of these equations, we must neglect all J_n with n greater than a certain number n_0, which depends on the number of equations available, on their degree of mutual independence, etc. This used to be a difficulty with this method, but it has been overcome long ago by least-squares collocation (Moritz 1980 a: Sect. 21). For details see Schwarz (1976).

From (7–23) it is seen that the coefficients of the J_n depend essentially on the inclination i. It is, therefore, important to use satellites with a wide variety of inclinations, in order to get equations with a high mutual independence.

Now the question arises which orbital elements are to be used for determining the coefficients J_n. The semimajor axis a clearly cannot be used at all. As for the other elements, we must distinguish between coefficients of even and of odd degree n. The even coefficients J_2, J_4, ... can be determined well from the regression of the node, $\Delta\Omega$, and the rotation of perigee, $\Delta\omega$. To see this, inspect (7–23). The even harmonics cause secular disturbances of Ω and ω, which are much larger than the long-periodic effects of the odd coefficients, since J_3, J_5, ... are multiplied by the small eccentricity e.

On the other hand, in Δe and Δi the odd coefficients J_3, J_5, ... have a much larger effect than the even coefficients, which here appear with the small factor e. Therefore, the odd coefficients are determined from Δe or Δi, or from the change of perigee distance $r_0 = FP$ (Fig. 7.3). Since r_0 is the radius vector for $v = 0$, we have from (7–5) and (7–6)

$$r_0 = \frac{p}{1+e} = a\,(1-e)\,, \qquad (7\text{--}25)$$

so that

$$\Delta r_0 = -a\,\Delta e \qquad (7\text{--}26)$$

because $\Delta a = 0$. Thus, the variation of perigee distance is proportional to the variation of eccentricity and may be used instead of Δe.

Numerical values

Helmert (1884: p. 472) used the regression of the node of the moon's orbit to determine J_2, which is the only coefficient to have an appreciable effect on it. Note that for $e \doteq 0$ and $p \doteq a \gg a_e$, the equation for $\Delta\Omega$ in (7–23) becomes

$$\Delta\Omega = -3\pi \left(\frac{a_e}{a}\right)^2 J_2 \cos i\,. \qquad (7\text{--}27)$$

Helmert found

$$J_2 = 1086.5 \cdot 10^{-6} \qquad (7\text{--}28)$$

by averaging two widely different values. This corresponds to a flattening of

$$1/f = 297.8 \pm 2.2\,. \qquad (7\text{--}29)$$

This value is quite close to the recent results but has a much larger uncertainty.

Reliable values by this method can only be obtained from close artificial satellites. Currently accepted values are, for example,

$$\begin{aligned} J_2 &= 1082.6359 \cdot 10^{-6}\,, \\ J_3 &= -2.5324 \cdot 10^{-6}\,, \\ J_4 &= -1.6198 \cdot 10^{-6}\,, \end{aligned} \qquad (7\text{--}30)$$

whose standard errors are assumed to be better than $\pm 0.01 \cdot 10^{-6}$. The value for J_2 has been taken from the report of the IAG by Groten (2004), accessible from www.gfy.ku.dk/~iag/HB2004/part5/51-groten.pdf. J_3 and J_4 are from the recent mission GRACE (see Sect. 7.5).

The most significant geodetic result is the reliable determination of J_2 and, therefore, of the flattening f, around $1/298.25$. Already in 1964, the International Astronomical Union (IAU) adopted the value 298.25 corresponding to $J_2 = 1082.7 \cdot 10^{-6}$ (see Sect. 2.11), followed by the IAG International Geodetic Reference Systems 1967 and then 1980, which in the slightly different form of the World Geodetic System 1984 (WGS 84) is standard even today (2005).

7.4   Rectangular coordinates of the satellite and perturbations

We now describe how the rectangular coordinates of the satellite are computed from the orbital elements. Then we will outline how they are affected by the irregularities of the gravity field. These considerations are necessary for the determination of tesseral harmonics from satellite observations.

We introduce an equatorial coordinate system $X^0 Y^0 Z^0$ that is at rest with respect to the stars. The origin is at the earth's center of mass. The Z^0-axis coincides with its axis of rotation; the $X^0 Y^0$-plane is the equatorial plane. The X^0-axis is the line of intersection of the equatorial plane and the ecliptic (the plane of the earth's orbit around the sun); according to astronomical terminology, it points to the *vernal equinox*. This coordinate system $X^0 Y^0 Z^0$ is fundamental in spherical astronomy. Note that the directions of the coordinate axes so defined are not completely constant in time. This fact requires certain refinements for which the reader is referred to Moritz and Mueller (1987: Chap. 7). In the present context, we consider the $X^0 Y^0 Z^0$-system as constant in time.

The relation between the rectangular coordinates of a satellite and the elements of its osculating ellipse (Sect. 7.2) at a certain time is found as follows. Consider Fig. 7.3 and the coordinate system $\mathbf{e}_1$, $\mathbf{e}_2$ defining the orbital plane. Assuming $\mathbf{e}_3$ orthogonal to this plane,

$$r \begin{bmatrix} \cos v \\ \sin v \\ 0 \end{bmatrix} \tag{7–31}$$

is the representation of the satellite in this system. This result may be transformed into the equatorial system $X^0 Y^0 Z^0$ by a rotation matrix $\mathbf{R}$ and results in a vector denoted as $\mathbf{X}^0 = [X^0, \ Y^0, \ Z^0]$. The transformation is obtained by

$$
\begin{bmatrix} X^0 \\ Y^0 \\ Z^0 \end{bmatrix} = \mathbf{R}\, r \begin{bmatrix} \cos v \\ \sin v \\ 0 \end{bmatrix} , \tag{7–32}
$$

where the matrix $\mathbf{R}$ is composed of three successive rotation matrices (see Figs. 7.2 and 7.3) and is given by

$$
\mathbf{R} = \mathbf{R}_3\{-\Omega\}\, \mathbf{R}_1\{-i\}\, \mathbf{R}_3\{-\omega\}
$$

$$
= \begin{bmatrix}
\cos\Omega\cos\omega & -\cos\Omega\sin\omega & \sin\Omega\sin i \\
-\sin\Omega\sin\omega\cos i & -\sin\Omega\cos\omega\cos i & \\[4pt]
\sin\Omega\cos\omega & -\sin\Omega\sin\omega & -\cos\Omega\sin i \\
+\cos\Omega\sin\omega\cos i & +\cos\Omega\cos\omega\cos i & \\[4pt]
\sin\omega\sin i & \cos\omega\sin i & \cos i
\end{bmatrix} , \tag{7–33}
$$

see Hofmann-Wellenhof et al. (2001: p. 43). The column vectors of the orthonormal matrix $\mathbf{R}$ are the axes of the orbital coordinate system represented in the equatorial system $\mathbf{X}_i^0$.

Substituting (7–33) into (7–32) and carrying out the multiplication (Montenbruck and Gill 2001: Eq. (2.51)) yields

$$
X^0 = r\left[\cos\Omega\,\cos(\omega + v) - \sin\Omega\,\sin(\omega + v)\cos i\right],
$$

$$
Y^0 = r\left[\sin\Omega\,\cos(\omega + v) + \cos\Omega\,\sin(\omega + v)\cos i\right], \tag{7–34}
$$

$$
Z^0 = r\,\sin(\omega + v)\,\sin i ,
$$

where, according to (7–5),

$$
r = \frac{a\,(1 - e^2)}{1 + e\,\cos v} . \tag{7–35}
$$

This expresses the rectangular coordinates of the satellite in terms of the elements of its osculating orbit, the true anomaly v fixing its position as a function of time.

Since the osculating ellipse does not remain constant, it is convenient to use a fixed *reference orbit* – for instance, the osculating ellipse E_0 at a certain instant t_0, having the elements a_0, e_0, i_0, Ω_0, ω_0, T_0. At a later instant t, the orbital elements will have changed to $a_0 + \Delta_t a$, $e_0 + \Delta_t e$, $i_0 + \Delta_t i$, $\Omega_0 + \Delta_t \Omega$, $\omega_0 + \Delta_t \omega$, $T_0 + \Delta_t T$, which corresponds to an osculating ellipse E_t.

The orbital elements in (7–34) refer to this instantaneous osculating ellipse, so that $a = a_0 + \Delta_t a$, etc. Therefore, the coordinates X^0, Y^0, Z^0 depend on the time in two ways: *explicitly*, through the true anomaly v, and *implicitly*, through the variable elements of the osculating orbit. We eliminate the implicit dependence in the following way. We evaluate (7–34) using the elements a_0, etc., of the fixed reference ellipse. Then the coordinates so obtained depend on the time only explicitly and correspond to a Keplerian motion in space along a fixed ellipse. To convert them into true coordinates X^0, Y^0, Z^0, they must be corrected by $\Delta_t X^0$, $\Delta_t Y^0$, $\Delta_t Z^0$, for which the linear terms of a Taylor expansion of (7–34) give

$$\Delta_t X^0 =$$

$$\frac{\partial X^0}{\partial a}\Delta_t a \frac{\partial X^0}{\partial e}\Delta_t e + \frac{\partial X^0}{\partial i}\Delta_t i + \frac{\partial X^0}{\partial \Omega}\Delta_t \Omega + \frac{\partial X^0}{\partial \omega}\Delta_t \omega + \frac{\partial X^0}{\partial v}\Delta_t v \,,$$

$$\Delta_t Y^0 =$$

$$\frac{\partial Y^0}{\partial a}\Delta_t a \frac{\partial Y^0}{\partial e}\Delta_t e + \frac{\partial Y^0}{\partial i}\Delta_t i + \frac{\partial Y^0}{\partial \Omega}\Delta_t \Omega + \frac{\partial Y^0}{\partial \omega}\Delta_t \omega + \frac{\partial Y^0}{\partial v}\Delta_t v \,,$$

$$\Delta_t Z^0 =$$

$$\frac{\partial Z^0}{\partial a}\Delta_t a \frac{\partial Z^0}{\partial e}\Delta_t e + \frac{\partial Z^0}{\partial i}\Delta_t i + \frac{\partial Z^0}{\partial \Omega}\Delta_t \Omega + \frac{\partial Z^0}{\partial \omega}\Delta_t \omega + \frac{\partial Z^0}{\partial v}\Delta_t v \,.$$

$$(7\text{–}36)$$

The partial derivatives are readily obtained by differentiating (7–34); note that r is a function of a, e, and v.

In these equations, we have used the perturbation of the true anomaly, $\Delta_t v$, instead of the perturbation of perigee epoch, $\Delta_t T$.

Perturbations expressed in terms of C_{nm} and S_{nm}
The perturbations of the orbital elements are found by integrating (7–10):

$$\Delta_t a = \int_{t_0}^{t} \dot{a}\, dt\,, \quad \Delta_t e = \int_{t_0}^{t} \dot{e}\, dt\,, \quad \ldots. \qquad (7\text{–}37)$$

A similar expression can be written for $\Delta_t v$. The components S, T, W of the perturbing force are expressed in terms of J_n, C_{nm}, and S_{nm} using equations (7–12), (7–13), and (7–17), where the perturbing potential

$$R = -\frac{GM}{a_e} \sum_{n=2}^{\infty} \left(\frac{a_e}{r}\right)^{n+1} \left[J_n P_n(\cos \vartheta) \right.$$

$$\left. - \sum_{m=1}^{n} (C_{nm} \cos m\lambda + S_{nm} \sin m\lambda)\, P_{nm}(\cos \vartheta) \right]$$

$$(7\text{–}38)$$

now also contains the tesseral harmonics.

By performing the integrations in (7–37), we obtain equations of the form

$$\Delta_t a = \sum_{n,m} \left(A_{nm} C_{nm} + \bar{A}_{nm} S_{nm} \right),$$

$$\Delta_t e = \sum_{n,m} \left(B_{nm} C_{nm} + \bar{B}_{nm} S_{nm} \right), \qquad (7\text{–}39)$$

$$\vdots \qquad \vdots$$

where the coefficients A_{nm}, etc., are functions of the time t and are, as a rule, periodic. Zonal and tesseral harmonics have been combined in (7–39) by setting $J_n = -C_{n0}$ and admitting the value $m = 0$; this practice will be continued in what follows.

The substitution of (7–39) into (7–36) gives the perturbation of the rectangular coordinates X^0, Y^0, Z^0 as functions of the harmonic coefficients $C_{n0} = -J_n$, C_{nm}, and S_{nm} in the form

$$\Delta_t X^0 = \sum_{n,m} \left(L_{nm} C_{nm} + \bar{L}_{nm} S_{nm} \right),$$

$$\Delta_t Y^0 = \sum_{n,m} \left(M_{nm} C_{nm} + \bar{M}_{nm} S_{nm} \right), \qquad (7\text{–}40)$$

$$\Delta_t Z^0 = \sum_{n,m} \left(N_{nm} C_{nm} + \bar{N}_{nm} S_{nm} \right),$$

where again L_{nm}, $\bar{L}_{nm}$, M_{nm}, etc., are functions of the time t.

These perturbations are added to the coordinates computed from (7–34) using the orbital elements of the reference ellipse E_0. In this way, we obtain the rectangular coordinates of the satellite in the form

$$X^0 = X^0(t;\, a_0, e_0, i_0, \Omega_0, \omega_0, T_0;\, C_{nm}, S_{nm}),$$

$$Y^0 = Y^0(t;\, a_0, e_0, i_0, \Omega_0, \omega_0, T_0;\, C_{nm}, S_{nm}), \qquad (7\text{–}41)$$

$$Z^0 = Z^0(t;\, a_0, e_0, i_0, \Omega_0, \omega_0, T_0;\, C_{nm}, S_{nm})$$

as *explicit* functions of the time t, containing as *constant* parameters the orbital element of the reference ellipse E_0 and the gravitational coefficients C_{nm} and S_{nm}. This is the advantage of (7–41) over the system (7–34), which formally is much simpler but depends on the *variable* orbital parameters of the osculating ellipse.

The actual expressions for (7–41) are very complicated. Therefore, we have been satisfied with outlining the procedure, referring the reader for details to the pioneering book by Kaula (1966 a) and to his papers given there.

7.5   Determination of tesseral harmonics and station positions

Zonal harmonics give rise to secular and long-periodic perturbations of the orbital elements a, e, etc. Therefore, their influence can be detected in changes of orbital parameters obtained by integrating over many revolutions of the satellite.

The perturbations due to tesseral harmonics have a much shorter period. The longest period of a harmonic of the order $m = 1$ is one day, for $m = 2$ it is only half a day, etc. Therefore, we must look for another method, which is sensitive enough to detect even short-periodic effects and extracts as much information as possible from the observations.

The observed elements are essentially spatial polar coordinates of the satellite with respect to the observing station: the distance s and the direction as determined by two angles. Corresponding to our coordinate system X^0, Y^0, Z^0 introduced in the preceding section, these two angles are the *right ascension* α and the *declination* δ, whose definition may be seen in Fig. 7.5. The angles α and δ are polar coordinates in three-dimensional space and were obtained by photographing the satellite against the background of stars, as outlined in Sect. 7.1. They are outdated nowadays but retained for geometrical intuition and symmetry. Most important are distances s measured by GPS, radar, or laser. Note that the measurement of the range rate ds/dt of the satellite by means of the Doppler effect is also important for the determination of tesseral harmonics and station positions.

Denoting in the equatorial system $X^0 Y^0 Z^0$ the rectangular coordinates of the terrestrial station P by X_P^0, Y_P^0, Z_P^0 and of the satellite S by X_S^0, Y_S^0, Z_S^0,

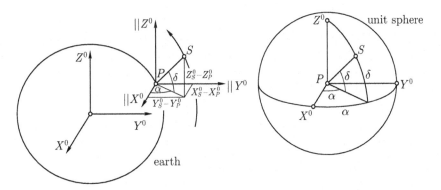

Fig. 7.5. Direction to the satellite defined by right ascension α and declination δ

we find by inspecting Fig. 7.5

$$X_S^0 - X_P^0 = s \cos \delta \cos \alpha \,,$$

$$Y_S^0 - Y_P^0 = s \cos \delta \sin \alpha \,, \tag{7-42}$$

$$Z_S^0 - Z_P^0 = s \sin \delta \,,$$

so that

$$\alpha = \tan^{-1} \frac{Y_S^0 - Y_P^0}{X_S^0 - X_P^0} \,,$$

$$\delta = \tan^{-1} \frac{Z_S^0 - Z_P^0}{\sqrt{(X_S^0 - X_P^0)^2 + (Y_S^0 - Y_P^0)^2}} \,, \tag{7-43}$$

$$s = \sqrt{(X_S^0 - X_P^0)^2 + (Y_S^0 - Y_P^0)^2 + (Z_S^0 - Z_P^0)^2} \,.$$

We now compute the rectangular coordinates X_P^0, Y_P^0, Z_P^0 of the observing station P. The system $X^0 Y^0 Z^0$, being fixed with respect to the stars, rotates with respect to the earth. The coordinates of P in this system are, therefore, functions of time. Let X_P, Y_P, Z_P be the coordinates of P in the usual geocentric coordinate system fixed with respect to the earth. In this system, the Z-axis, coinciding with the Z^0-axis, is the earth's axis of rotation; the X-axis lies in the mean meridian plane of Greenwich, corresponding to the longitude $\lambda = 0°$; and the Y-axis points to $\lambda = 90°$ east. Figure 7.6 shows that

$$X_P^0 = X_P \cos \theta_0 - Y_P \sin \theta_0 \,,$$

$$Y_P^0 = X_P \sin \theta_0 + Y_P \cos \theta_0 \,, \tag{7-44}$$

$$Z_P^0 = Z_P \,.$$

The angle θ_0 is called *Greenwich sidereal time*; its value is

$$\theta_0 = \omega t \,, \tag{7-45}$$

where ω is the angular velocity of the earth's rotation. It is proportional to the time t and, in appropriate units, measures it. Thus, absolute Greenwich time is needed to convert the terrestrial coordinates X_P, Y_P, Z_P to the celestial coordinates X_P^0, Y_P^0, Z_P^0 that are required in (7–42) and (7–43).

As a final step, we substitute the station coordinates, as given by (7–44), and the satellite coordinates, as symbolized by (7–41), into (7–43), obtaining expressions of the form

$$\alpha = \alpha(X_P, Y_P, Z_P; \; t; \; a_0, e_0, i_0, \Omega_0, \omega_0, T_0; \; C_{nm}, S_{nm}) \,,$$

$$\delta = \delta(X_P, Y_P, Z_P; \; t; \; a_0, e_0, i_0, \Omega_0, \omega_0, T_0; \; C_{nm}, S_{nm}) \,, \tag{7-46}$$

$$s = s(X_P, Y_P, Z_P; \; t; \; a_0, e_0, i_0, \Omega_0, \omega_0, T_0; \; C_{nm}, S_{nm}) \,.$$

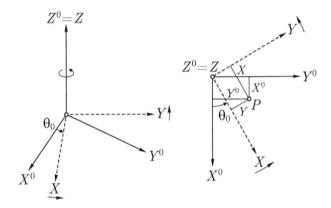

Fig. 7.6. Geocentric coordinate systems $X^0 Y^0 Z^0$ (celestial) and XYZ (terrestrial)

Besides depending on the station coordinates and the time, they also contain the orbital and gravitational parameters.

Every observation furnishes an equation of type (7–46). Provided we have a sufficient number of such observation equations, we can solve them for the station coordinates X_P, Y_P, Z_P, for the orbital parameters a_0, e_0, etc., of the reference ellipse, and for a certain number of gravitational parameters C_{nm} and S_{nm}. This is the principle of the *orbital method*. In practice, differential formulas will be applied to determine corrections to assume approximate values by means of a least-squares adjustment. Therefore, the actual analytical developments are from the outset directed toward obtaining differential formulas corresponding to (7–46). The substitutions indicated above are, thus, consistently performed in terms of the corresponding differential expressions. In this way we are able to operate with linear equations and to employ that efficient tool of linear analysis, matrix calculus. Simple though the principle of this procedure is, the details when written out are nevertheless so complicated that the reader must again be referred to the literature, e.g., Kaula (1966 a), Montenbruck and Gill (2001). Computer formula manipulation is also used.

Besides these analytical problems, which have been satisfactorily solved, the geodetic application of (7–46) raises difficulties similar in principle to those involved in the determination of zonal harmonics by means of (7–24), but even more serious in practice. Strictly speaking, an infinite number of unknowns, C_{nm}, S_{nm}, etc., are to be determined from a finite number of observations. In order to get a definite solution, it must be assumed that the effect of higher-degree terms is negligibly small. But even then there are very many unknowns: coordinates of the observing stations, parameters of the reference orbit, and gravitational parameters; in addition, other unknowns must be included to take into account nongravitational forces acting on the satellite, such as air drag. An appropriate computational tool is least-squares collocation with parameters (Moritz 1980 a: Sect. 16).

To get a strong solution, observations should be evenly distributed both in space (with respect to the inclination of the satellites used) and in time.

Present results

At present (2005), several determinations of tesseral harmonics up to the degree 360 are available from a combination of satellite and terrestrial data. Soon the degree 1800 will be achieved. These coefficients represent the large-scale features of the disturbing potential T and, hence, of the geoid, since the geoidal height is given by $N = T/\gamma$. There is a general agreement between the essential aspects of these determinations as expressed in geoidal maps, although the details of these maps, and even more so the individual coefficients, are rather different.

As an example we take the first nonzonal coefficients, C_{22} and S_{22}, which, according to Sect. 2.6, Eq. (2–95), express the inequality of the earth's principal equatorial moments of inertia or, somewhat loosely speaking, its triaxiality. According to Groten (2004), we have $C_{22} = (1574.5 \pm 0.7) \cdot 10^{-9}$ and $S_{22} = (-903.9 \pm 0.7) \cdot 10^{-9}$.

Concerning the order of magnitude, J_2 is on the order of magnitude of 10^{-3}, where all the other coefficients are of order 10^{-6}. This is why the earth can be approximated by an ellipsoid so well.

7.6 New satellite gravity missions

7.6.1 Motivation and introductory considerations

Accuracy requirements in geodesy, geophysics, and oceanography for detailed gravity field information amount to 1 mgal for gravity anomalies. The related accuracy for the geoid ranges from 1 to 2 cm. In the presatellite era, the earth's gravity field was known with high accuracy only in a few regions of the world. Primarily, the available accurate gravity field information was based on terrestrial and airborne measurements. This implied that in large parts of the world there were virtually no gravity data available.

Why do we need the earth's gravity field at all? Following Pail (2003), first, the gravity field reflects the mass inhomogeneities in the earth's interior and on the earth's surface. Second, it is fundamental for the determination of the geoid (see Chap. 11) which, in its turn, may be regarded as a physical reference surface for a number of geodynamic processes (subject to continents, oceans, ice masses, atmosphere, etc.) and their interaction. The mass inhomogeneities are a necessary prerequisite to understand convection motions in the earth's mantle which are responsible for plate tectonics. Some large and many small lithospheric plates with a thickness of some 100–200 km move with a relative velocity of some centimeters per year. At the edges of the plates, seismic zones and volcanoes are situated.

Many time-dependent earth-related processes can be regarded as changes of the mass distribution and, thus, influence the gravity field, e.g., ocean circulation, ice mass variations, sea level change, tides, volcanism, post-glacial rebound. These variations may be categorized according to their periodicity. Some of these effects are extremely long-periodic or secular, e.g., plate tectonics with about 100 million years. In contrast, changes of the ice masses may amount to some 10 years only; even immediate events like earthquakes may occur.

These variations are referred to a global physical reference surface, the geoid. Therefore, the more accurately we know the geoid, the better we accurately understand the previously mentioned effects. Referring to various disciplines, the earth's gravity field is important for, e.g., geodesy, geophysics, oceanography, and climatology.

Geodesy

As mentioned in Sect. 5.3, GPS has revolutionized geodesy in many respects. Despite the tremendous importance of GPS, in Sect. 5.4 it was shown that the user of GPS gets only *geometric* quantities: WGS 84 coordinates, i.e., geocentric rectangular coordinates X, Y, Z or, computed from them, ellipsoidal coordinates φ, λ, h (see Sect. 5.6.1). Therefore, the height obtained by GPS, i.e., the ellipsoidal height h, is purely geometric. To transform these heights into orthometric heights H by $H = h - N$, the geoidal undulation N is required. Using satellites to determine the earth's gravity field, a globally uniform height system will result.

Additionally, an accurate knowledge of the earth's gravity field improves the orbit determination of satellites.

Oceanography

The sea surface topography (SST), i.e., the difference between the geoid and the mean sea surface, can be determined when combining satellite altimetry data and the earth's gravity field data. From Fig. 7.7 we obtain the relation

$$h = N + \text{SST} + \Delta H + a \,, \tag{7–47}$$

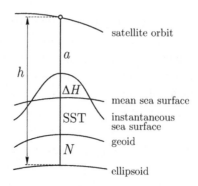

Fig. 7.7. Satellite altimetry

where h is the ellipsoidal height of the altimeter satellite (based on orbit computations), N is the geoidal height, SST is the sea surface topography to be derived, ΔH is caused by the instantaneous tidal effect, and a is the altimeter measurement. Note that (7–47) is a simplified representation, since, e.g., usually SST is split into a dynamic and a constant part. Refer to Seeber (2003: Sect. 9.3.1) for more details.

Knowing the sea surface topography, ocean currents and circulations may be explained, which is highly interesting for our understanding of the global energy transport. Ocean currents together with their time variations are an important indicator for climatic changes.

This method suffers from different accuracy influences in the results: when referring the mean sea surface to the ellipsoid, centimeter accuracy could be achieved. Involving the gravity model and referring the sea surface topography to the geoid as in Fig. 7.7, an improved geoid is required for a consistent accuracy level.

Geophysics

As mentioned earlier, the earth's gravity field reflects the mass inhomogeneities in the interior of the earth. Knowing gravity values on the earth's surface and, in addition, complementary data (e.g., magnetic and seismic data), improved models for the structure and processes in the earth's interior may be obtained. These processes may cause the movement of tectonic plates which are responsible for earthquakes. Thus, we see that the gravity field is the fundamental link in a chain of interactive processes. Using more descriptive terms, an improved knowledge on the gravity field may yield more accurate methods to predict earthquakes. This justifies any effort on the determination of the earth's gravity field.

7.6.2   Measurement concepts

From the introduction above, the need for an accurate determination of the earth's gravity field becomes evident. Three different measurement concepts evolved, leading to three different gravity field satellite missions:

- satellite-to-satellite (SST) tracking in high-low mode being realized by the "Challenging Minisatellite Payload" (CHAMP) mission,
- satellite-to-satellite tracking in low-low mode being realized by the "Gravity Recovery and Climate Experiment" (GRACE), and
- satellite gravity gradiometry, the objective of the "Gravity Field and Steady State Ocean Circulation Explorer" (GOCE) mission.

Before giving some details on the objectives and payloads of the missions, the different concepts are briefly described.

Satellite-to-satellite tracking in high-low mode

The principle is shown in Fig. 7.8. The orbit of the low earth orbit (LEO) satellite is continuously determined by satellites of global systems such as GPS, GLONASS or, in the future, Galileo. Note that the term "high-low mode" is not really appropriate because the satellites of GPS, GLONASS, and Galileo belong to the mean earth orbit (MEO) satellites and not to the high earth orbit (HEO) satellites. However, we keep the notation as used in Seeber (2003: Sect. 10.1). Apart from satellite-to-satellite tracking, the LEO satellite uses an accelerometer. In principle, three-dimensional perturbing accelerations caused by the earth's gravity field are measured. These accelerations correspond to first derivatives of the gravitational potential V.

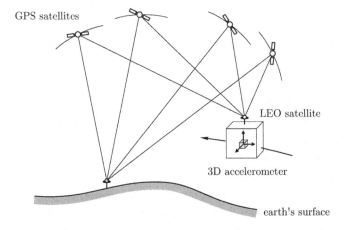

Fig. 7.8. Satellite-to-satellite tracking in high-low mode

The gravity field is derived by inverting (in the sense of inverse problems, cf. the remark on inverse problems at the end of Sect. 1.13) the information obtained from the satellite orbit

Satellite-to-satellite tracking in low-low mode
The principle is shown in Fig. 7.9. Two LEO satellites are placed in the same orbit but separated by some hundreds of kilometers (about 220 km in the case of GRACE). Ranges and range rates between the satellites are measured to utmost accuracy. Individually, the orbit of each LEO satellite is affected by perturbing accelerations which correspond to the first derivatives of the gravitational potential. In combination, differences of accelerations result. In addition, the position of the LEOs is determined by GPS satellites. This means that inherently satellite-to-satellite tracking in high-low mode is also implied. The effect of nongravitational forces on the satellite, e.g., due to air drag, must either be compensated or measured by an accelerometer.

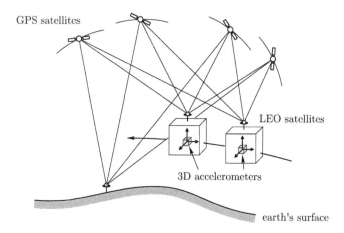

Fig. 7.9. Satellite-to-satellite tracking in low-low mode

Satellite gravity gradiometry
Compared to the just decribed low-low mode of satellite-to-satellite tracking with a long baseline between the two LEOs, the baseline between the accelerometer units tends to zero in case of satellite gravity gradiometry. This is achieved by placing both units into a single satellite (Fig. 7.10). Therefore, satellite gradiometry is the measurement of acceleration differences in three spatial orthogonal directions between the test masses of the six accelerometer units (two on each of the three axes) inside the satellite. In other words, the measured signal is the difference in gravitational acceleration at the satellite,

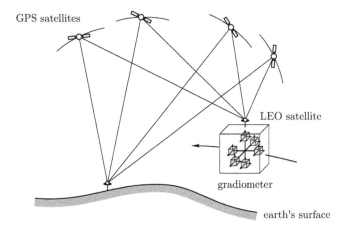

Fig. 7.10. Satellite gravity gradiometry with a three-axis gradiometer

where the gravitational signal arises from the attracting masses of the earth. Thus, the measured signal corresponds to the gradients of the component of the gravity acceleration, i.e., the second derivatives of the gravitational potential. For instance, in obvious notation we read from Fig. 7.11

$$\frac{V_{x2} - V_{x1}}{\Delta z} = \frac{\Delta V_x}{\Delta z} \doteq \frac{\partial V_x}{\partial z} = V_{xz}\,. \tag{7–48}$$

Summarizing the briefly described three methods, satellite-to-satellite tracking in high-low mode, satellite-to-satellite tracking in low-low mode, and satellite gravity gradiometry, we may say that the basic observable is gravitational acceleration. Following Rummel et al. (2002), the case of satellite-to-satellite tracking in high-low mode corresponds to a three-dimensional position, velocity or acceleration determination of a LEO satellite. The three-dimensional accelerometry corresponds to gravity acceleration. Mathematically, this is expressed by the first derivatives of the gravitational potential.

Considering the low-low mode, the principle corresponds to the line-of-

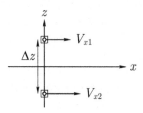

Fig. 7.11. Measuring the second derivative V_{xz}

sight measurement of the range, range rate or acceleration difference between the two low-orbiting satellites. The intersatellite link corresponds to acceleration differences between the two LEO satellites. Mathematically, this is expressed by the difference of first derivatives of the gravitational potential over a long baseline (i.e., the distance between the two LEO satellites).

In the case of satellite gradiometry, three-dimensional acceleration differences referring to the very short baseline realized by the gradiometer are measured. The gradient of gravity components corresponds to the acceleration gradient. Mathematically, this is expressed by the second derivatives of the gravitational potential.

Another feature inherent to satellite gravity missions should be kept in mind: the amplification of errors by the factor $(r/R)^{n+1}$ when transferring the measurement comprising the signal and noise from satellite altitude to the earth's surface. The factor $(r/R)^{n+1}$ describes the field attenuation with altitude. This error amplification effect is minimized by using an orbit as low as possible and by not measuring the potential V itself or its gradient but rather its second-order derivatives as in gravity gradiometry.

7.6.3 The CHAMP mission

The information on the challenging minisatellite payload (CHAMP) mission has been extracted primarily from http://op.gfz-potsdam.de/champ.

The Geoforschungszentrum Potsdam initiated the CHAMP idea and has the main responsibility. The primary CHAMP objectives are the following:

- mapping of the global gravity field, or, more specifically, to accurately determine the long-wavelength features of the static earth gravity field and its temporal variations (caused, e.g., by atmospheric mass redistributions, ocean circulation, sea level changes resulting from polar ice melting);
- mapping of the global magnetic field, or, more specifically, to accurately determine the main and crustal magnetic field of the earth and its space-time variations;
- profiling of the ionosphere and the troposphere, or, more specifically, to derive from GPS signal refraction data information on the temperature, water vapor, and electronic content of the atmosphere.

The CHAMP mission was launched on July 15, 2000 from the Russian Plesetsk cosmodrome. The main mission parameters of the respective satellite are the following:

- almost circular (eccentricity $e < 0.004$) and near-polar ($i = 87°$) orbit,

- initial altitude of 454 km,
- designed lifetime of five years for the mission (but the life expectation is much higher!),
- weight of 522 kg, length of 8.3 m (including a "boom" of 4 m length), width of 1.6 m, height of 0.75 m.

This initial altitude may be regarded as a compromise between gravity and magnetic field measurements. Considering the gravity field, a lower altitude would be desirable. Primarily due to atmospheric drag, the altitude will decrease to about 300 km and even less which is important because of an increasing sensitivity with respect to gravity field coefficient determination.

The reason for the curious 4 m boom is that the magnetometry assembly must be separated from the main body of the satellite ("magnetic cleanliness reasons", see http://op.gfz-potsdam.de/champ).

To achieve the mission goals, the following payload is on board of the satellite:

- dual-frequency GPS receiver connected to a multiple antenna system to determine the orbit of the CHAMP satellite using code and phase pseudoranges;
- three-axis accelerometer to measure the nongravitational accelerations acting on the spacecraft (air drag, solar radiation pressure, albedo, etc.);
- laser retroreflector for backup tracking to measure two-way ranges between ground stations and the satellites with 1–2 cm accuracy; these measurements support the precise orbit determination;
- fluxgate magnetometer to measure the vector components of the magnetic field of the earth (this instrument is supported by a scalar magnetometer to provide a calibration capability of the fluxgate magnetometer);
- equipment to determine the ion density and temperature, the drift velocity, and the electric field;
- two advanced star trackers to provide high-precision attitude information as required for the three-axis accelerometer, the digital ion drift meter, but also for the attitude control of the satellite.

Typical other equipment required for a proper operation of the satellite but with no specific relation to the scientific objectives of the mission is not detailed here, such as the cold gas propulsion system, the thermal control system, the power generation, the data handling, the telemetry, tracking and command system. Furthermore, we do not list items of the control segment of the CHAMP mission, but refer the reader to the previously mentioned homepage.

As explained before, the measuring principle for CHAMP is satellite-to-satellite tracking in high-low mode. The gravity field of the earth perturbes the CHAMP satellite orbit. These perturbing accelerations correspond to first derivatives of the gravitational potential V. This implies that the gravity field of the earth may be derived from observed gravitational satellite orbit perturbations applying numerical orbit integration (Montenbruck and Gill 2001) or using the energy balance principle (Ilk 1999, Jekeli 1999, Sneeuw et al. 2002).

For further reading see Reigber et al. (2003), Seeber (2003: Sect. 10.2.2).

7.6.4 The GRACE mission

The information on the gravity recovery and climate experiment (GRACE) mission has been extracted primarily from http://op.gfz-potsdam.de/grace.

The GRACE mission is a joint project between the U.S. National Aeronautics and Space Administration (NASA) and the Deutsches Zentrum für Luft- und Raumfahrt (DLR). The primary objectives of the mission are the following:

- determination of the global high-resolution gravity field of the earth,
- temporal gravity variations.

In addition, another task is the determination of the total electron content by GPS measurements to get knowledge on the refractivity in the ionosphere and troposphere. The two satellites of this mission were launched simultaneously on March 17, 2002 from the Russian Plesetsk cosmodrome. The main mission parameters of the two satellites are the following:

- almost circular (eccentricity $e < 0.005$) and near-polar ($i = 89°$) orbit,
- initial altitude between 485 km and 500 km,
- the two satellites are some 220 km apart (this requires orbit maneuvers every one or two months to maintain the separation between the two spacecraft),
- design lifetime of the mission is five years (but extended operation is envisaged),
- the weight of each satellite is about 480 kg and the length about 3 m.

As with CHAMP, also the altitude of the GRACE satellites will decrease in the course of their lifetime primarily because of atmospheric drag. The amount of this decrease depends on the solar activity cycle and may accumulate in the mission lifetime to some 50 km on low activity, and up to 200 km on high activity, see http://op.gfz-potsdam.de/grace.

The range between the two satellites must be determined extremely accurately. Its range rate must be known to better than $1\,\mu\mathrm{m\,s}^{-1}$, which is achieved by intersatellite microwave measurements. The basic idea is that variations in the gravity field cause variations in the range between the two satellites; areas of stronger gravity will affect the lead satellite first and, therefore, accelerate it away from the following satellite (Seeber 2003: p. 479).

GRACE will not only provide a static global gravity field but also its temporal variations.

To achieve the mission goals, the following payload is on board of the two satellites:

- The K-band ranging system is the key instrument of GRACE to measure the range changes between both satellites using dual-band microwave signals (i.e., two one-way ranges) with a precision of about $1\,\mu\mathrm{m\,s}^{-1}$. The ranges are obtained at a sampling rate of 10 Hz.

- The GPS receiver serves for the precise orbit determination of the GRACE spacecraft and provides data for atmospheric and ionospheric profiling. To achieve this, satellite-to-satellite tracking between the GRACE satellites and the GPS satellites is realized. A navigation solution comprising position, velocity, and a time mark is derived on board. The navigation solution is required for the attitude control system. The precise orbit based on code and carrier pseudoranges is determined on ground.

- The attitude and orbit control system comprises a cold gas propulsion system, three magnetic torque rods, star trackers, a three-axis inertial reference unit to measure angular rates, and a three-axis magnetometer.

- The accelerometer measures all nongravitational accelerations on the GRACE spacecraft, e.g., due to air drag or solar radiation pressure.

- The laser retroreflector is a passive payload instrument used to reflect short laser pulses transmitted by ground stations. The distance between a ground station and a GRACE satellite can be measured with an accuracy of 1–2 cm. The laser retroreflector data are primarily used together with the GPS receiver data for the precise orbit determination.

In 2004, the GRACE science team released to the public a first version of a new earth gravity field model complete to degree and order 150. The resulting improved geoid together with satellite altimetry will advance the knowledge on oceanographic, geodetic, and solid earth issues such as oceanic heat flux, change of sea level, ocean currents, precise positioning, orbit determination, and leveling.

The GRACE concept can be regarded as a one-dimensional gradiometer with a very long baseline of 220 km (Seeber 2003: p. 480). In contrast to this concept, GOCE uses very short baselines (50 cm) in three directions.

7.6.5 The GOCE mission

The main sources of this section are www.esa.int/export/esaLP/goce.html, ESA (1999), Müller (2001), Drinkwater et al. (2003), and Pail (2003).

The gravity field and steady-state ocean circulation explorer (GOCE) mission is a Core Mission of the ESA Living Planet Programme. The primary objectives of the GOCE mission are to measure the earth's stationary gravity field and to model the geoid with extremely high accuracy. More specifically:

- to determine the gravity anomalies with an accuracy of 1 mgal,
- to determine the geoid with an accuracy of 1–2 cm,
- to achieve these results at a spatial resolution better than 100 km.

According to the above mission requirements, GOCE is intended for a representation of the gravity potential by spherical harmonics complete at least to degree and order 200 (corresponding to the spatial resolution of 100 km), but 250 is envisaged.

From the *geodetic point of view*, a global geoid of 1–2 cm accuracy and a gravity field model accurate to 1 mgal at about 100 km spatial resolution may be used – among many other important applications – for the following purposes:

- Control (or replacement) of traditional leveling by leveling with GPS. In Sect. 4.6 we have learned the basic equation (4–72), $H = h - N$, relating the orthometric height H (above the geoid), the ellipsoidal height h (above the ellipsoid), and the geoidal undulation N. With N accurately known from GOCE and h measured by GPS (Sects. 5.5, 5.6.1), the orthometric height H is readily obtained.
- Worldwide unification of height systems so as to refer to one height datum which allows for comparison of different sea levels (e.g., in the North Sea and in the Mediterranean) and sea-level changes (which may be caused by melting continental ice sheets). Remember that the geoid is defined as an equipotential surface which follows a hypothetical ocean surface at rest (in the absence of tides and currents and other smaller influences). Consequently, a precise geoid is crucial in deriving accurate measurements of ocean currents and sea-level changes.

- Providing a significant improvement in satellite orbit determination and prediction. This especially applies to low-orbiting satellites. The highly accurate gravity field will enable a better separation of the perturbations caused by the static gravity field and other perturbing forces (not only the nongravitational forces caused by air drag and solar radiation pressure but also perturbations caused by the solid earth and ocean tides).

The duration of the mission is scheduled with nominally 20 months, including a 3-month commissioning and calibration phase and two measurement phases, each lasting six months and separated by a long eclipse period. The other main mission parameters are the following:

- due for launch in 2007 from Plesetsk in Russia,
- sun-synchronous orbit, inclination 96.5°,
- measurement altitudes: approximately 250 km,
- single ground station in Kiruna, Sweden, to exchange data and commands; the European Space Operations Center (ESOC) at Darmstadt will be used for mission and satellite control.

The main payload components are the following:

- three-axis gravity gradiometer based on three pairs of electrostatic servo-controlled accelerometers to measure gravity gradients in three spatial orthogonal directions: the desired signal is the difference in gravitational acceleration (between a pair of accelerometers separated by 0.5 m) at the test mass location inside the spacecraft caused by gravity anomalies from attracting masses of the earth;
- geodetic dual-frequency (to compensate for ionospheric delays) multichannel GPS receiver with codeless tracking capability to (1) determine the orbit of the GOCE satellite and (2) derive gravity information from this orbit (the first task is performed by satellite-to-satellite tracking in the high-low mode: this provides knowledge of the precise position of the [low] spacecraft relative to [high] reference satellites such as the GPS satellites; the second task is solved by orbit perturbation analysis yielding gravity information);
- laser retroreflector to enable tracking by ground-based laser stations;
- attitude control accomplished by actuators comprising an ion thruster assembly, star trackers, a three-axis magnetic torquer, and some other sensors;
- length of the satellite about 5 m, cross section of $1\,\text{m}^2$, weight about 1000 kg.

Referring to the results, the main output of this mission will be the following:

- spherical-harmonic coefficients for the gravitational potential, see, e.g., (2–80),
- corresponding variance-covariance matrix.

Derived products from this main output are geoidal heights, gravity anomalies, and also oceanographic data.

It is important to mention that the GPS orbit analysis of GOCE will rather yield long-wavelength information of the gravity field, while the satellite gravity gradiometry will yield the short-wavelength information.

GOCE is the first "drag-free" mission, which implies that the satellite moves in free fall around the earth. Therefore, a drag compensation and attitude control system is required to compensate for drag forces and torques.

This and more information may be found in Rebhahn et al. (2000), Drinkwater et al. (2003), Pail (2003), www.esa.int/livingplanet/goce.

Measurements

The basic principle of gradiometry in GOCE is the measurement of acceleration differences for a very short baseline. Considering two accelerometers separated by 50 cm on one axis, Müller (2001) and Pail (2003) write the two observation equations as

$$\begin{aligned} \mathbf{a}_1 &= \left[\mathbf{M} + \dot{\boldsymbol{\Omega}} + \boldsymbol{\Omega}\boldsymbol{\Omega}\right]\Delta\mathbf{x} + \mathbf{f}_{\mathrm{ng}}, \\ \mathbf{a}_2 &= -\left[\mathbf{M} + \dot{\boldsymbol{\Omega}} + \boldsymbol{\Omega}\boldsymbol{\Omega}\right]\Delta\mathbf{x} + \mathbf{f}_{\mathrm{ng}}, \end{aligned} \tag{7–49}$$

where $\mathbf{a}_1$ and $\mathbf{a}_2$ are the measured accelerations of the two accelerometers on the axis, and $\mathbf{M}$ is the Marussi tensor,

$$\mathbf{M} = \begin{bmatrix} \dfrac{\partial^2 V}{\partial x^2} & \dfrac{\partial^2 V}{\partial x\,\partial y} & \dfrac{\partial^2 V}{\partial x\,\partial z} \\[2ex] \dfrac{\partial^2 V}{\partial x\,\partial y} & \dfrac{\partial^2 V}{\partial y^2} & \dfrac{\partial^2 V}{\partial y\,\partial z} \\[2ex] \dfrac{\partial^2 V}{\partial x\,\partial z} & \dfrac{\partial^2 V}{\partial y\,\partial z} & \dfrac{\partial^2 V}{\partial z^2} \end{bmatrix}, \tag{7–50}$$

which comprises the second derivatives of the gravitational potential (our target quantity!). Furthermore, the skewsymmetric matrix

$$\boldsymbol{\Omega} = \begin{bmatrix} 0 & \omega_3 & -\omega_2 \\ -\omega_3 & 0 & \omega_1 \\ \omega_2 & -\omega_1 & 0 \end{bmatrix} \tag{7–51}$$

comprises the components of the angular velocity and is used to describe the orientation of the gradiometer. Since $\boldsymbol{\Omega}$ is skewsymmetric, the tensor $\boldsymbol{\Omega\Omega}$ is symmetric. Finally, $\boldsymbol{\Delta x}$ in (7–49) is the vector from the intersection of the three coordinate axes to the respective accelerometer (where the same length is assumed), and $\mathbf{f}_{ng}$ comprises all nongravitational effects (air drag, solar radiation pressure, etc.).

Now we once add ("common mode") and once subtract ("differential mode") the two accelerations in (7–49) and obtain

$$(\mathbf{a}_1 + \mathbf{a}_2)/2 = \mathbf{f}_{ng} \,,$$
$$(\mathbf{a}_1 - \mathbf{a}_2)/2 = \left[\mathbf{M} + \dot{\boldsymbol{\Omega}} + \boldsymbol{\Omega\Omega}\right]\boldsymbol{\Delta x} \,, \tag{7–52}$$

where we can extract the nongravitational effects $\mathbf{f}_{ng}$ in the common mode. Introducing the quantity

$$\boldsymbol{\Gamma} = \mathbf{M} + \dot{\boldsymbol{\Omega}} + \boldsymbol{\Omega\Omega} \tag{7–53}$$

and assuming a known geometry of the gradiometer, i.e., $\boldsymbol{\Delta x}$ may safely assumed to be known, then the remaining task is to extract the gravity gradient tensor $\mathbf{M}$ from $\boldsymbol{\Gamma}$. This can be achieved by the two relations

$$(\boldsymbol{\Gamma} - \boldsymbol{\Gamma}^T)/2 = \dot{\boldsymbol{\Omega}} \,,$$
$$(\boldsymbol{\Gamma} + \boldsymbol{\Gamma}^T)/2 = \mathbf{M} + \boldsymbol{\Omega\Omega} \,, \tag{7–54}$$

where the superscript T denotes transposition. To verify these relations, a little matrix calculus is needed. If, generally, $\mathbf{K}$ is a symmetric matrix, then we have $\mathbf{K} = \mathbf{K}^T$. If $\mathbf{K}$ is a skewsymmetric matrix, then we have $\mathbf{K} = -\mathbf{K}^T$.

Referring now to (7–53), we know that $\mathbf{M}$ is symmetric, $\boldsymbol{\Omega}$ is skewsymmetric, and $\boldsymbol{\Omega\Omega}$ is symmetric. Therefore, transposing (7–53) yields

$$\boldsymbol{\Gamma}^T = \mathbf{M} - \dot{\boldsymbol{\Omega}} + \boldsymbol{\Omega\Omega} \,. \tag{7–55}$$

Using now (7–53) and (7–55), we get immediately

$$\boldsymbol{\Gamma} - \boldsymbol{\Gamma}^T = 2\dot{\boldsymbol{\Omega}} \tag{7–56}$$

and, finally,

$$(\boldsymbol{\Gamma} - \boldsymbol{\Gamma}^T)/2 = \dot{\boldsymbol{\Omega}} \,, \tag{7–57}$$

which completes our proof for the first relation of (7–54). To prove the second relation of (7–54), we add Eqs. (7–53) and (7–55):

$$\boldsymbol{\Gamma} + \boldsymbol{\Gamma}^T = 2\mathbf{M} + 2\boldsymbol{\Omega\Omega} \tag{7–58}$$

or

$$(\boldsymbol{\Gamma} + \boldsymbol{\Gamma}^T)/2 = \mathbf{M} + \boldsymbol{\Omega}\boldsymbol{\Omega}\,, \qquad\qquad (7\text{--}59)$$

which concludes our proof.

Since we have determined $\dot{\boldsymbol{\Omega}}$ in (7–57), we can get $\boldsymbol{\Omega}$ by integration:

$$\boldsymbol{\Omega}(t) = \boldsymbol{\Omega}(t_0) + \int_{t_0}^{t} \dot{\boldsymbol{\Omega}}\, dt\,, \qquad\qquad (7\text{--}60)$$

where the initial orientation $\boldsymbol{\Omega}(t_0)$ is obtained from the star trackers. Squaring the result for $\boldsymbol{\Omega}(t)$ yields $\boldsymbol{\Omega}\boldsymbol{\Omega}$, which is needed in (7–59) so that we find

$$\mathbf{M} = (\boldsymbol{\Gamma} + \boldsymbol{\Gamma}^T)/2 - \boldsymbol{\Omega}\boldsymbol{\Omega} \qquad\qquad (7\text{--}61)$$

as final result for the desired Marussi tensor $\mathbf{M}$. Many more details may be found in Rummel (1986).

8 Modern views on the determination of the figure of the earth

8.1 Introduction

In the preceding chapters we have usually followed what might be called the conservative approach to the problems of physical geodesy using classical observations. The geodetic measurements – astronomical coordinates and azimuths, horizontal angles, gravity observations, etc. – are reduced to the geoid, and the "geodetic boundary-value problem" is solved for the geoid by means of Stokes' integral and similar formulas. The geoid then serves as a basis for establishing the position of points of the earth's surface.

The advantage of this approach is that the geoid is a level surface, capable of a simple definition in terms of the physically meaningful and geodetically important potential W. The geoid represents the most obvious mathematical formulation of a horizontal surface at sea level. This is why the use of the geoid simplifies geodetic problems and makes them accessible to geometrical intuition.

The disadvantage is that the potential W inside the earth, and hence the geoid $W = $ constant, depends on the density ϱ because of Poisson's Eq. (2–9),

$$\Delta W = -4\pi \, G \, \varrho + 2\omega^2 \, . \qquad (8\text{–}1)$$

Therefore, in order to determine or to use the geoid, the density of the masses at every point between the geoid and the ground must be known, at least theoretically. This is clearly impossible, and therefore some assumptions concerning the density must be made, which is unsatisfactory theoretically, even though the practical influence of these assumptions is usually rather small.

For this reason it is of basic importance that *M.S. Molodensky* in 1945 was able to show that the physical surface of the earth can be determined from geodetic measurements alone, without using the density of the earth's crust. This requires that the concept of the geoid be abandoned. The mathematical formulation becomes more abstract and more difficult. Both the gravimetric method and the astrogeodetic method can be modified for this purpose. The gravity anomalies and the deflections of the vertical now refer to the ground, and no longer to sea level; the "height anomalies" at ground level take the place of the geoidal undulations.

These developments have considerably broadened our insight into the principles of physical geodesy and have also introduced powerful new methods for tackling classical problems. Hence their basic theoretical significance is by no means lessened by the fact that many scientists prefer to retain the geoid because of its conceptual and practical advantages.

In this chapter, we first give a concise survey of the conventional determination of the geoid by means of gravity reductions, in order to understand better the modern ideas. After an exposition of Molodensky's theory, we show how the new methods may be applied to classical problems such as gravity reduction or the determination of the geoid by gravimetric and astrogeodetic methods. It should be mentioned that the terms "modern" and "conventional" merely serve as convenient labels; they do not imply any connotation of value or preferability.

Part I: Gravimetric methods

8.2 Gravity reductions and the geoid

The integrals of Stokes and of Vening Meinesz and similar formulas presuppose that the disturbing potential T is harmonic on the geoid, which implies that there are no masses outside the geoid. This assumption – no masses outside the bounding surface – is necessary if we wish to treat any problem of physical geodesy as a boundary-value problem in the sense of potential theory. The reason is that the boundary-value problems of potential theory always involve harmonic functions, that is, solutions of Laplace's equation

$$\Delta T = 0. \qquad (8\text{–}2)$$

This is equivalent to $\Delta V = 0$. Proof: $T = W - U$ (U is the normal potential), $\Delta W = 2\omega^2$ outside the earth (density zero, only rotation, $\Delta U = 2\omega^2$ for the same reason, hence $\Delta T = \Delta W - \Delta U = 2\omega^2 - 2\omega^2 = 0$). Since then $\Delta W = 2\omega^2$ rather than zero by Eq. (2–9), it is not quite correct to call the external gravity potential W harmonic as well, but we may nevertheless do so for simplicity. No misunderstanding is possible.

We know, for instance, that the determination of T or N from gravity anomalies Δg may be considered as a third boundary-value problem (see Sect. 1.13).

Since there are masses outside the geoid, they must be moved inside the geoid or completely removed before we can apply Stokes' integral or related

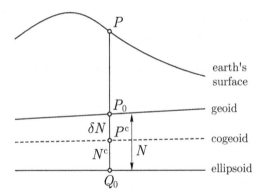

Fig. 8.1. Geoid and cogeoid

formulas. *This is the purpose of the various gravity reductions.* They were considered extensively in Chap. 3; we therefore can limit ourselves to pointing out those theoretical features that are relevant to our present problem.

If the external masses, the masses outside the geoid, are removed or moved inside the geoid, then gravity changes. Furthermore, gravity is observed at ground level but is needed at sea level. Thus, the reduction of gravity involves the consideration of these two effects, in order to obtain boundary values on the geoid.

This *regularization* of the geoid by removing the external masses unfortunately also changes the level surfaces and hence, in general, the geoid. This is the *indirect effect*; the changed geoid is called the *cogeoid* or the regularized geoid.

The principle of this method may be described as follows (Jung 1956: p. 578); see Fig. 8.1.

1. The masses outside the geoid are, by computation, either removed entirely or else moved inside the geoid. The effect of this procedure on the value of gravity g at the station P is considered.

2. The gravity station is moved from P down to the geoid, to the point P_0. Again, the corresponding effect on the gravity is considered.

3. The indirect effect, the distance $\delta N = P_0 P^{\mathrm{c}}$, is obtained by dividing the change in potential at the geoid, δW, by normal gravity (Bruns' theorem):

$$\delta N = \frac{\delta W}{\gamma}.$$

(8–3)

4. The gravity station is now moved from the geoidal point P_0 to the

cogeoid, to the point P^c (hence the notation with upper index c). This gives the boundary value of gravity at the cogeoid, g^c.

5. The shape of the cogeoid is computed from the reduced gravity anomalies

$$\Delta g^c = g^c - \gamma \qquad (8\text{--}4)$$

by Stokes' formula, which gives $N^c = QP^c$.

6. Finally, the geoid is determined by considering the indirect effect. The geoidal undulation N is thus obtained as

$$N = N^c + \delta N . \qquad (8\text{--}5)$$

Remark. At first sight it may seem that the masses between the geoid and the cogeoid should be removed if the cogeoid happens to be below the geoid, because Stokes' formula is applied to the cogeoid. However, this is not necessary, and therefore we need not be concerned with a "secondary indirect effect". The argument is a little too technical to be presented here; see Moritz (1965: p. 26).

In principle, every gravity reduction that gives boundary values at the geoid is equally suited for the determination of the geoid, provided the indirect effect is properly taken into account. Thus, the selection of a good reduction method should be made from other points of view, such as the geophysical meaning of the reduced gravity anomalies, the simplicity of computation, the feasibility of interpolation between the gravity stations, the smallness or even absence of the indirect effect, etc. (see Sect. 3.7).

The *Bouguer reduction* corresponds to a complete removal of the external masses. In the *isostatic reduction*, these masses are shifted vertically downward according to some theory of isostasy. In Helmert's *condensation reduction*, the external masses are compressed to form a surface layer on the geoid. The Bouguer reduction and especially the isostatic reduction (in modern terminology *topographic-isostatic reduction*) are used as auxiliary quantities for computational purposes, especially to facilitate interpolation.

The *free-air anomaly* is nowadays used in three senses:

1. at ground level (on the physical surface of the earth) it is simply the gravity anomaly in the sense of Molodensky (Sect. 8.4);

2. at sea level it may be identified with the analytical continuation of the Molodensky anomaly from ground down to sea level. This will be considered in detail in Sect. 8.6. A final review will be found in Sect. 8.15.

3. The free-air anomaly can be theoretically interpreted as an approximation of the classical condensation anomaly in the sense of Helmert (Sect. 3.9). This is one of the interpretations of the frequent practice to simply apply Stokes' formula to the classical free-air anomaly, where only the standard normal free-air reduction is applied to measured gravity g, see Eq. (8–6) below.

This is pretty rigorously the gravity anomaly in the sense of Molodensky (item 1 above), so there is another interpretation of this frequent practice: it is a (conscious or unconscious) use of Molodensky's method in the zero approximation (i.e., only Stokes' formula without Molodensky correction g_1, see Sect. 8.6). Of course, this works only in a reasonably flat terrain.

Important remark. Curiously enough, it helps if the terrain correction (Sect. 3.4) is applied; this is explained in Moritz (1980 a: Sect. 48) as some kind of Molodensky correction g_1 and in Moritz (1990: p. 244) by isostatic reduction.

Also the *Poincaré–Prey reduction* is quite different (Sect. 3.5). It gives the actual gravity inside the earth. It does not give boundary values but is used for orthometric heights (Chap. 4).

In all reduction methods it is necessary to know the density of the masses above the geoid. In practice, this involves some kind of an assumption – for instance, putting $\varrho = 2.67 \text{ g cm}^{-3}$. A second assumption is usually made in the free-air reduction, which is part of the reduction of gravity to the geoid: the actual free-air gravity gradient is assumed to be equal to the normal gradient

$$\frac{\partial \gamma}{\partial h} \doteq -0.3086 \text{ mgal m}^{-1}. \qquad (8\text{--}6)$$

These two assumptions falsify our results, at least theoretically.

The second assumption can be avoided by using the actual free-air gradient as computed by the methods of Sect. 2.20. The anomalies Δg to be used in formula (2–394) must be gravity anomalies reduced to the geoid: gravity g after steps 1 and 2 of the above description, minus normal gravity γ on the ellipsoid. This presupposes that in step 2 a preliminary free-air reduction using the normal gradient has been applied first.

Deflections of the vertical

The indirect effect affects the deflection of the vertical as well as the geoidal height. We have found

$$N = N^c + \delta N, \qquad (8\text{--}7)$$

where N^c is the undulation of the cogeoid, the immediate result of Stokes' formula, and δN is the indirect effect. By differentiating N in a horizontal direction, we get the deflection component along this direction:

$$\varepsilon = -\frac{\partial N}{\partial s} = -\frac{\partial N^c}{\partial s} - \frac{\partial(\delta N)}{\partial s}\,. \qquad (8\text{--}8)$$

This means that we must add to the immediate result of Vening Meinesz' formula, $-\partial N^c/\partial s$, a term representing the horizontal derivative of δN (see also Sect. 3.7).

To repeat, the main purpose is to obtain a simple boundary surface. The geoid approximated by an ellipsoid or even a sphere is a much easier boundary surface than the physical surface of the earth, to which we turn now.

8.3 Geodetic boundary-value problems

It is, however, quite easy to understand the general principles. In space we have the well-known fact that the gravity vector **g** and the gravity potential (geopotential) W are related by

$$\mathbf{g} = \operatorname{grad} W \equiv \left[\frac{\partial W}{\partial x}, \frac{\partial W}{\partial y}, \frac{\partial W}{\partial z}\right], \qquad (8\text{--}9)$$

which shows that the force **g** is the gradient vector of the potential.

Let S be the earth's topographic surface and let W and **g** be the geopotential and the gravity vector on this surface. Then there exists a relation

$$\mathbf{g} = f(S, W)\,, \qquad (8\text{--}10)$$

the gravity vector **g** on S is a function of the surface S and the geopotential W on it. This can be seen in the following way. Let the surface S and the geopotential W on S be given. The gravitational potential V is obtained by subtracting the potential of the centrifugal force Φ, which is simple and perfectly known (Sect. 2.1):

$$V = W - \Phi\,. \qquad (8\text{--}11)$$

The potential V outside the earth is a solution of Laplace's equation $\Delta V = 0$ and consequently *harmonic* (Sect. 1.3). Thus, knowing V *on* S, we can obtain V outside S by solving Dirichlet's boundary-value problem, the first boundary-value problem of potential theory, which is practically always uniquely solvable (Sect. 1.12) at least if V is sufficiently smooth on S. After having found V as a function in space outside S, we obtain the gravitational force $\operatorname{grad} V$. Adding the well-known and simple vector of the centrifugal force, we obtain the gravity vector $\mathbf{g}$ outside and, by continuity, on S.

This is precisely what (8–10) means. The modern general concept of a *function* can be explained as a *rule of computation*, indicating that given S and W on S, we can uniquely calculate $\mathbf{g}$ on S. Note that f is not a function in the elementary sense but rather a "nonlinear operator", but we disregard this for the moment. Therefore we may formulate:

(1) *Molodensky's boundary-value problem* is the task to determine S, the earth's surface, if $\mathbf{g}$ and W on it are given. Formally, we have to solve (8–10) for S:

$$S = F_1(\mathbf{g}, W), \tag{8–12}$$

that is, we get *geometry from gravity*.

(2) *GPS boundary-value problem.* Since we have GPS at our disposal, we can consider S as known, or at least determinable by GPS. In this case, the geometry S is known, and we can solve (8–10) for W:

$$W = F_2(S, \mathbf{g}), \tag{8–13}$$

that is, we get *potential from gravity*. As we shall see, this is far from being trivial: we have now a method to *replace leveling*, a tedious and time-consuming old-fashioned method, *by GPS leveling*, a fast and modern technique (Sect. 4.6).

In spite of all similarities, we should bear in mind a fundamental difference: (8–13) solves a *fixed-boundary problem* (boundary S given), whereas (8–12) solves a *free-boundary problem*: the boundary S is a priori unknown ("free"). Fixed-boundary problems are usually simpler than free ones.

This is only the principle of both solutions. The formulation is quite easy to understand. The direct implementation of these formulas is difficult, however, because that would imply the solution of "hard inverse function theorems" of nonlinear functional analysis. For numerical computations, we know series solutions, in the form of "Molodensky series", which are sufficient for all present purposes and which can, furthermore, be derived in an elementary fashion, without needing integral equations (Molodenski 1958; Molodenskii et al. 1962; Moritz 1980 a: Sect. 45). Here we shall outline the known elementary solution for Molodensky's problem and immediately extend it to the GPS problem. Both problems will be solved by very similar Molodensky series.

The simplest possible example

Let the boundary surface S be a sphere of radius R. The earth is represented by this sphere which is considered homogeneous and nonrotating. The potential W is identical to the gravitational potential V, so that on the surface S we have constant values

$$W = \frac{GM}{R},$$

$$g = \frac{GM}{R^2}. \tag{8-14}$$

Knowing W and g, we have

$$R = \frac{W}{g}, \tag{8-15}$$

the radius of the sphere S. Thus, we have solved Molodensky's problem in this trivial but instructive example. We have indeed got geometry (i.e., R) from physics (i.e., g and W)!

8.4   Molodensky's approach and linearization

We have just seen that the reduction of gravity to sea level necessarily involves assumptions concerning the density of the masses above the geoid. This is equally true of other geodetic computations when performed in the conventional way.

To see this, consider the problem of computing the ellipsoidal coordinates φ, λ, h from the natural coordinates Φ, Λ, H, as described in Chap. 5. The geometric ellipsoidal height h above the ellipsoid is obtained from the orthometric height H above the geoid and the geoidal undulation N by

$$h = H + N. \tag{8-16}$$

The determination of N was considered in Chap. 2 and elsewhere in this book. To compute H from the results of leveling, we need the mean gravity $\bar{g}$ along the plumb line between the geoid and the ground (Sect. 4.3). Since gravity g cannot be measured inside the earth, we compute it by Prey's reduction, for which we must know the density of the masses above the geoid.

The ellipsoidal coordinates φ and λ are obtained from the astronomical coordinates Φ and Λ and the deflection components ξ and η by

$$\varphi = \Phi - \xi, \quad \lambda = \Lambda - \eta \sec \varphi. \tag{8-17}$$

The coordinates Φ and Λ are measured on the ground; ξ and η can be computed for the geoid by Vening Meinesz' formula, the indirect effect being taken into account according to Sect. 8.2. To apply the above formulas, either Φ and Λ must be reduced down to the geoid or ξ and η must be reduced up to the ground. In both cases this involves the reduction for the curvature of the plumb line (Sect. 5.15), which also depends on the mean value $\bar{g}$ through its horizontal derivatives. Hence Prey's reduction enters here too.

Thus we see that in the conventional approach to the problems of physical geodesy we must know the density of the outer masses or make assumptions concerning it. To avoid this, Molodensky proposed a different approach in 1945.

Figure 8.2 shows the geometrical principles of this method, which is essentially a linearization of Eq. (8–10). The ground point P (i.e., point on the earth's surface S) is again projected onto the ellipsoid according to Helmert. However, the ellipsoidal height h is now determined by

$$h = H^* + \zeta, \tag{8–18}$$

the *normal height* H^* replacing the orthometric height H, and the *height anomaly* ζ replacing the geoidal undulation N.

This will be clear if one considers the surface whose normal potential U at every point Q is equal to the actual potential W at the corresponding point P, so that $U_Q = W_P$, corresponding points P and Q being situated on the same ellipsoidal normal. This surface is called the *telluroid* (Hirvonen 1960, 1961). The vertical distance from the ellipsoid to the telluroid is the normal height H^* (Sect. 4.4), whereas the ellipsoidal height h is the vertical distance from the ellipsoid to the earth's surface. Thus, the difference between these two heights is the height anomaly

$$\zeta = h - H^*, \tag{8–19}$$

closely corresponding to the geoidal undulation $N = h - H$, which is the difference between the ellipsoidal and the orthometric height.

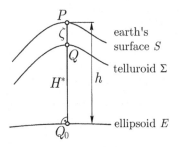

Fig. 8.2. Telluroid, normal height H^*, and height anomaly ζ

The normal height H^*, and hence the telluroid Σ, can be determined by leveling combined with gravity measurements, according to Sect. 4.4. First the geopotential number of P, $C = W_0 - W_P$, is computed by

$$C = \int_0^P g \, dn \,, \tag{8-20}$$

where g is the measured gravity and dn is the leveling increment. The normal height H^* is then related to C by an *analytical* expression such as (4–63),

$$H^* = \frac{C}{\gamma_{Q_0}} \left[1 + (1 + f + m - 2f \sin^2\varphi) \frac{C}{a\,\gamma_{Q_0}} + \left(\frac{C}{a\,\gamma_{Q_0}} \right)^2 \right], \tag{8-21}$$

where γ_{Q_0} is the normal gravity at the ellipsoidal point Q_0. Note that H^* is independent of the density.

The normal height H^* of a ground point P is identical with the ellipsoidal height h, the height above the ellipsoid, of the corresponding telluroid point Q. If the geopotential function W were equal to the normal potential function U at every point, then Q would coincide with P, the telluroid would coincide with the physical surface of the earth, and the normal height of every point would be equal to its ellipsoidal height. Actually, however, $W_P \neq U_P$; hence the difference

$$\zeta_P = h_P - H_P^* = h_P - h_Q \tag{8-22}$$

is not zero. This explains the term "height anomaly" for ζ.

The gravity anomaly is now defined as

$$\Delta g = g_P - \gamma_Q \,; \tag{8-23}$$

it is the difference between the actual gravity as measured on the ground and the normal gravity on the telluroid. The normal gravity on the telluroid, which we shall briefly denote by γ, is computed from the normal gravity at the ellipsoid, γ_{Q_0}, by the normal free-air reduction, but now applied *upward*:

$$\gamma \equiv \gamma_Q = \gamma_{Q_0} + \frac{\partial\gamma}{\partial h} H^* + \frac{1}{2!} \frac{\partial^2\gamma}{\partial h^2} H^{*2} + \cdots . \tag{8-24}$$

For this reason, the new gravity anomalies (8–23) are called *free-air anomalies*. They are *referred to ground level*, whereas the conventional gravity anomalies have been referred to sea level. Therefore, the new free-air anomalies have nothing in common with a free-air reduction of actual gravity to sea level, except the name. This distinction should be carefully kept in mind.

A direct formula for computing γ at Q is (2–215),

$$\gamma = \gamma_{Q_0} \left[1 - 2(1 + f + m - 2f \sin^2\varphi) \frac{H^*}{a} + 3 \left(\frac{H^*}{a} \right)^2 \right], \tag{8-25}$$

where γ_{Q_0} is the corresponding value on the ellipsoid.

The height anomaly ζ may be considered as the distance between the geopotential surface $W = W_P =$ constant and the corresponding spheropotential surface $U = W_P =$ constant at the point P. In Sect. 2.14 (Fig. 2.15), we have denoted this distance by N_P and have found that Bruns' formula (2–237) also applies to this quantity. Hence, for $\zeta = N_P$ we have

$$\zeta = \frac{T}{\gamma} , \qquad (8\text{–}26)$$

where $T = W_P - U_P$ is the disturbing potential at ground level, and γ the normal gravity at the telluroid.

It may be expected that ζ is connected with the ground-level anomalies Δg by an expression analogous to Stokes' formula for the geoidal height N. This is indeed true. However, the telluroid is not a level surface, and to every point P on the earth's surface corresponds in general a different geopotential surface $W = W_P$. Therefore, the relation between Δg and ζ in the new theory is considerably more complicated than for the geoid. In Molodensky's original formulation, the problem involves an integral equation, which may be solved by an iteration, the first term of which is given by Stokes' formula. We shall use an equivalent but much simpler approach without integral equation.

Finally, we remark that we may also plot the height anomalies ζ above the ellipsoid. In this way we get a surface that is identical with the geoid over the oceans, because there $\zeta = N$, and is very close to the geoid anywhere else. This surface has been called the *quasigeoid* by Molodensky. However, the quasigeoid is not a level surface and has no physical meaning whatever. It must be considered as a concession to conventional conceptions that call for a geoidlike surface. From this point of view, the normal height of a point is its elevation above the quasigeoid, just as the orthometric height is its elevation above the geoid.

Gravity disturbance

As usual, the gravity disturbance is defined by

$$\delta g = g_P - \gamma_P . \qquad (8\text{–}27)$$

It is a typical new feature introduced into the practice of physial geodesy by GPS, because GPS determines the ellipsoidal coordinates φ, λ, h directly at the surface point P, so that now δg can be considered observational data instead of Δg.

Linearization

The linearization applies equally well for the Molodensky problem and the GPS problem. The geometry is familiar (Fig. 8.2).

We recall the surface Σ, the *telluroid*, which is defined by the condition

$$U(Q) = W(P) \,. \tag{8-28}$$

We note that (8–28) is the surface equivalent to the classical relation for sea level (Fig. 8.3)

$$U(Q_0) = W(P_0) \,. \tag{8-29}$$

Equation (8–28) would apply with

$$W(P_0) = W_0 = \text{ constant} \tag{8-30}$$

if S were an equipotential surface, the geoid, which is the case only over the oceans with the usual simplifying assumption that the surface of the ocean is an equipotential surface not changing with time (Fig. 8.3).

Molodensky's theory does not use the geoid directly but the physical earth's surface. We repeat once more that this is Molodensky's epochal idea which radically changed the course of physical geodesy since 1945.

We shall, however, use the fictitious case of S being an equipotential surface, but only as a first (or zero-order) assumption in a perturbation approach for the real earth's surface (Molodensky series). This first approximation is the spherical case to be considered in the next section.

Now we consider the linearization in more detail. The ellipsoidal height h is directly determined by GPS. It may be decomposed into

$$h = H^* + \zeta \,. \tag{8-31}$$

Here, H^* is the normal height and ζ is the height anomaly, whose definitions are seen from Fig. 8.2. In the GPS case we do know the earth's surface S directly, but the telluroid Σ and the height anomalies ζ are still required for formulating the boundary condition, just as the knowledge of the geoid does not make superfluous the reference ellipsoid.

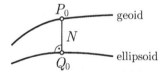

Fig. 8.3. Geoid and ellipsoid

The definition of the *gravity anomaly* Δg and the *gravity disturbance* δg has, on the earth's surface, the same form as in the classical case of geoid and sea level:

$$\Delta g = g_P - \gamma_Q = -\frac{\partial T}{\partial h} + \frac{1}{\gamma}\frac{\partial \gamma}{\partial h}\,T\,, \qquad (8\text{--}32)$$

$$\delta g = g_P - \gamma_P = -\frac{\partial T}{\partial h}\,. \qquad (8\text{--}33)$$

The gravity disturbance δg has become practically important only through GPS, since h, the ellipsoidal height of P, can be measured using GPS and hence γ_P, the normal gravity γ at P, can be determined.

As usual, Bruns' formula applies at P_0 (classical geoid height N) and P (Molodensky height anomaly ζ) as well:

$$N = \frac{T(P_0)}{\gamma}\,, \qquad (8\text{--}34)$$

$$\zeta = \frac{T(P)}{\gamma}\,, \qquad (8\text{--}35)$$

with some approximate value for γ such as $\gamma_{45°}$. Equation (8–32) can be reformulated as the *boundary conditions for the Molodensky problem*

$$\frac{\partial T}{\partial h} - \frac{1}{\gamma}\frac{\partial \gamma}{\partial h}\,T + \Delta g = 0\,, \qquad (8\text{--}36)$$

cf. (2–251), and *for the GPS problem*, cf. (2–252),

$$\frac{\partial T}{\partial h} + \delta g = 0\,. \qquad (8\text{--}37)$$

These two boundary conditions apply at the surface S (Molodensky) and at sea level as well.

Finally we introduce the spherical approximation, disregarding the flattening f in the equations (which are linear relations between small quantities).

Note: The spherical approximation is a formal operation (disregarding f in small *ellipsoidal* quantities) and does not mean using a "reference sphere" instead of a reference ellipsoid in any geometrical sense (Moritz 1980 a: p. 15). This would imply geoidal heights on the order of 20 km!

Then (8–36) and (8–37) reduce to

$$\frac{\partial T}{\partial r} + \frac{2}{r}\,T + \Delta g = 0\,, \qquad (8\text{--}38)$$

$$\frac{\partial T}{\partial r} + \delta g = 0\,. \qquad (8\text{--}39)$$

These equations, for the Molodensky and the GPS problem, are valid both at sea level (classical) and at S (Molodensky).

8.5   The spherical case

As we have agreed, we work formally with a sphere (the reference ellipsoid stays at its geometric place!). This means putting $r = R = \text{constant}$. Furthermore, we assume (fictitiously!) that S is a level surface.

Expanding T and Δg into a series of Laplace spherical harmonics, see (2–322) and (2–320), we find

$$T(\vartheta, \lambda) = \sum_2^\infty T_n(\vartheta, \lambda), \tag{8–40}$$

$$\Delta g(\vartheta, \lambda) = \sum_2^\infty \Delta g_n(\vartheta, \lambda) \tag{8–41}$$

on the surface of the sphere, whence by (8–38) and (2–321) with $r = R$,

$$T = R \sum_{n=2}^\infty \frac{\Delta g_n}{n - 1}. \tag{8–42}$$

The summation starts conventionally with $n = 2$, rather than $n = 0$, for several reasons, one of them being that $n = 1$ would lead to a zero denominator in (8–42).

Using (2–325) and (2–326) leads to the well-known Stokes' formula

$$T = \frac{R}{4\pi} \iint_\sigma S(\psi) \, \Delta g \, d\sigma, \tag{8–43}$$

where

$$S(\psi) = \sum_{n=2}^\infty \frac{2n + 1}{n - 1} P_n(\cos \psi), \tag{8–44}$$

where $P(\cos \psi)$ are Legendre polynomials. Here ψ denotes the spherical distance from the point at which T is to be computed.

In exactly the same way, we obtain for the gravity disturbance with the boundary condition (8–39), summarizing the derivation in Sect. 2.18,

$$\delta g(\vartheta, \lambda) = \sum_0^\infty \delta g_n(\vartheta, \lambda), \tag{8–45}$$

$$T(\vartheta, \lambda) = R \sum_{n=0}^\infty \frac{\delta g_n}{n + 1}, \tag{8–46}$$

and the formula of Neumann–Koch

$$T = \frac{R}{4\pi} \iint\limits_{\sigma} K(\psi)\,\delta g\,d\sigma\,, \tag{8–47}$$

where

$$K(\psi) = \sum_{n=0}^{\infty} \frac{2n+1}{n+1} P_n(\cos\psi) \tag{8–48}$$

and, by summation of this series,

$$K(\psi) = \frac{1}{\sin(\psi/2)} - \ln\left(1 + \frac{1}{\sin(\psi/2)}\right) \tag{8–49}$$

being the Neumann–Koch function.

So in the GPS boundary problem on the sphere, the solution (8–47) is completely analogous to the formula of Stokes (8–43) for the classical problem.

The fact that the GPS problem is conceptually simpler (fixed-boundary surface) than Molodensky's problem (free-boundary surface) is expressed by the fact that Stokes' function must start with $n = 2$, since $n = 1$ gives a zero denominator, whereas Neumann–Koch's function (8–48) is regular for all n.

In both cases, the height anomaly ζ (here the geoidal height) is given by Bruns' formula

$$\zeta = \frac{T}{\gamma} \doteq \frac{T}{\gamma_0}\,. \tag{8–50}$$

In the spherical approximation, γ may be, in formulas of Bruns' and Stokes' type, replaced by our usual mean value $\gamma_0 = \gamma_{45°}$.

We will see that these spherical solutions form the base for an elementary solution of Molodensky's problem and the GPS problem for the earth's surface. We only mention the well-known fact that, for the earth's surface S, these two problems are *oblique-derivative problems*, since the direction of the plumb line does not coincide with the normal to the earth's surface, at least on land. Thus the GPS boundary problem *for S* is not a spherical Neumann problem, which always involves the normal derivative!

8.6 Solution by analytical continuation

8.6.1 The idea

The idea is very simple (Fig. 8.4). Our observations Δg or δg, given on the earth's surface S, are "reduced", or rather "analytically continued" (upward or downward, see below and Fig. 8.5), to a level surface (or normal level

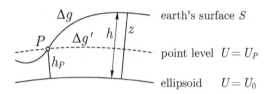

Fig. 8.4. Analytical continuation from the earth's surface to point level

surface $U = U_P$, which for our purpose is the same). In the spherical approximation, both surfaces $U = U_P$ and $U = U_0$ are concentric spheres, but only in the precise sense of the spherical approximation as explained above.

We also use the term "harmonic continuation" because the analytically continued function satisfies Laplace's equation. This will be explained in detail later.

An expansion into a Taylor series gives immediately

$$\Delta g = \Delta g^* + z\,\frac{\partial \Delta g^*}{\partial z} + \frac{1}{2!}\,z^2\,\frac{\partial^2 \Delta g^*}{\partial z^2} + \frac{1}{3!}\,z^3\,\frac{\partial^3 \Delta g^*}{\partial z^3} + \cdots$$

$$= \Delta g^* + \sum_{n=1}^{\infty} \frac{1}{n!}\,z^n\,\frac{\partial^n \Delta g^*}{\partial z^n}\,,$$

$$(8\text{--}51)$$

where

$$z = h - h_P \qquad\qquad (8\text{--}52)$$

is the elevation difference with respect to the computation point P. For the present, we assume the series (8–51) to be convergent. Here Δg^* is the gravity anomaly at point level (Fig. 8.4). The use of a Taylor series is typical for *analytical continuation*. For instance, Taylor series are a standard tool for analytical continuation of functions of a complex variable.

8.6.2   First-order solution

It is particularly easy to give a solution as a first approximation. With γ_0 from (8–50) we have

$$\zeta = \frac{R}{4\pi\,\gamma_0} \iint_\sigma \left(\Delta g - \frac{\partial \Delta g}{\partial h}\,h \right) S(\psi)\,d\sigma + \frac{\partial \zeta}{\partial h}\,h\,. \qquad (8\text{--}53)$$

This follows from the geometrical interpretation of this equation which is evident from Fig. 8.5 a. We see that the free-air anomalies Δg at ground level are "reduced" downward to sea level to become

$$\Delta g^{\text{harmonic}} = \Delta g - \frac{\partial \Delta g}{\partial h}\,h \qquad\qquad (8\text{--}54)$$

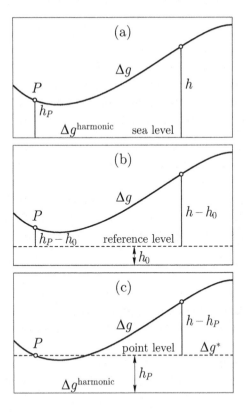

Fig. 8.5. Harmonic continuation to sea level (a), to an arbitrary level (b), and to the level of point P (c)

(the superscript "harmonic" denotes harmonic continuation to sea level; see Fig. 8.5 and the paragraph "A note on terminology" below); then Stokes' integral gives height anomalies at sea level which are reduced upward to ground level by adding the term $\frac{\partial \zeta}{\partial h} h$.

Harmonic continuation to point level

The elevation h in (8–53) is taken above sea level (see Fig. 8.5 a). If we examine the arguments leading to this equation, we will find that the sea level is not distinguished from any other level. If we reckon the elevation above some other reference level, which has the elevation h_0 above sea level, we must replace h by $h - h_0$ (see Fig. 8.5 b). Thus (8–53) is equivalent to

$$\zeta = \frac{R}{4\pi\,\gamma_0} \iint_\sigma \left[\Delta g - \frac{\partial \Delta g}{\partial h} (h - h_0) \right] S(\psi)\, d\sigma + \frac{\partial \zeta}{\partial h} (h - h_0) . \qquad (8\text{--}55)$$

In particular we may take as reference level the level of the point P itself, so that

$$h_0 = h_P \, , \tag{8–56}$$

where P is the point at which the height anomaly ζ is computed. If this choice is made, the last term in the above expression will be zero, because outside the integral h always means h_P, so that $h - h_0 = h_P - h_P = 0$. Thus we have

$$\zeta = \frac{R}{4\pi\,\gamma_0} \iint\limits_{\sigma} \left[\Delta g - \frac{\partial \Delta g}{\partial h}(h - h_P)\right] S(\psi)\,d\sigma \, . \tag{8–57}$$

This formula is particularly simple; geometrically it means that the free-air anomalies are "reduced" (in the sense of "analytically or harmonically continued") from the ground to the level of the computation point P (see Fig. 8.5 b). Thus, the reference level is different for different computation points.

As we have already indicated at the beginning of Sect. 8.6.1, Fig. 8.5 c shows that harmonic continuation by Eq. (8–57) is *upward* for surface points *below* the level of P and *downward* for surface points *above* the level of P.

Important remark

Equation (8–57) is really a genuine spherical Stokes formula applied to a "reference sphere", namely, to the spherical "point level"! An immediate consequence: this formula can be simply differentiated *horizontally* to give a genuine Vening Meinesz formula in the sense of Sect. 2.19 for the *deflections of the vertical*. This remark is relevant for Sect. 8.7.

Vertical derivative

The vertical derivative $\partial/\partial r$ can be expressed in terms of surface values by the well-known spherical formula (Sect. 1.14)

$$\frac{\partial f}{\partial r} = -\frac{1}{R}f + \frac{R^2}{2\pi} \iint\limits_{\sigma} \frac{f - f_Q}{l_0^3}\,d\sigma \, . \tag{8–58}$$

Q is the surface point where $\partial f/\partial r$ is computed and to which f in the first term on the right-hand side refers, σ denotes the unit sphere, and

$$l_0 = 2R \sin\frac{\psi}{2} \, . \tag{8–59}$$

This gives $\partial \Delta g/\partial r$ if we put $f = \Delta g$ in (8–58). We may also introduce the linear gradient operator L by

$$L(f) = \frac{R^2}{2\pi} \iint\limits_{\sigma} \frac{f - f_Q}{l_0^3}\,d\sigma \, . \tag{8–60}$$

(The first term on the right-hand side of (8–58) is much smaller and can be omitted.)

The term $\partial\zeta/\partial r$ no longer occurs in (8–57) as it did in (8–53) and (8–55), and will not be needed.

Computational formulas; the Molodensky correction

Our computational formula is (8–57). We split it up as follows: The free-air anomaly Δg is continued (downward or upward) from ground level to the level of point P, obtaining

$$\Delta g^* = \Delta g + g_1 \,, \tag{8-61}$$

where the Molodensky correction is

$$g_1 = -\frac{\partial \Delta g}{\partial h}(h - h_P) = -\frac{\partial \Delta g}{\partial r}(h - h_P) \tag{8-62}$$

(in spherical approximation) with

$$\frac{\partial \Delta g}{\partial r} = \frac{R^2}{2\pi}\iint\limits_{\sigma}\frac{\Delta g - \Delta g_Q}{l_0^3}\,d\sigma \,. \tag{8-63}$$

Then we finally get

$$\zeta = \zeta_0 + \zeta_1 \,, \tag{8-64}$$

where

$$\zeta_0 = \frac{R}{4\pi\,\gamma_0}\iint\limits_{\sigma}\Delta g\,S(\psi)\,d\sigma \tag{8-65}$$

is the simple Stokes formula applied to ground-level free-air anomalies Δg, and the Molodenski correction for ζ is

$$\zeta_1 = \frac{R}{4\pi\,\gamma_0}\iint\limits_{\sigma}g_1\,S(\psi)\,d\sigma \,. \tag{8-66}$$

This is the first-order solution, or linear solution.

Important remark

Please note carefully that we are using "linear", or "first-order", in two very different senses:

- *general linearization*, linear in quantities of the anomalous potential, such as N or ζ, as introduced in Sect. 2.12 and Sect. 8.4 and implied everywhere throughout the book, and

- *linear approximation in h* used very generally in first-order "Molodensky corrections" such as (8–62) or (8–66) but not in (8–67) or (8–68).

In fact, to a higher approximation

$$\Delta g^* = \Delta g + g_1 + g_2 + g_3 + \cdots \tag{8–67}$$

and

$$\zeta = \zeta_0 + \zeta_1 + \zeta_2 + \zeta_3 + \cdots . \tag{8–68}$$

Generalizing (8–66), we have

$$\zeta_i = \frac{R}{4\pi\,\gamma_0} \iint_\sigma g_i \, S(\psi) \, d\sigma , \tag{8–69}$$

where $i = 1, 2, 3, \ldots$. For the deflection of the vertical we have similar expressions, see Sect. 8.7; compare also (8–75) and (8–76).

8.6.3   Higher-order solution

The following recursion formulas are somewhat advanced and may be omitted. From Moritz (1980 a: Sect. 45) we may take the recursion formula for the correction terms g_n, which are evaluated recursively by

$$g_n = -\sum_{r=1}^{n} z^r \, L_r(g_{n-r}) , \tag{8–70}$$

starting from

$$g_0 = \Delta g . \tag{8–71}$$

Here the operator L_n is also defined recursively:

$$L_n(\Delta g) = n^{-1} L_1[L_{n-1}(\Delta g)] \tag{8–72}$$

starting with

$$L_1 = L \tag{8–73}$$

with the gradient operator L defined above, (8–60), and z given by (8–52).

8.6.4   Problems of analytical continuation

Analytical continuation comes from the theory of complex variables and means extending the domain, on which the function is defined, by the use of Taylor series. Complex functions always satisfy Laplace's equations in two dimensions and are therefore harmonic.

Also in three dimensions, functions satisfying Laplace's equation are called harmonic, as we know well. Analytical continuation is again best defined by Taylor series, and analytical continuation is frequently called *harmonic continuation* (Kellogg 1929: Chap. X).

Above we have been misusing the all-round word "reduced" in the sense of "analytical" or "harmonic" continuation and will continue to do so for brevity. As we have seen in Sect. 8.2 and will see in Sect. 8.9, it is not a gravity reduction in the standard sense of explicit mass removal. The Taylor series whose first term is (8–54) is an *analytical* operation performed on the external potential directly at ground level, preserving the Laplace equation $\Delta W = 0$. (In fact, $\Delta W = 2\omega^2$, but let us, as we did in (8–2), for a while forget earth rotation, which implies $\omega = 0$ and $W = V$.) Thus, it is a harmonic function and our "reduction" is really *analytical continuation* as a harmonic function or briefly *harmonic continuation*. **Harmonic continuation is the key notion in modern physical geodesy, from Molodensky's problem to least-squares collocation.** Its full meaning will gradually emerge in what follows, as a notion which is surprisingly simple and general. Symbols like $\Delta g^{\text{harmonic}}$ will relate to harmonic continuation. In what follows, we shall sometimes continue to use "reduce downward" or "continue downward" instead of "harmonically continue downward" and use "reduce upward" in a similar sense. We also use "continue upward". Only in doubt, the clumsy expression "harmonically continue upward" should be employed. Also "analytical continuation" is used. It all means the same. In the present context, confusion is hardly possible.

Hence we see why gravity anomalies Δg *at ground level* may be used for f in (8–58), whereas the equivalent expression (2–394) was originally derived for gravity anomalies *at sea level*. Since $\Delta g^{\text{harmonic}}$ and Δg differ only by terms of the order of h, the difference between using $\Delta g^{\text{harmonic}}$ or Δg in (8–62) causes only an error of the order of h^2, which is negligible in the linear approximation.

Analytical continuation: historical remarks

The use of analytical continuation has an interesting history. It was first considered as a possibility by Molodensky himself, already before 1945, but he soon rejected this method! Molodensky was a profound mathematician, with a high regard for mathematical rigor. He would not be satisfied with intuitive heuristic approaches so common in mathematical physics, also in the present book.

In fact, the analytical continuation of the external gravitational potential into the interior of the earth's masses is very likely to become singular at some points. As a serious mathematician, Molodensky rejected the use of singular functions for regular purposes.

Still, analytical continuation continued to exert an irresistable fascination because its use is so easy. It was rediscovered around 1960 by A. Bjerhammar. At the General Assembly of the International Union of Geodesy and Geophysics in Berkeley, California, in 1963, one of the authors (H.M.) talked to Bjerhammar about these difficulties, but Bjerhammar refused to take them seriously. After a long discussion he convinced H.M. that analytical continuation was rigorously possible for discrete boundary data (all our terrestrial gravity measurements are discrete) and approximately possible for continuous boundary data.

This admittedly intuitive thinking was made rigorous by the idea of Krarup (1969) that Runge's theorem, well known for approximation of analytical functions of a complex variable, should be applied to the problem of analytical continuation of harmonic functions in space. Runge's theorem, in the form of Krarup, loosely speaking says that, even if the external geopotential cannot be regularly continued from the earth surface S into its interior, it can be made continuable by an arbitrarily small change of the geopotential at S. Another historical remark: the Krarup–Runge theorem for harmonic functions in space goes back at least to Szegö and to Walsh (both around 1929), cf. Frank and Mises (1930: pp. 760–762). It is always dangerous to talk about priorities! A detailed discussion will be found in Moritz (1980 a: Sects. 6 to 8).

More on the validity of this method

Let us summarize. The presupposition of this method is that the earth's external gravitational potential can be continued, as a regular harmonic function, analytically down to sea level. This is the case if and only if it is possible to shift the masses outside the ellipsoid into its interior in such a way that the potential outside the earth remains unchanged or, in other words, if the analytical continuation of the disturbing potential T is a regular function everywhere between the earth's surface and the ellipsoid. Thus, the question arises whether the external potential can be analytically continued down to sea level.

Rigorously, as we have just remarked, the answer must be in the negative, in view of the irregularities of topography (Molodenski et al. 1962: p. 120; Moritz 1965: Sect. 6.4). This fact is also related to the divergence at the earth's surface of the spherical-harmonic expansion for the external potential (Sect. 2.5).

However, by Krarup–Runge's theorem, the analytical continuation of the external potential down to sea level is possible *with sufficient accuracy for all practical purposes*. Actually it is possible with any accuracy you wish; if you are not satisfied with $1\,\mathrm{mgal}$, prescribe $10^{-3}\,\mathrm{mgal}$ or $10^{-1000}\,\mathrm{mgal}$!

Bjerhammar has pointed out that the assumption of a complete continuous gravity coverage at every point of the earth's surface, from which the above negative answer follows, is unrealistic because we can measure gravity only at discrete points. If the purpose of physical geodesy is understood as the *determination of a gravity field that is compatible with the given discrete observations*, then it is always possible to find a potential that can be analytically continued down to the ellipsoid. This is the theoretical basis for *least-squares collocation*.

Here we need only one result: **Do not worry about analytical continuation! It is always possible with an arbitrarily small error being not equal to** 0 (though not for one being 0).

So, in the same year 1969, Marych and Moritz independently found an elementary solution by analytical continuation in the form of an infinite series denoted as "Molodensky series". Details can be found in Moritz (1980 a): The original form of Molodensky's series obtained by solving an integral equation is found in Sect. 45. Pellinen's equivalence proof that *the simple "analytical continuation solution" and Molodensky's integral equation solution are equivalent* (that means, the series are termwise equal!) is found in Sect. 46.

We remark that analytical continuation is a purely mathematical concept independent of the density of the topographic masses. Thus, it is not an "introduction of gravity reduction through the backdoor", which would be contrary to the spirit of Molodensky's theory.

8.6.5   Another perspective

Consider Fig. 8.6. Let us assume that the analytical downward continuation of Δg to the sea level surface has been performed, obtaining $\Delta g^{\text{harmonic}}$. The sea-level anomalies $\Delta g^{\text{harmonic}}$ then generate, on the physical surface of the earth, a field of gravity anomalies Δg that is identical with the actual gravity anomalies on the earth's surface as obtained from observation. Therefore, the gravity anomalies that they generate outside the earth must also be identical

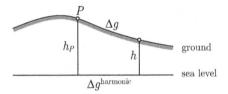

Fig. 8.6. Free-air anomalies at ground level, Δg, and at sea level, $\Delta g^{\text{harmonic}}$

with the actual gravity anomalies outside the earth, since the function $r\,\Delta g$ is harmonic according to Sect. 2.14.

(Remark: we are consistently using the notation Δg for ground level, $\Delta g^{\text{harmonic}}$ for sea level, and Δg^* for point level; see Fig. 8.5.)

It follows that the harmonic function T that is produced by $\Delta g^{\text{harmonic}}$ according to Pizzetti's generalization (2–302) of Stokes' formula

$$T(r, \vartheta, \lambda) = \frac{R}{4\pi} \iint_{\sigma} \Delta g^{\text{harmonic}}\, S(r, \psi)\, d\sigma \qquad (8\text{–}74)$$

is identical with the actual disturbing potential of the earth *outside and on its surface.*

Applications

Assume that we got in some way (e.g., by the Taylor series mentioned above or by collocation to be treated in Chap. 10 or by a high-resolution gravitational field from satellite observations) the downward continuation $\Delta g^{\text{harmonic}}$ to sea level. Then we can compute the external gravity field, its spherical harmonics, etc., rigorously by means of the conventional formulas of Chaps. 2 and 6, provided we use $\Delta g^{\text{harmonic}}$ rather than Δg in the relevant formulas. For instance, the coefficients of the spherical harmonics of the gravitational potential may be obtained by expanding the function $\Delta g^{\text{harmonic}}$ according to Sect. 1.9 together with Sect. 1.6. If we wish to compute the height anomaly ζ at a point P at ground level, we must remember that P lies *above* the ellipsoid, so that the formulas for the external gravity field are to be applied. By Bruns' formula $\zeta = T/\gamma_0$ (8–50), we get

$$\zeta = \frac{R}{4\pi\,\gamma_0} \iint_{\sigma} \Delta g^{\text{harmonic}}\, S(r, \psi)\, d\sigma \,, \qquad (8\text{–}75)$$

where $r = R + h$ and h is the topographic height of P in some sense of Chap. 4. (We do not need it very accurately, but it means that h is *formally* added to the constant radius R of the mean terrestrial sphere, which has no real-world geometric interpretation!) Cf. Eq. (6–57). The function $S(r, \psi)$ is expressed by (2–303), (6–22) or (6–35). Similarly, ξ and η, being deflections of the vertical above sea level, must be computed by (6–41) and the second and third equation of (6–30). This gives

$$\xi = \frac{t}{4\pi\,\gamma_0} \iint_{\sigma} \Delta g^{\text{harmonic}}\, \frac{\partial S(r, \psi)}{\partial \psi}\, \cos\alpha\, d\sigma \,,$$

$$\eta = \frac{t}{4\pi\,\gamma_0} \iint_{\sigma} \Delta g^{\text{harmonic}}\, \frac{\partial S(r, \psi)}{\partial \psi}\, \sin\alpha\, d\sigma \,, \qquad (8\text{–}76)$$

where $\partial S(r, \psi)/\partial \psi$ is expressed by the second equation of (6–32) or the second equation of (6–36). The linear approximation of (8–74) is evidently equivalent to (8–53).

This indirect procedure, downward continuation to sea level and again upward continuation to ground level or above, has the advantage that only the conventional spherical formulas are needed; yet at the same time the irregularities of the earth's topography are fully taken into account. The downward continuation of Δg need be performed only once; the resulting anomalies $\Delta g^{\mathrm{harmonic}}$ may be stored and used for all further computations.

Just as Δg is related to $\Delta g^{\mathrm{harmonic}}$ by analytical continuation, so are ζ and N^{harmonic}, the height of a "harmonic geoid". A final and hopefully instructive and not too difficult review will be found in Sect. 8.15.

An elementary explanation from daily life

Generally, "analytical continuation" means continuation by the same mathematical formula: Taylor series, Laplace equation, or even an elementary explicit equation.

Let us illustrate the meaning of analytical continuation by means of an almost trivial example from everyday life (Fig. 8.7). A person is driving a car along a road which at first is completely straight; at point B, however, it suddenly turns into a circular curve. Thus, our person first drives along the straight segment of the road. Unfortunately, he is tired and sleepy just when the straight road suddenly turns into a circular curve. Thus, our sleepy driver fails to turn the steering wheel and goes straight ahead, the car leaving the road. Fortunately, the slope is mild, the driver immediately takes control again and manages to bring the car to a stop at C' without major damages. The driver (one of the authors of this book) has even found the experience an excellent example to illustrate analytical continuation in his courses!

The gravitational potential corresponds to the road ABC, which, after some idealization, can be considered *"piecewise analytic"*, consisting of the straight line AB and the circular arc BC. The transition from the straight line to the circle is continuous and differentiable at B, but the curvature

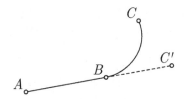

Fig. 8.7. An illustration of analytical continuation

changes discontinuously from 0 to $1/R$, where R is the radius of the circular arc. Therefore, the function "road" is continuous and continuously differentiable but has a discontinuous second derivative at point B, just as the function "gravitational potential" is everywhere continuous and continuously differentiable but has discontinuous second derivatives at the earth surface as we have seen in Sect. 1.2. The straight line has the equation $y'' = 0$ (which is the "one-dimensional Laplace equation"!), corresponding to the external potential satisfying $\Delta V = 0$. Thus, neither the "function road" nor the "function gravitational potential" may be considered an everywhere analytical function, but each may be said to consist of a "linear" or "harmonic" piece ($y'' = 0$ or $\Delta V = 0$: Laplace, respectively) and a "nonlinear" piece ($y'' \neq 0$ or $\Delta V = -4\pi G \varrho$: Poisson). For the road, the analytical continuation is the straight line for which $y'' = 0$ even beyond point B, the path followed by the car without action of the sleepy driver towards C', and for the potential it is a function satisfying $\Delta V_{\text{analytical continuation}} = 0$ even in the interior of the earth.

8.7 Deflections of the vertical

The effect of Molodensky-type corrections is even much more important on the deflections of the vertical ξ, η than on the height anomalies ζ. This is shown by their order of magnitude in high mountains: the Molodensky correction for the height anomaly might be of the order of 0.3 m, whereas for vertical deflections they may be on the order of 0.3 arc seconds, which corresponds to 10 m (1 arc second corresponds to 30 m in position). *The difference is more than one order of magnitude!*

The consideration of a Molodensky type of correction to the deflections of the vertical is easiest by using analytical continuation to point level (Sect. 8.6). Differentiating (8–57) in north-south and east-west direction, we get the corresponding Vening Meinesz equations

$$\xi = \frac{1}{4\pi\,\gamma_0} \iint_{\sigma} \left[\Delta g - \frac{\partial \Delta g}{\partial h}\,(h - h_P)\right] \frac{dS}{d\psi}\,\cos\alpha\,d\sigma\,,$$

$$\eta = \frac{1}{4\pi\,\gamma_0} \iint_{\sigma} \left[\Delta g - \frac{\partial \Delta g}{\partial h}\,(h - h_P)\right] \frac{dS}{d\psi}\,\sin\alpha\,d\sigma\,. \tag{8–77}$$

Its geometrical interpretation is analogous to that of (8–57). The gravity anomalies Δg are "reduced" to the level of point P so that we obtain

$$\Delta g^{\text{harmonic}} = \Delta g - \frac{\partial \Delta g}{\partial h}\,(h - h_P)\,. \tag{8–78}$$

Since these anomalies *refer to a level surface,* Vening Meinesz' formula can now be applied directly and gives (8–77).

Relation with the ellipsoidal geodetic coordinates

The deflection components ξ and η as given by the above expressions represent the deviation of the actual plumb line from the normal plumb line at the ground point P. Therefore, they are defined by

$$\xi = \Phi - \varphi^* ,$$
$$\eta = (\Lambda - \lambda^*) \cos \varphi .$$

$(8–79)$

The symbols Φ and Λ represent the astronomical coordinates of P referred to the ground. The symbols φ^* and λ^* represent the *"normal coordinates"* of P, defining the direction of the normal plumb line at P; they are not identical with the ellipsoidal coordinates φ and λ of P, which are the coordinates of the foot point Q_0 of the straight perpendicular to the ellipsoid (Fig. 8.8).

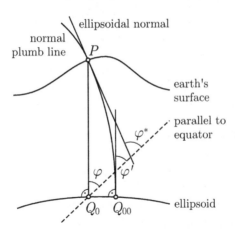

Fig. 8.8. Normal latitude φ^* and ellipsoidal latitude φ

The normal coordinates of P, φ^* and λ^*, differ from the normal coordinates of Q_{00}, φ' and λ', by the correction for the normal curvature of the plumb line (see Sect. 5.15). Formula (5–147) gives

$$\varphi^* = \varphi' + f^* \frac{h}{R} \sin 2\varphi ,$$
$$\lambda^* = \lambda' .$$

$(8–80)$

Because of the rotational symmetry, we have rigorously

$$\lambda' = \lambda ,$$

$(8–81)$

since Q_0 and Q_{00} lie on the same ellipsoidal meridian. Furthermore, even in extreme cases the distance between Q_0 and Q_{00} can never exceed a few centimeters. For this reason, we may also set

$$\varphi' = \varphi \qquad (8\text{–}82)$$

without introducing a perceptible error. Hence, we can identify φ' and λ' with φ and λ, which are the *ellipsoidal coordinates* of P according to Helmert's projection (Sect. 5.5). Therefore, we may replace the above equations for φ^* and λ^* by

$$\varphi^* = \varphi + f^* \frac{h}{R} \sin 2\varphi\,,$$
$$\lambda^* = \lambda\,. \qquad (8\text{–}83)$$

Introducing the deflection components according to Helmert's projection, defined as

$$\xi_{\text{Helmert}} = \Phi - \varphi\,,$$
$$\eta_{\text{Helmert}} = (\Lambda - \lambda)\,\cos\varphi\,, \qquad (8\text{–}84)$$

we see that they are related to ξ and η by the equations

$$\xi_{\text{Helmert}} = \xi + f^* \frac{h}{R} \sin 2\varphi\,,$$
$$\eta_{\text{Helmert}} = \eta\,. \qquad (8\text{–}85)$$

Therefore, ξ and ξ_{Helmert} differ by the normal reduction for the curvature of the plumb line,

$$-\delta\varphi_{\text{normal}} = f^* \frac{h}{R} \sin 2\varphi\,. \qquad (8\text{–}86)$$

The deflection components ξ_{Helmert} and η_{Helmert} are used in astrogeodetic computations; ξ and η are those obtained gravimetrically from formulas such as (8–77) and (8–88) below.

These relations are *mathematically* quite analogous to the corresponding equations (5–138) for the conventional method using the geoid, but now, with the use of the *normal* curvature, the once formidable obstacle of the correction for plumb-line curvature *practically* belongs to the past.

Remark on accuracy

With Molodensky's theory, the accuracy problem mentioned at the end of Sect. 2.21 even aggravates, because in a mountainous terrain it is almost impossible to compute the Molodensky corrections with an accuracy of $0.03''$ (say), so that these observations cannot be directly used for precise horizontal positions.

Astronomical field observations for latitude, longitude, and azimuth have an accuracy around $0.3''$, which is sufficient for classical trigonometric net computation and astrogeodetic observation of the geoid (Sect. 5.14).

8.8 Gravity disturbances: the GPS case

The basic fact is that for gravity disturbances the derivation of "Molodensky corrections" g_n is identical to the Δg case. The reason is that the gravity disturbance δg has exactly the same analytical behavior as the gravity anomaly Δg since $r\,\delta g$, as a function in space, is harmonic together with $r\,\Delta g$. Thus, the arguments are literally the same, only Δg has to be replaced by δg, and Stokes' formula must be replaced by the Neumann–Koch formula (8–47) and similarly for Vening Meinesz' formula.

Therefore, we obtain

$$\zeta = \frac{R}{4\pi\,\gamma_0} \iint\limits_{\sigma} \delta g\, K(\psi)\, d\sigma + \sum_{n=1}^{\infty} \frac{R}{4\pi\,\gamma_0} \iint\limits_{\sigma} g_n\, K(\psi)\, d\sigma\,, \tag{8–87}$$

$$\xi = \frac{1}{4\pi\,\gamma_0} \iint\limits_{\sigma} \delta g\, \frac{dK}{d\psi}\, \cos\alpha\, d\sigma + \sum_{n=1}^{\infty} \frac{1}{4\pi\,\gamma_0} \iint\limits_{\sigma} g_n\, \frac{dK}{d\psi}\, \cos\alpha\, d\sigma\,,$$

$$\eta = \frac{1}{4\pi\,\gamma_0} \iint\limits_{\sigma} \delta g\, \frac{dK}{d\psi}\, \sin\alpha\, d\sigma + \sum_{n=1}^{\infty} \frac{1}{4\pi\,\gamma_0} \iint\limits_{\sigma} g_n\, \frac{dK}{d\psi}\, \sin\alpha\, d\sigma\,. \tag{8–88}$$

For the "Vening Meinesz GPS formula" (8–88), we find by differentiation of (8–49):

$$\frac{dK}{d\psi} = -\frac{1}{2} \frac{\cos(\psi/2)}{\sin^2(\psi/2)} \frac{1}{1+\sin(\psi/2)}\,. \tag{8–89}$$

The correction terms g_n are evaluated recursively by

$$g_n = -\sum_{r=1}^{n} z^r\, L_r(g_{n-r})\,, \tag{8–90}$$

but now we start from

$$g_0 = \delta g\,. \tag{8–91}$$

We only have to replace Δg by δg and $S(\psi)$ by $K(\psi)$. The operators L remain the same.

Let us summarize again our trick for solving the modern boundary-value problems (Molodensky and Koch). It is difficult to directly work with the complicated earth's surface S. Therefore, by analytical continuation of Δg or

δg, respectively, we reduce these complicated problems to the corresponding spherical problems, for which the solution is simple and well known.

The similarity of the Molodensky series for the Molodensky problem, on the one hand, and for the GPS boundary problem, on the other hand, is very clear because Δg and δg have the same analytical and geometric structure.

At the same time, this similarity is very surprising since the two underlying boundary problems are mathematically quite different, as we have seen in Sect. 8.3 (compare Eqs. (8–12) and (8–13)). Nonetheless, (8–87) does give the potential as (8–13) requires: by Bruns' theorem, which is the omnipresent link between geometry and physics, we have

$$T = \gamma \zeta \,. \tag{8–92}$$

Then

$$W = U + T \tag{8–93}$$

is the geopotential required by (8–13), and

$$C = W_0 - W \tag{8–94}$$

is the geopotential number, the physical measure of height above sea level, conventionally obtained by the cumbersome method of leveling, but now computed in a direct way from gravity data. This is the physical, more general, equivalent of the geometric determination of the normal height by $H^* = h - \zeta$, according to Eq. (8–31).

It can be shown that, in the linear approximation, the Molodensky correction for the gravity disturbance has the same form as for the gravity anomaly and can for each quantity be computed using either Δg or δg.

The formulas for the Molodensky corrections and their numerical values are the same to the linear approximation.

All this shows the power of Molodensky's approach even in problems he never treated himself.

8.9   Gravity reduction in the modern theory

In Sect. 8.2, we have considered gravity reductions from the point of view of the determination of the geoid. It is quite remarkable that these reductions, such as the Bouguer or the isostatic reduction, can also be incorporated into the new method of direct determination of the earth's physical surface, although with essentially changed meaning (Pellinen 1962; Moritz 1965: Sect. 5.2).

Let the masses outside the geoid be removed or moved inside the geoid, as described in Sect. 8.2, and consider the effect of this procedure on quantities *referred to the ground.*

We denote the changes in potential and in gravity by δW and δg; then the new values at ground will be

$$W^c = W - \delta W \,,$$

$$g^c = g - \delta g \,. \tag{8-95}$$

(It is clear that δg here is not the gravity disturbance!) The disturbing potential $T = W - U$ becomes

$$T^c = T - \delta W \,. \tag{8-96}$$

The physical surface S as such will remain unchanged, but the telluroid Σ will change, because its points Q are defined by $U_Q = W_P$, and the potential W at any surface point P will be affected by the mass displacements according to (8–95). The distance $Q\,Q^c$ between the original telluroid Σ and the changed telluroid Σ^c (Fig. 8.9) is given by

$$Q\,Q^c = \frac{\delta W}{\gamma} \tag{8-97}$$

according to Bruns' theorem. This is identical with the variation of the height anomaly ζ, so that

$$\delta\zeta = \zeta - \zeta^c = \frac{\delta W}{\gamma} \,. \tag{8-98}$$

Normal gravity γ on the telluroid Σ becomes on the changed telluroid Σ^c

$$\gamma^c = \gamma + \frac{\partial\gamma}{\partial h}\,\delta\zeta = \gamma + \frac{1}{\gamma}\frac{\partial\gamma}{\partial h}\,\delta W \,, \tag{8-99}$$

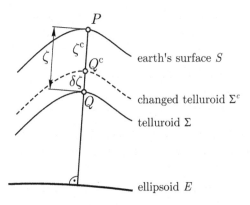

earth's surface S

changed telluroid Σ^c

telluroid Σ

ellipsoid E

Fig. 8.9. Telluroid before and after gravity reduction, Σ and Σ^c

so that the new gravity anomaly will be

$$\Delta g^c = g^c - \gamma^c = (g - \delta g) - \left(\gamma + \frac{1}{\gamma} \frac{\partial \gamma}{\partial h} \delta W \right) \qquad (8\text{--}100)$$

or

$$\Delta g^c = \Delta g - \delta g - \frac{1}{\gamma} \frac{\partial \gamma}{\partial h} \delta W . \qquad (8\text{--}101)$$

The reduced gravity anomaly Δg^c consists of the free-air anomaly (in the Molodensky sense) Δg and two reductions:

1. the direct effect, $-\delta g$, of the shift of the outer masses on g; and

2. the "indirect effect",

$$-\frac{1}{\gamma} \frac{\partial \gamma}{\partial h} \delta W , \qquad (8\text{--}102)$$

of this shift on γ, because of the change of the telluroid to which γ refers.

Let us repeat once more that all these anomalies Δg^c refer to the physical surface of the earth, to "ground level"!

If the masses outside the geoid are completely removed, then Δg^c is a Bouguer anomaly; if the outer masses are shifted vertically downward according to some isostatic hypothesis, then Δg^c is an isostatic anomaly, etc. In this way we may get a "ground equivalent" for each conventional gravity reduction. The two are always related by analytical continuation. See below for the isostatic anomalies; for analytical continuation see Sect. 8.6.

Now we may describe the determination of the height anomalies ζ in a way that is similar to the corresponding procedure for the geoidal undulations N of Sect. 8.2:

1. The masses outside the geoid are, by computation, removed entirely or else moved inside the geoid; W and g change to W^c and g^c according to (8–95).

2. The point at which normal gravity is computed is moved from the ellipsoid upward to the telluroid point Q.

3. The indirect effect, the distance $Q\,Q^c = \delta\zeta$, is computed by (8–98).

4. The point to which normal gravity refers is now moved from the point Q of the telluroid Σ to the point Q^c of the changed telluroid Σ^c, according to (8–99).

5. The changed height anomalies ζ^c are computed from the "reduced" gravity anomalies Δg^c (8–101) by any solution of Molodensky's problem, such as Eq. (8–57) or (8–68).

6. Finally, the original height anomalies ζ are obtained by considering the indirect effect according to

$$\zeta = \zeta^c + \delta\zeta \, . \tag{8--103}$$

The purpose of this somewhat complicated procedure is to make use of the well-known advantages of Bouguer and isostatic anomalies. The Bouguer anomalies, and even more so the isostatic anomalies, are *smoother and more representative* than the free-air anomalies and can, therefore, be interpolated more easily and more accurately.

The isostatic gravity anomalies Δg^c in the new sense are thus quite analogous to the conventional isostatic anomalies; accordingly for any other type of gravity reduction. The difference is that now the isostatic anomalies, etc., *refer to the physical surface of the earth* as well as the free-air anomalies. If the isostatic anomalies in this new sense are analytically continued from the earth's surface down to the geoid, then isostatic anomalies in the conventional sense are obtained. Nowadays, in view of the "remove-restore principle", one speaks usually of *topographic-isostatic reduction* while continuing to speak of isostatic anomalies.

Hence, the isostatic anomalies according to the conventional definition (at sea level) and those according to the new definition (at ground level) are related through analytical continuation. This fact leads to two conclusions. First, the difference between the isostatic anomalies according to these two definitions will be small, because the distance along which this analytical continuation is made is only the height above sea level and because the isostatic reduction achieves a strong smoothing of the anomalous gravity field. This difference is considerably smaller than the corresponding difference between free-air anomalies at ground level and at sea level. This fact clearly provides a computational advantage if isostatic anomalies are used in a formula such as (8--74).

Second, we obtain a relation between the conventional and the modern use of gravity reduction if the method of downward continuation, as discussed in the preceding section, is applied for obtaining the height anomalies. As we have just seen, the gravity anomalies Δg^{c*} at sea level, obtained by downward continuation of the isostatic ground-level anomalies Δg^c, are identical with the isostatic anomalies in the conventional sense. Hence, we obtain on the one hand the height anomalies by

$$\zeta = \frac{R}{4\pi\gamma} \iint_{\sigma} \Delta g^{c*}\, S(R+h, \psi)\, d\sigma + \left(\frac{\delta W}{\gamma}\right)_{\text{ground}} \tag{8--104}$$

according to (8–75) and (8–103), and on the other hand the geoidal undulations by

$$N = \frac{R}{4\pi\,\gamma_0} \iint\limits_{\sigma} \Delta g^{c*}\, S(\psi)\, d\sigma + \left(\frac{\delta W}{\gamma}\right)_{\text{geoid}} \qquad (8\text{--}105)$$

according to the ordinary Stokes formula applied to Δg^{c*} and (8–5). Since the height anomalies refer to the elevation h, the function $S(R+h, \psi)$ replaces in (8–104) the original function of Stokes $S(\psi) \equiv S(R, \psi)$, which occurs in (8–105) because the geoidal undulation refers to zero elevation. We could use γ_0 in (8–104) as well. Summarizing, we have the following steps:

1. Computation of the free-air anomaly at ground level, Δg, according to (8–23).
2. Computation of the isostatic anomaly at ground level, Δg^c, according to (8–101).
3. Downward continuation of Δg^c by (8–54), where Δg and $\Delta g^{\text{harmonic}}$ are replaced by Δg^c and Δg^{c*}. The resulting isostatic anomalies at sea level, Δg^{c*}, may now be used for two purposes: either for
4a. the determination of the physical surface of the earth according to (8–104), or for
4b. the determination of the geoid according to (8–105).

An error in the assumed density of the masses below the earth's surface affects the geoidal undulations as determined from (8–105) but does not influence the height anomalies resulting from (8–104). This is clear because a wrong guess of the density means only that the masses above sea level are not completely removed, which is no worse than not removing them at all when using free-air anomalies.

This method is of particular interest for practical computations, as we will see later. It has become popular by the name *"remove-restore method"*, invented by K. Colic and others, see Sect. 11.1.

An almost final remark on free-air reduction

The apparently so simple topic of free-air reduction in reality is formidably complex and complicated. Therefore, it is not possible to treat it in one block. The problem is rather like a mountain which can only be investigated by accessing it from various sides. An initial glance has been given as early as in Chap. 3, and the reader is asked to return to the paragraph "The many facets of free-air reduction" in Sect. 3.9. Now it is much easier to understand the remarks made there. What we now understand as harmonic continuation offers a possibility to *interpret free-air reduction as a mass-transporting gravity reduction*: the topographic masses are transported into the interior of the earth in such a way that the exterior potential remains unchanged. This is not unlike the Rudzki reduction, where *the geoid* remains unchanged. Whereas the Rudzki reduction is, however, "constructive" in the sense that a way of performing it can be described, our present interpretation of free-air reduction as harmonic continuation is nonconstructive, it is an "improperly posed" inverse problem; cf. Anger and Moritz (2003) and www.inas.tugraz.at/forschung/InverseProblems/AngerMoritz.html, as well as Fig. 8.10.

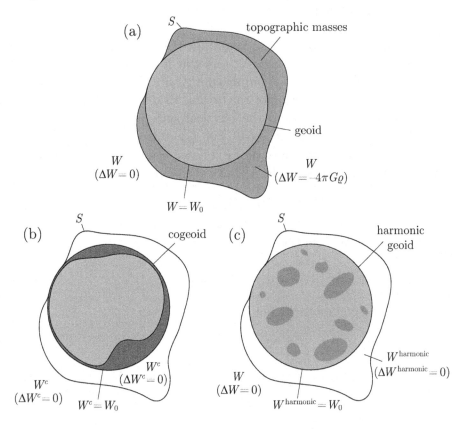

Fig. 8.10. (a) Geoid and topographic masses, (b) mass displacement in gravity reduction, (c) "ill-defined" mass displacement in free-air reduction as harmonic continuation

Important remark
The isostatic gravity anomalies and the topographic-isostatically reduced deflections of the vertical (Sect. 8.14) are fundamental for least-squares collocation in mountain areas (Sect. 11.2). **Thus, the spatial approach due to Molodensky is basic even for least-squares collocation!**

Exercise

Collecting all these remarks into a separately readable paper on the various aspects of free-air reduction would be a nice task for a seminar work. The present authors offer a prize of Euro 500, the "Molodensky Prize", to the first excellent review paper on this topic.

8.10   Determination of the geoid from ground-level anomalies

We have seen that it is possible to determine the physical surface of the earth by means of the height anomalies ζ, and the direction of the plumb line on it by means of the deflection components ξ and η, from free-air anomalies referred to the ground. If we plot the orthometric height H downward along the plumb line, starting from the physical surface, then the locus of the points so obtained will be the geoid (Fig. 8.11).

This geometrical idea may be formulated analytically in the following way. Conventionally, the height h above the ellipsoid is given by

$$h = H + N \,;\tag{8--106}$$

according to the modern theory, by

$$h = H^* + \zeta \,.\tag{8--107}$$

From these two equations we get

$$N - \zeta = H^* - H \,.\tag{8--108}$$

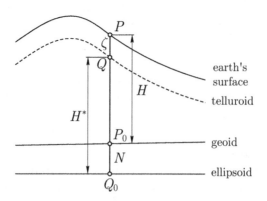

Fig. 8.11. Geoid at a depth H below the earth's surface

This means that the difference between the geoidal undulation N and the height anomaly ζ is equal to the difference between the normal height H^* and the orthometric height H. Since ζ is also the undulation of the quasigeoid, this difference is also the distance between geoid and quasigeoid.

According to Sect. 4.5, the two heights are defined by

$$H = \frac{C}{\bar{g}} \,, \qquad H^* = \frac{C}{\bar{\gamma}} \,, \tag{8–109}$$

where C is the geopotential number, $\bar{g}$ is the mean gravity along the plumb line between geoid and ground, and $\bar{\gamma}$ is the mean normal gravity along the normal plumb line between ellipsoid and telluroid. By eliminating C between these two equations, we readily find

$$H^* - H = \frac{\bar{g} - \bar{\gamma}}{\bar{\gamma}} \, H \,, \tag{8–110}$$

which is also the distance between the geoid and the quasigeoid, see (8–108); hence

$$N = \zeta + \frac{\bar{g} - \bar{\gamma}}{\bar{\gamma}} \, H \,. \tag{8–111}$$

The height anomaly ζ may be expressed, for instance, by Molodensky's formula (8–57). Then we obtain

$$N = \frac{R}{4\pi \, \gamma_0} \iint\limits_{\sigma} \Delta g \, S(\psi) \, d\sigma + \frac{R}{4\pi \, \gamma_0} \iint\limits_{\sigma} g_1 \, S(\psi) \, d\sigma + \frac{\bar{g} - \bar{\gamma}}{\bar{\gamma}} \, H \,, \tag{8–112}$$

where g_1 is the term (8–62). Thus N is given by Stokes' integral, *applied to free-air anomalies at ground level*, and two small corrections, where

1. the term containing g_1 represents the effect of topography;

2. the term containing $\bar{g} - \bar{\gamma}$ represents the distance between the geoid and the quasigeoid.

If we neglect these two corrections, then the geoidal undulations N are given by Stokes' integral using free-air anomalies. This was first noted by Stokes in 1849. A new approach by Jeffreys (1931) by means of Green's identities started several developments which culminated in the work of Molodensky and others.

The advantage of this method for the determination of N is that the density of the masses above sea level enters only indirectly, as an effect on the orthometric height H through the mean gravity $\bar{g}$, which must be computed by a Prey reduction (Sect. 3.5). Hence, as far as errors in the density are concerned, the geoidal undulation N as obtained by this method is as accurate as the orthometric height.

As a matter of fact, the gravity anomaly Δg in this method refers to ground level; it is the difference between gravity at ground and normal gravity at the telluroid. Instead of using directly this free-air anomaly, we may also use other gravity anomalies – for instance, the isostatic anomaly in the sense of Sect. 8.9.

To repeat a simple but fundamental principle: Δg, δg, ξ, η, ζ as obtained by Molodensky's theory **primarily** *always refer to the physical earth's surface and not to sea level!*

8.11   A first balance

The new methods described in this chapter are primarily intended for the determination of the physical surface of the earth, but they are also well suited for the determination of the geoid (Sect. 8.10). Their essential feature is that the gravity anomalies *now* refer to the ground, whether we deal with free-air anomalies or with isostatic or other similarly reduced gravity anomalies (Sect. 8.9).

The immediate result is the height anomaly ζ, the separation between the geopotential and the corresponding spheropotential surface at ground level. By plotting the height anomalies above the ellipsoid, we get the quasigeoid. This geoid-like surface has no physical significance, but it furnishes a convenient visualization of the height anomalies. By plotting the orthometric height from the earth's surface vertically downward, we obtain the geoid.

It is instructive to compare the geoid and the quasigeoid. The geoidal undulation N and ζ, the undulation of the quasigeoid, are related by (8–111), or

$$N - \zeta = \frac{\bar{g} - \bar{\gamma}}{\bar{\gamma}} H = H^* - H \,. \tag{8–113}$$

The term $\bar{g} - \bar{\gamma}$ is approximately equal to the Bouguer anomaly; this may be seen by using (4–32) for γ together with

$$\bar{\gamma} \doteq \gamma - \frac{1}{2} \frac{\partial \gamma}{\partial h} H \,. \tag{8–114}$$

The quantity $\bar{\gamma}$ in the denominator can be replaced by our usual constant γ_0. Since the Bouguer anomaly is rather insensitive to local topographic irregularities, the coefficient is locally constant so that there is approximately a linear relation between ζ and the local irregularities of the height H. In other words, the *quasigeoid mirrors the topography* (Fig. 8.12).

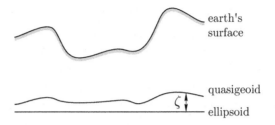

Fig. 8.12. Quasigeoid

To get a quantitative estimate of the difference $N - \zeta$, we again use the fact that

$$\frac{\bar{g} - \bar{\gamma}}{\bar{\gamma}} \doteq \frac{\Delta g_B}{981 \text{ gal}} \doteq 10^{-3} \Delta g_B \,, \qquad (8\text{--}115)$$

where Δg_B is the Bouguer anomaly in gal, so that

$$(\zeta - N)_{[m]} \doteq -\Delta g_{B\,[\text{gal}]} \cdot H_{[\text{km}]} \,. \qquad (8\text{--}116)$$

Since Δg_B is usually negative on the continents, the differences $\zeta - N$ are usually positive there. In other words, the height anomaly ζ is in general greater than the corresponding geoidal undulation N on land. We have $\zeta = N$ on the oceans. If $\Delta g_B = -100 \text{ mgal} = -0.1 \text{ gal}$ and $H = 1 \text{ km}$, then

$$\zeta - N = 0.1 \text{ m} \,. \qquad (8\text{--}117)$$

Furthermore, the Bouguer anomaly depends on the *mean* elevation of the terrain, decreasing approximately by 0.1 gal per 1 km average elevation. Assuming as a rough estimate, which may be verified by inspecting maps of Bouguer anomalies,

$$\Delta g_{B\,[\text{gal}]} = -0.1\, H^{\text{av}}_{[\text{km}]} \,, \qquad (8\text{--}118)$$

we obtain

$$(\zeta - N)_{[m]} \doteq +0.1\, H^{\text{av}}_{[\text{km}]}\, H_{[\text{km}]} \,, \qquad (8\text{--}119)$$

where H is the height of the station and H^{av} is an average height of the area considered. We see that the difference $\zeta - N$ increases faster than the elevation, almost as the square of the elevation. As a matter of fact, this formula is suited only to give an idea of the order of magnitude (see also Sect. 11.3).

Since $\zeta - N = H - H^*$, the approximate formulas given above may also be used to estimate the differences between the orthometric height H and the normal height H^*.

A theoretically important point is that the quasigeoid can be determined without hypothetical assumptions concerning the density, but not so the geoid. The avoidance of such assumptions has been the guiding idea of Molodensky's research. However, orthometric heights are but little affected by errors in density. The error in H due to the imperfect knowledge of the density hardly ever exceeds 1–2 decimeters even in extreme cases (Sect. 4.3). It is presumably smaller than the inaccuracy of the corresponding ζ even with very good gravity coverage, because of inevitable errors of interpolation, etc. If, therefore, the method of Sect. 8.10 is used, the geoid can be determined with virtually the same accuracy as the quasigeoid. Note that it is theoretically even possible to eliminate completely the errors arising from the use of the geoid (Moritz 1962, 1964). Thus, we may well retain the geoid with its physical significance and its other advantages.

How much do Molodensky's formulas differ from the corresponding equations of Stokes and Vening Meinesz? The deviation of ζ from the result of the original Stokes formula is given by the equivalent expressions

$$\zeta_1 = \frac{R}{4\pi\,\gamma_0} \iint\limits_{\sigma} g_1\, S(\psi)\, d\sigma \quad \text{or} \quad \zeta_1 = -\frac{R}{4\pi\,\gamma_0} \iint\limits_{\sigma} \frac{\partial \Delta g}{\partial h}\, (h - h_P)\, S(\psi)\, d\sigma$$

$$(8\text{--}120)$$

according to Eqs. (8–62) and (8–66). This correction may even be smaller than the difference $\zeta - N$ (see Sect. 11.3).

It is appropriate again to point out that the deflection of the vertical is relatively more affected by the Molodensky correction than is the height anomaly. In extreme cases, this correction may attain values of a few seconds, as studies of models by Molodensky (Molodenski et al. 1962: pp. 217–225) indicate. This is considerable, since $1''$ in the deflection corresponds to 30 m in position. Numerical estimates will be found in Chap. 11.

We may summarize the result of applying Stokes' and Vening Meinesz' formulas to free-air anomalies directly, without any corrections. Stokes' formula yields height anomalies ζ with high accuracy; for many practical purposes, we may, in addition, identify these height anomalies with the corresponding geoidal undulations N. Vening Meinesz' formula gives deflections of the vertical at ground level that are relatively less accurate but often acceptable.

An advantage of the modern theory is its direct relation to the external gravity field of the earth, which is particularly important nowadays for the computation of the effect of gravitational disturbances on spacecraft trajectories and satellite orbits. It is immediately clear that ground-level quantities, such as free-air gravity anomalies, are better suited for this purpose than the corresponding quantities referred to the geoid, which is separated from the external field by the outer masses. For the computation of the external field and of spherical harmonics, the method described in Sect. 8.6.5

is particularly appropriate (see also Sect. 6.5).

Practically it is usually adequate to consider only the linear approximation by using (8–57). In many cases it is even possible to neglect the correction $-(\partial \Delta g/\partial h)\, h$, identifying the sea-level free-air anomalies $\Delta g^{\text{harmonic}}$ with the corresponding ground-level anomalies Δg. In agreement with Sect. 3.9, these free-air anomalies $\Delta g^{\text{harmonic}} = \Delta g$ may also be considered approximations to condensation anomalies in the sense of Helmert. This approximation is particularly sufficient for the external gravity field, spherical harmonics, and geoidal undulations or height anomalies. For deflections of the vertical, it is often necessary to use a more careful approach, such as the consideration of the indirect effect with mass-transporting gravity reductions (Sect. 8.2) or the modern methods of Sect. 8.9.

In high and steep mountains, the approach of Molodensky and others through free-air anomalies encounters practical difficulties, such as unreliability of interpolation, large corrections, and other computational problems. To avoid this, isostatic reduction in the modern sense shoud be used. Thus the clash between "conventional" (geoid) and "modern" (Molodensky-type) ideas gives way to an important synthesis. For another synthesis, see least-squares collocation in Sects. 10.2 and 11.2.

For further study, especially of the historic aspects, the reader is referred to the book by Molodenski et al. (1962) and the M.S. Molodensky Anniversary Volume edited by Moritz and Yurkina (2000).

Part II: Astrogeodetic methods according to Molodensky

8.12   Some background

The computation of a detailed geoid, or of a detailed gravity potential field, in limited areas, especially in mountainous regions, has not been very much in the focus of attention recently. There may be various reasons for this.

For decades now, global geoid determinations, either from satellite data or from a combination of satellite and gravimetric data have been in the center of interest (Lerch et al. 1979, Reigber et al. 1983, Rapp 1981). Even (almost) purely gravimetric global and local geoids have been successfully computed (March and Chang 1979), between the classic Heiskanen (1957) and the modern local geoid (Kühtreiber 2002 b). An excellent recent reference volume is that by Tsiavos (2002).

Over the oceans, the geoid is now known to an accuracy of perhaps a few centimeters, due to satellite altimetry. Unfortunately, satellite altimetry does not work over land areas. The classical method for a detailed geoid determination on the continents has been the gravimetric method, in spite of the fact that it is severely handicapped by lack of an adequate gravity coverage (or lack of information on such a coverage). Thus, we have the paradoxical situation that on the oceans, long a stepchild of geodesy, the geoid is now in general known much better than on the continents.

Still, the gravimetric method has continued to fascinate theoreticians because it gives rise to very interesting and deep mathematical problems, related to the geodetic boundary-value problem discussed above in this chapter.

These enormous practical and theoretical developments concerning global satellite and gravimetric gravity field determination have somewhat overshadowed the determination of detailed geoids in smaller areas, in particular, astrogeodetic geoids. Especially in mountainous regions, local geoid determinations are difficult. The gravimetric method does not work very well in high mountains. The astrogeodetic method, using astronomical observations of latitude and longitude, does work well there but is considered time-consuming and somewhat old-fashioned, perhaps also because working during the night is not very popular nowadays. An appropriate use of gravity and astrogeodetic data in high mountains must involve some topographic-isostatic reduction. Furthermore, the theory behind the astrogeodetic method is not nearly as attractively difficult as the theory of Molodensky's problem. Last but not least, high-mountain areas are exceptional and, apart from such countries as Switzerland and Austria, are frequently regions of little economic interest. For these and similar reasons, the mainstream of geodetic practice and theory has flown with grand indifference around high mountains, ignoring such trivial obstacles.

Still, a country such as Switzerland has made a virtue out of necessity and has traditionally been very active in local astrogeodetic geoid determination (Elmiger 1969, Gurtner 1978, Gurtner and Elmiger 1983). Austria has followed up (Österreichische Kommission für die Internationale Erdmessung 1983). It has been found that, even besides the problem of getting the required observations, the underlying theory is not so trivial as one might think and shows quite interesting features.

Concerning measurements, astronomical observations have again proved very feasible in mountains; see the articles by Erker, Bretterbauer and Gerstbach, Lichtenegger and Chesi in Chap. 2 of Österreichische Kommission für die Internationale Erdmessung (1983), followed by Sünkel et al. (1987). The main advantages of astrogeodetic versus gravimetric data for local geoid determination in mountainous regions may be summarized as follows:

1. It is sufficient to have astrogeodetic deflections of the vertical in the

region of geoid determination; no data are needed outside that region as they would be in the gravimetric method.

2. Errors in the topographic height have significantly less influence on deflections than on gravity data. Thus, a relatively crude terrain model will be sufficient for the use of astrogeodetic data.

As a matter of fact, the two types of data are not mutually exclusive; an optimal geoid determination will combine astrogeodetic deflections of the vertical, gravity anomalies, and possibly data of other type. A suitable technique for this purpose is least-squares collocation to be discussed in Chap. 10.

From the observational point of view it is interesting to note that *inertial surveying techniques* will be able to furnish deflections of the vertical and gravity anomalies rapidly and with sufficient accuracy for many purposes.

Let us finally try to give a list of various methods of geoid determination:

- conventional satellite techniques (Doppler, laser, etc.),
- satellite-to-satellite tracking,
- satellite gradiometry,
- satellite altimetry,
- gravimetry,
- astrogeodesy, and,
- most directly, GPS leveling (Sect. 4.6).

As a general rule, these methods are listed in such a way as to start with the most global and end up with the most local method, that is, according to decreasing globality or increasing locality. In general, going down the list also corresponds to increasing resolution and accuracy.

Again it should be stressed that these methods complement each other and should be combined for best results.

New satellite gravity missions have been discussed in Sect. 7.6.

Astrogeodetic method according to Molodensky

The remaining part of this chapter deals primarily with the lower end of the list, providing a detailed theory of astrogeodetic local geoid determination in areas with difficult topography. The role (and necessity) of topographic-isostatic reduction is investigated in detail. The computations for Austria give concrete numerical results for questions which have been much discussed theoretically, such as the difference between geoidal heights and height anomalies according to Molodensky (quasigeoidal heights), or the numerical effect of analytical continuation from the earth's surface to sea level (Österreichische Kommission für die Internationale Erdmessung 1983).

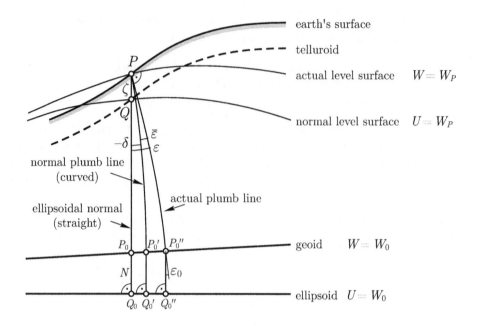

Fig. 8.13. The basic geometry

As a warm-up, let us return to basics and remember some main principles of Molodensky's geometry. Figure 8.13 illustrates the basic quantities. In the classical theory, the geoid is defined by its deviation N from a reference ellipsoid; N is the geoidal height. The geoid is a level surface $W = W_0 = $ constant of the gravity potential W; the ellipsoid is defined to be the level surface $U = U_0 = $ constant of a normal gravity potential U; the constants W_0 and U_0 are usually assumed to be equal (Sect. 2.12).

For the modern theory according to Molodensky (Sect. 8.4), to each point P of the earth's surface we associate a point Q in such a way that Q lies on the straight ellipsoidal normal through P and that

$$U(Q) = W(P). \tag{8-121}$$

That is, Q is defined such that its normal potential U equals the actual potential W of P.

This corresponds to the classical relation

$$U_0 = U(Q_0) = W(P_0) = W_0 \tag{8-122}$$

mentioned above, by which U_0 is taken to be equal to W_0 (Fig. 8.13). By the same correspondence, the height anomaly according to Molodensky,

$$\zeta = QP, \tag{8-123}$$

is the modern equivalent of the classical geoidal height,

$$N = Q_0 P_0 \,. \tag{8-124}$$

Using the anomalous potential

$$T = W - U \,, \tag{8-125}$$

we have according to Bruns' theorem

$$\zeta = \left(\frac{T}{\gamma}\right)_Q \,, \qquad N = \left(\frac{T}{\gamma}\right)_{Q_0} \,, \tag{8-126}$$

where γ denotes the ellipsoidal normal gravity.

The points P_0 form the geoid, and the points Q_0 constitute the ellipsoid, both being level surfaces (of W and U, respectively). On the other hand, the points P form the earth's surface, and the set of points Q defines an auxiliary surface, denoted as telluroid according to R.A. Hirvonen. As a matter of fact, neither the earth's surface nor the telluroid are level surfaces, which makes matters more complicated than in the classical situation, where we deal with level surfaces.

Following a suggestion of Molodensky, one could plot the height anomalies ζ as vertical distances from the reference ellipsoid. Thus one obtains a geoid-like surface, the quasigeoid, and ζ could be considered as quasigeoidal heights. In contrast to the geoid, however, the quasigeoid is not a level surface and does not admit of a natural physical interpretation. Therefore, working with height anomalies ζ, it is best to consistently consider them quantities referred to the earth's surface (vertical distances between earth surface and telluroid), rather than using the quasigeoidal concept. A summary will be given in Sect. 8.15.

The classical gravity anomaly Δg_0 at sea level is defined as

$$\Delta g_0 = g(P_0) - \gamma(Q_0) \,, \tag{8-127}$$

where g denotes gravity and γ normal gravity. So far, $g(P_0)$ denotes the actual gravity on the geoid; we are not yet here considering mass-transporting gravity reductions.

Analogously we have according to Molodensky:

$$\Delta g = g(P) - \gamma(Q) \,. \tag{8-128}$$

Generally we will, as far as feasible, use the subscript "0" to designate quantities referred to sea level, to distinguish them from quantities referred to the earth's surface, which do not carry such a subscript. For instance, Δg_0

refers to sea level and Δg to the earth's surface. With GPS we have gravity disturbances

$$\Delta g = g(P) - \gamma(P) \,. \tag{8--129}$$

Regarding plumb line definition, we must distinguish three lines (Fig. 8.13):

1. the straight ellipsoidal normal $Q_0\, P$,
2. the actual plumb line $P_0''\, P$,
3. the normal plumb line $P_0'\, P$.

Geometrically, the ellipsoidal normal is defined as the straight line through P perpendicular to the ellipsoid. The (actual) plumb line is defined by the condition that, at each point of the line, the tangent coincides with the gravity vector $\mathbf{g}$ at that point; the plumb line is very slightly curved, but its curvature is irregular, being determined by the irregularities of topographic masses. The normal plumb line, at each of its points, is tangent to the normal gravity vector $\boldsymbol{\gamma}$; it possesses a curvature that is even smaller and completely regular.

The points P_0, P_0', and P_0'' coincide within a few decimeters, and we will not distinguish them in what follows. The reason is that the distance, in arc seconds, between P_0 and P_0'' is much smaller than the effect of plumb line curvature (Sect. 5.15). The same applies for $Q_0, Q_0', $ and Q_0''.

The direction of the gravity vector $\mathbf{g}$ is the direction of (the tangent to) the plumb line. It is determined by two angles, the astronomical latitude Φ and the astronomical longitude Λ. Let Φ, Λ be referred to the earth's surface (to point P) and Φ_0, Λ_0 to the geoid (strictly speaking, to point P_0''). The differences

$$\delta\varphi = \Phi_0 - \Phi \,, \qquad \delta\lambda = \Lambda_0 - \Lambda \tag{8--130}$$

express the effect of plumb line curvature (Fig. 8.14). You may also wish to refer back to Fig. 5.18. Hence, we have

$$\Phi_0 = \Phi + \delta\varphi \,, \qquad \Lambda_0 = \Lambda + \delta\lambda \,. \tag{8--131}$$

Knowing the plumb line curvature $\delta\Phi, \delta\Lambda$, we could use these simple formulas to compute the sea-level values Φ_0, Λ_0 from the observed surface values Φ, Λ.

In the same way as Φ, Λ are related to the actual plumb line, the ellipsoidal latitude φ and the ellipsoidal longitude λ refer to the straight ellipsoidal normal. The quantities

$$\xi = \Phi - \varphi \,, \qquad \eta = (\Lambda - \lambda)\cos\varphi \tag{8--132}$$

are the components of the deflection of the vertical in a north-south and an east-west direction. For an arbitrary azimuth α, the vertical deflection ε is given by

$$\varepsilon = \xi\cos\alpha + \eta\sin\alpha \,. \tag{8--133}$$

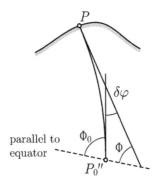

Fig. 8.14. Curvature of the plumb line along a north-south profile

These quantities ξ, η, ε refer to the earth's surface. Figure 8.13 shows ε.
  Similarly, we have for the geoid

$$\xi_0 = \Phi_0 - \varphi, \quad \eta_0 = (\Lambda_0 - \lambda)\cos\varphi, \tag{8-134}$$

$$\varepsilon_0 = \xi_0 \cos\alpha + \eta_0 \sin\alpha. \tag{8-135}$$

See again Fig. 8.13 for ε_0, noting that we do not distinguish the normals in
Q_0 and Q_0'' as we have mentioned above.
  In addition, we need the normal direction of the plumb line at the surface
point P; it is defined as the tangent to the normal plumb line at P; the
corresponding latitude and longitude will be denoted by $\bar\varphi, \bar\lambda$. In this "local"
notation, there is no danger of confusion with the spherical coordinate $\bar\varphi$
used in earlier chapters. Hence, we have

$$\varphi = \bar\varphi + \delta\varphi_{\text{normal}}, \quad \lambda = \bar\lambda + \delta\lambda_{\text{normal}}, \tag{8-136}$$

where $\delta\varphi, \delta\lambda$ express the normal plumb line curvature. These equations are
the "normal equivalent" to (8-131): the "normal surface values" $\bar\varphi, \bar\lambda$ correspond to the "actual surface values" Φ, Λ and the ellipsoidal values φ, λ
correspond to the geoidal values Φ_0, Λ_0. To make the analogy complete, we
should replace $\varphi = \varphi(P_0)$ by $\varphi(P_0')$, but we have consistently neglected such
differences.
  In contrast to the actual plumb line curvature, it is very easy to compute
the normal curvature of the plumb line: from (5-147) we have

$$\delta\varphi_{\text{normal}} = -0.17'' \, h_{\text{[km]}} \sin 2\varphi, \quad \delta\lambda_{\text{normal}} = 0, \tag{8-137}$$

where $h_{\text{[km]}}$ denotes elevation in kilometers.
  Since the ellipsoidal normal and hence φ, λ are geometrically defined, we
may call the quantities (8-132) "geometric deflections of the vertical" at the

earth's surface. On the other hand, the normal plumb line is physically (or dynamically) defined by means of the external gravity field of an equipotential ellipsoid. Hence also $\bar{\varphi}, \bar{\lambda}$ are dynamically defined. The quantities obtained by replacing φ, λ by $\bar{\varphi}, \bar{\lambda}$ so that

$$\bar{\xi} = \Phi - \bar{\varphi}, \quad \bar{\eta} = (\Lambda - \bar{\lambda})\cos\varphi, \tag{8–138}$$

are called "dynamical deflections of the vertical" at the earth's surface. By (8–136) and (8–137) we have

$$\bar{\xi} = \xi + \delta\varphi_{\text{normal}}, \quad \bar{\eta} = \eta, \tag{8–139}$$

since $\delta\lambda_{\text{normal}} = 0$. For an azimuth α we accordingly have

$$\bar{\varepsilon} = \bar{\xi}\cos\alpha + \bar{\eta}\sin\alpha. \tag{8–140}$$

Compare ε and $\bar{\varepsilon}$ in Fig. 8.13 and note that in this figure δ denotes the curvature of the normal plumb line for the azimuth α given by the analogous formula

$$\delta = \delta\varphi_{\text{normal}}\cos\alpha + (\delta\lambda_{\text{normal}}\cos\varphi)\sin\alpha = \delta\varphi_{\text{normal}}\cos\alpha. \tag{8–141}$$

8.13   Astronomical leveling revisited

From Fig. 8.15 we take the well-known differential relation

$$dN = -\varepsilon_0\,ds, \tag{8–142}$$

where ε_0 denotes the deflection of the vertical at the geoid. Integration between two points A and B yields the difference between their geoidal heights:

$$N_B - N_A = -\int_A^B \varepsilon_0\,ds, \tag{8–143}$$

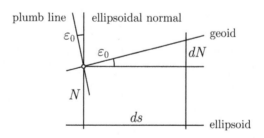

Fig. 8.15. Astronomical leveling according to Helmert

or, approximately,

$$N_B - N_A = -\frac{\varepsilon_{0A} + \varepsilon_{0B}}{2} s_{AB}, \qquad (8\text{–}144)$$

where s_{AB} denotes the horizontal distance between A and B. The minus sign is conventional. Cf. Sect. 5.14.

A corresponding relation to height anomalies according to Molodensky is found as follows (Molodensky et al. 1962: p. 125):

$$d\zeta = \frac{\partial \zeta}{\partial s} ds + \frac{\partial \zeta}{\partial h} dh, \qquad (8\text{–}145)$$

notations following Fig. 8.16. Since the earth's surface is not a level surface, we also have a vertical part $(\partial \zeta/\partial h)\,h$ in addition to the usual horizontal part $(\partial \zeta/\partial s)\,ds$. The vertical part arises from change in height and is usually smaller than the horizontal part.

In analogy to (8–142), the horizontal part is given by

$$\frac{\partial \zeta}{\partial s} = -\bar{\varepsilon}, \qquad (8\text{–}146)$$

where $\bar{\varepsilon}$ denotes the dynamical deflection of the vertical at the earth's surface; cf. (8–140) and Fig. 8.13. For the vertical part we have from (8–126):

$$\frac{\partial \zeta}{\partial h} = \frac{\partial}{\partial h}\left(\frac{T}{\gamma}\right) = \frac{1}{\gamma}\left(\frac{\partial T}{\partial h} - \frac{1}{\gamma}\frac{\partial \gamma}{\partial h} T\right) \qquad (8\text{–}147)$$

or

$$\frac{\partial \zeta}{\partial h} = -\frac{\Delta g}{\gamma} = -\frac{g - \gamma}{\gamma} \qquad (8\text{–}148)$$

according to the fundamental equation of physical geodesy (8–36).

Hence (8–145) becomes

$$d\zeta = -\bar{\varepsilon}\,ds - \frac{g - \gamma}{\gamma}\,dh. \qquad (8\text{–}149)$$

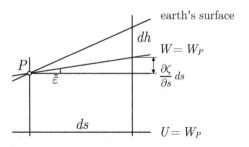

Fig. 8.16. Astronomical leveling according to Molodensky

Integrating this relation yields the difference of the height anomaly

$$\zeta_B - \zeta_A = - \int_A^B \bar{\varepsilon} \, ds - \int_A^B \frac{\Delta g}{\gamma} \, dh \, ; \tag{8-150}$$

the gravity anomaly Δg refers to the earth's surface according to (8–128). The first term on the right-hand side represents the Helmert integral (8–143) of the surface deflection $\bar{\varepsilon}$, and the second term is Molodensky's correction to the Helmert integral, necessary to obtain height anomalies. This correction depends on the gravity g at the earth's surface.

8.14 Topographic-isostatic reduction of vertical deflections

For the reasons mentioned at the end of the preceding section, it is natural to try and find a way which makes use of the clear advantages of the topographic-isostatic reduction but avoids the problems inherent in a free-air reduction from the surface point P to the geoidal point P_0.

In Sect. 8.9, we have treated the reduction of gravity from the modern point of view. The second formula of (8–95) is

$$g^c = g - \delta g \, . \tag{8-151}$$

Everything is referred to the ground point P, and $\delta g = \delta g_{\mathrm{TI}}$ is the effect of gravity reduction on g, also at P. In the *topographic-isostatic reduction which we use here exclusively*, it is the gravitational attraction of the topography minus the gravitational attraction of the compensating isostatic masses, topography minus isostasy.

To get the topographic-isostatic gravity anomaly, we subtract normal gravity γ, also referred to ground level, more precisely, to the corresponding telluroid point Q. Thus,

$$\Delta g^c = \Delta g - \delta g_{\mathrm{TI}}. \tag{8-152}$$

The explanation is trivial: you are standing at point P and watch how the topography is removed to fill the isostatic mass deficits, but by a miracle you are still hovering at P, now in "free air".

Application to deflections of the vertical
The gravity anomaly is only one component of the anomalous gravity vector, the other two being the vertical components ξ and η, both, of course, multiplied by γ to get the dimensions right. *Thus, ξ and η can be isostatically reduced in exactly the same way.*

For ξ and η, (8–152) becomes

$$\xi^c = \xi - \xi_{TI} + \delta\varphi_{\text{normal}}\,, \qquad \eta^c = \eta - \eta_{TI}\,. \tag{8–153}$$

By means of (8–139) this may be written

$$\xi^c = \bar{\xi} - \xi_{TI}\,, \qquad \eta^c = \bar{\eta} - \eta_{TI}\,. \tag{8–154}$$

The interpretation of (8–154), however, is clear, simple, and rigorous: from the dynamic deflections of the vertical at P, which are the very quantities $\bar{\xi}$ and $\bar{\eta}$, we subtract the effect of the topographic-isostatic masses, ξ_{TI} and η_{TI} likewise at P. The vertical deflections so obtained, ξ^c and η^c, thus do not really refer to the (co-)geoid; in reality, *they refer to the earth's surface!*

But what, then, means the normal plumb line curvature $\delta\varphi_{\text{normal}}$ in (8–153)? Does it not mean a reduction from the earth's surface to sea level? No, in Eqs. (8–139) it only denotes the transformation between the geometrical and the dynamical deflection of the vertical, both referred to the point P of the earth's surface. This is also clear from Fig. 8.13, which illustrates the formula

$$\bar{\varepsilon} = \varepsilon + \delta\,, \tag{8–155}$$

extending (8–139) to an arbitrary azimuth, δ being defined by (8–141).

This interpretation of (8–153) or (8–154) as isostatically reduced deflections of the vertical at the earth's surface is exact, whereas the interpretation of (8–8) as deflections at the cogeoid was only approximate. This is the desired rigorous interpretation of our isostatically reduced vertical deflections.

This interpretation exactly corresponds to the modern view of gravity reduction according to the theory of Molodensky. According to this view, the isostatically (or in some other way) reduced gravity anomalies continue to refer to the earth's surface. The classical gravity reduction (Sect. 8.2) had comprised two procedures: mass transport and shift $P \to P_0$; the new view of gravity reduction only considers the mass transport; the problematic shift $P \to P_0$ is avoided.

Formally, a "normal free-air reduction"

$$F = -\frac{\partial\gamma}{\partial h}\,h \tag{8–156}$$

may be said to occur also in Molodensky's theory: normal gravity γ in the new definition (8–128) of the gravity anomaly, where it refers to the telluroid point Q, is computed by

$$\gamma = \gamma_{Q_0} + \frac{\partial\gamma}{\partial h}\,h\,, \tag{8–157}$$

with $h = Q_0 Q$ denoting the normal height of P. But instead of reducing actual gravity g downward, from P to P_0, now normal gravity is reduced upward from Q_0 to Q. Whereas for the first process the use of the normal gradient $\partial\gamma/\partial h$ is problematic, it is fully justified for the second process.

In a similar way, we might interpret $\delta\varphi_{\mathrm{normal}}$ as a reduction of φ for normal curvature of the plumb line upwards, say, from P_0 to P. This is possible because in (8–136) φ could be said to refer to P_0' (because P_0 and P_0' practically coincide), and because $\bar{\varphi}$ denotes the latitude of the tangent to the normal plumb line at P. This interpretation is instructive because of the analogy with gravity reduction, though regarding φ and $\bar{\varphi}$ as ellipsoidal and dynamic latitude of the same point P appears more natural. Refer again to our key figure (Fig. 8.13).

As pointed out above, the present interpretation of ξ^c and η^c as isostatically reduced deflections of the vertical at the earth's surface is conceptually rigorous and therefore also practically more accurate, but this decisive advantage implies a computational drawback if integration along a profile is used: Since this integration must now be performed along the earth's surface and not along a level surface such as the geoid, computation will be more complicated. Instead of the simple Helmert formula (8–143), we now must use the Molodensky formula (8–150):

$$\zeta_B^c - \zeta_A^c = -\int_A^B \varepsilon^c \, ds - \int_A^B \frac{g^c - \gamma}{\gamma} \, dh \qquad (8\text{–}158)$$

with

$$\varepsilon^c = \xi^c \cos\alpha + \eta^c \sin\alpha \,, \qquad (8\text{–}159)$$

and $\Delta g^c = g^c - \gamma$, where g^c is the isostatically reduced surface value of gravity (measured value g minus attraction of the topographic-isostatic masses).

From the isostatic height anomalies ξ^c obtained in this way, we then get the actual height anomalies ζ by applying the indirect effect:

$$\zeta = \zeta^c + \delta\zeta \qquad (8\text{–}160)$$

with

$$\delta\zeta = \frac{T_{\mathrm{TI}}}{\gamma} \,. \qquad (8\text{–}161)$$

This is completely analogous to (8–5) and (8–3), but now T_{TI} is the potential of the topographic-isostatic masses at the surface point P. As a matter of fact, normal gravity in (8–3) refers to the ellipsoid, and in (8–161) to the telluroid, but the difference is generally small.

For higher mountains, the isostatic reduction procedure described in the present section is preferable in practice to a direct application of Molodensky's formula (8–150) because the isostatically reduced vertical deflections are much smoother and easier to interpolate. It is, however, extremely laborious from a computational point of view since the integration must be performed along the earth's surface (or, what is practically the same, along the telluroid).

We remark that the computational drawback of the present method, the Molodensky integration along the earth's surface, can be completely avoided if we perform our computations in space: instead of integrating along a surface, we perform collocation in space. This modern procedure, to be described in the next chapter, permits a simple and computationally convenient use of surface deflections and also their combination with gravimetric and other data. Still, the present developments are necessary for a full understanding of the collocation approach.

Final remarks

In these last sections we tried to apply the same principle for topographic-isostatic reduction (the "remove-restore method") *at point level* to all terrestrial data related to the gravity vector: gravity anomalies and disturbances (Sect. 8.9) and deflections of the vertical (Sect. 8.14). This unified view of isostatically reduced data thus makes them directly suitable for combined solutions by least-squares collocation to be treated in Chap. 10.

8.15   The meaning of the geoid

We now review the geoid and some surfaces that might be able to replace it. We will again confirm the unique role of the geoid as a standard surface of physical geodesy.

The meaning of the geoid is very simple. It is defined in Sect. 2.2 as one of the equipotential surfaces (level surfaces, surfaces of constant gravity potential)

$$W(x, y, z) = \text{constant.} \qquad (8\text{--}162)$$

The constant is chosen so that, on the oceans, the geoid coincides with mean sea level:

$$W(x, y, z) = W_0. \qquad (8\text{--}163)$$

This is the usual classical equation of the geoid. So what is the problem? Well, theory and practice are different, in geodesy as well as in daily life. First, we must disregard small tidal effects (on the order of 50 cm). This is done by applying a suitable tidal model and is not too problematic. In fact, we have numerous geoids determined from satellite observations. Second, they are usually expressed in terms of a series of spherical harmonics. If taken at sea level, such a series may diverge (this is related to the difficulties of downward continuation, cf. Sect. 8.6). Such a possible divergence may concern mathematicians, but it should not concern geodesists, for several reasons:

1. Our spherical-harmonic expansions are not infinite series but finite polynomials, by their very determination and computations. So divergence problems do not exist; the question is only good approximation.

2. Such approximating polynomials of spherical harmonics always exist for arbitrary accuracy requirements (Frank and Mises 1930: p. 760). In geodesy we usually speak of Runge's theorem. The whole subject is thorougly discussed in Moritz (1980 a: Sects. 6 to 8).

3. If you use spatial collocation, the behavior (harmonic or not, convergent or divergent, ...) of the solution is completely determined by the covariance function used. One always uses "good" covariance functions, which are harmonic and analytic down to a sphere completely inside the earth.

So forget all about the convergence problem. *It is practically solved.* Further discussions beyond the results obtained so far would have to be made at a very high mathematical level. The question can be made as complicated as desired; if looked at it from the right angle, it is simple.

Geoid and downward continuation

Therefore, and by the reasoning at the end of Sect. 10.1, the geoid computed by (harmonic!) spherical-harmonic expressions and by collocation is not a level surface of the actual geopotential W but a *level surface of a harmonic downward continuation* of W, for the simple reason that the base functions both of spherical harmonics and of collocation satisfy Laplace's equation (8–2). We may speak of a "harmonic geoid". This again emphasizes the importance of analytical continuation (Sect. 8.6). We have deliberately used the indefinite article "a" in the italicized expression above, because harmonic downward continuation is an inverse problem and thus has no unique solution (see below).

The application of collocation to ξ, η, Δg *without gravity reduction* gives height anomalies ζ and undulations of the harmonic geoid, N^{harmonic}, by simply varying the elevation parameter (h and zero, respectively) in the collocation program. A completely analogous fact was remarked at the end of the last section for the case of height anomalies ζ^c and cogeoidal heights N^c. In the case of Molodensky's problem (without or with gravity reduction), we have seen a completely similar behavior with the application of the generalized Stokes and Vening Meinesz formulas, (8–75) and (8–76).

Geoid, harmonic geoids, and quasigeoid

The *geoid in the usual sense* of Eqs. (2–18) or (8–163) is defined purely by nature and is independent of geodetic observations (except for the tidal corrections). Its disadvantage is that it depends on the "topographic masses" above the geoid whose density is unknown, at least in principle. This drawback seems to be theoretical rather than practical.

The *harmonic geoids* are equipotential surfaces of an analytical downward continuation. We shall be careful to denote the harmonically continued potential by W^{harmonic} so that

$$W^{\mathrm{harmonic}} = W_0 = \text{constant} \qquad (8\text{–}164)$$

denote harmonic geoid(s).

To repeat, analytical downward continuation based on discrete data at the earth's surface is an *inverse problem* (Sect. 1.13; for more information see www.inas.tugraz.at/forschung/InverseProblems/AngerMoritz.html) which has infinitely many possible solutions. For collocation, e.g., each solution corresponds to the choice of a different covariance function.

Thus, the "harmonic geoid" is not uniquely defined. It is a product not only of nature but also of the computational method used. It cannot, therefore, replace the real geoid as a standard surface.

The "cogeoids" of the various gravity reductions (Sect. 8.2) are intermediate computational concepts and should never be used in place of the geoid. The topographic-isostatic height anomalies at point level, ζ^{c}, and the heights of the topographic-isostatic cogeoid, N^{c}, are related to each other by *analytical continuation*. The same collocation formula applies if the height anomaly $f(P)$ is computed at sea level with elevation parameter 0 to give N^{c}, or, if $f(P)$ is computed at point level with elevation parameter h, to give ζ^{c}. (The elevation parameter h is a height above sea level in any of the definitions of Chap. 4.) See item 5 at the end of Sect. 10.2.

For the limiting case of Fig. 8.5 c, take the question: "How is the undulation N^{harmonic} of a 'harmonic geoid' related to the height anomaly ζ above it on the ground and on the same vertical?" Answer: "By analytic continuation!"

Another special question to which the answer is also easy: "Which gravity reduction leaves the geoid unchanged?" Answer: "The Rudzki reduction" (Sect. 3.8). So why not use it? It changes the external potential, which today is of paramount importance.

"What is the difference between the Rudzki reduction and the harmonic downward continuation?" Answer: "The Rudzki reduction leaves the geoid unchanged but changes the external geopotential: there is $W^{\mathrm{c}} = W = W_0$ *only on the geoid*, but $W^{\mathrm{c}} \neq W$ outside the earth, which is inadmissible.

The harmonic continuation leaves the external geopotential unchanged but changes the geoid: $W^{\text{harmonic}} = W$ outside the earth and on the earth's surface, but $W^{\text{harmonic}} \neq W$ at sea level.

Height anomalies and quasigeoid

The height anomalies ζ refer to the physical earth's surface. They find their natural physical interpretation in Hirvonen's *telluroid*. Molodensky proposed to plot ζ above the reference ellipsoid and get the "quasigeoid". Thus, ζ gives the quasigeoid in exactly the same way as the geoidal height N gives the geoid. However, this analogy is purely formal. There is no way to interpret the quasigeoid as a surface of constant potential or find any other physical interpretation for it. Again, it cannot replace the real geoid as a standard surface.

Thus, in spite of all modern developments, the geoid retains its role as a standard reference for physical geodesy. However, the reader must have a clear view of all the concepts reviewed in this section, see Forsberg and Tscherning (1997).

A final remark on the many facets of free-air reduction

Now, dear reader, having struggled through almost the whole book, you will be able to understand the *disjecta membra* on free-air reduction strewn all over it, such as Sects. 3.3, 3.9, 8.2, 8.6, 8.9, and the present section. Have a couple of nice mountaineering tours!

9 Statistical methods in physical geodesy

9.1 Introduction

Some of the most important problems of gravimetric geodesy are formulated and solved in terms of integrals extended over the whole earth. An example is Stokes' formula. Thus, in principle, we need the gravity g at every point of the earth's surface. As a matter of fact, even in the densest gravity net we measure g only at relatively few points so that we must estimate g at other points by *interpolation*. In large parts of the oceans we have made no observations at all; these gaps must be filled by some kind of *extrapolation*.

Mathematically, there is no difference between interpolation and extrapolation; therefore they are denoted by the same term, *prediction*.

Prediction (i.e., interpolation or extrapolation) cannot give exact values; hence, the problem is to estimate the errors that are to be expected in the gravity g or in the gravity anomaly Δg. As usual, gravity disturbances δg are appropriately comprised whenever we speak of gravity anomalies.

Since Δg is further used to compute other quantities, such as the geoidal undulation N or the deflection components ξ and η, we must also investigate the influence of the prediction errors of Δg on N, ξ, η, etc. This is called *error propagation*, which will play a basic role.

It is also important to know which prediction method gives highest accuracy, either in Δg or in derived quantities N, ξ, η, etc. To be able to find these "best" prediction methods, it is necessary to have solved the previous problem, to know the prediction error of Δg and its influence on the derived quantities.

Summarizing, we have the following problems:

1. estimation of interpolation and extrapolation errors of Δg (or δg);

2. estimation of the effect of these errors on derived quantities (N, ξ, η, etc.);

3. determination of the best prediction method.

Since we are interested in the average rather than the individual errors, we are led to a statistical treatment. This will be the topic of the present chapter.

9.2   The covariance function

It is quite remarkable that all the problems mentioned above can be solved
by means of only one function of one variable, without any other information.
This is the *covariance function* of the gravity anomalies.

First we need a measure of the average size of the gravity anomalies Δg.
If we form the average of Δg over the whole earth, we get the value zero:

$$M\{\Delta g\} \equiv \frac{1}{4\pi} \iint_{\sigma} \Delta g \, d\sigma = 0 . \tag{9–1}$$

The symbol M stands for the average over the whole earth (over the unit
sphere); this average is equal to the integral over the unit sphere divided
by its area 4π. The integral is zero if there is no term of degree zero in the
expansion of the gravity anomalies Δg into spherical harmonics, that is, if a
reference ellipsoid of the same mass as the earth and of the same potential
as the geoid is used. This will be assumed throughout this chapter.

Note that if this is not the case, that is, if $M\{\Delta g\} = m \neq 0$, then we
may form new gravity anomalies $\Delta g^* = \Delta g - m$ by subtracting the average
value m. Then $M\{\Delta g^*\} = 0$ and all the following developments apply to the
"centered" anomalies Δg^*.

Clearly, the quantity $M\{\Delta g\}$, which is zero, cannot be used to characterize the average size of the gravity anomalies. Consider then the average
square of Δg,

$$\text{var}\{\Delta g\} \equiv M\{\Delta g^2\} = \frac{1}{4\pi} \iint_{\sigma} \Delta g^2 \, d\sigma . \tag{9–2}$$

It is called the *variance* of the gravity anomalies. Its square root is the *root
mean square (rms) anomaly*:

$$\text{rms}\{\Delta g\} \equiv \sqrt{\text{var}\{\Delta g\}} = \sqrt{M\{\Delta g^2\}} . \tag{9–3}$$

The rms anomaly is a very useful measure of the average size of the gravity
anomalies; it is usually given in the form

$$\text{rms}\{\Delta g\} = \pm 35 \text{ mgal} ; \tag{9–4}$$

the sign $\pm$ expresses the ambiguity of the sign of the square root and sym-
bolizes that Δg may be either positive or negative. The rms anomaly is very
intuitive; but the variance of Δg is more convenient to handle mathematically and admits an important generalization.

Instead of the average square of Δg, consider the average product of the gravity anomalies $\Delta g \, \Delta g'$ at each pair of points P and P' that are at a constant distance s apart. This average product is called the *covariance* of the gravity anomalies for the distance s and is defined by

$$\text{cov}_s\{\Delta g\} \equiv M\{\Delta g \, \Delta g'\} \,. \tag{9-5}$$

The average is to be extended over all pairs of points P and P' for which $PP' = s = \text{constant}$.

The covariances characterize the *statistical correlation* of the gravity anomalies Δg and $\Delta g'$, which is their tendency to have about the same size and sign. If the covariance is zero, then the anomalies Δg and $\Delta g'$ are uncorrelated or independent of one another (note that in the precise language of mathematical statistics, zero correlation and independence are not quite the same, but we may neglect the difference here!); in other words, the size or sign of Δg has no influence on the size or sign of $\Delta g'$. Gravity anomalies at points that are far apart may be considered uncorrelated or independent because the local disturbances that cause Δg have almost no influence on $\Delta g'$ and vice versa.

If we consider the covariance as a function of distance $s = PP'$, then we get the *covariance function* $C(s)$ mentioned at the beginning:

$$C(s) \equiv \text{cov}_s\{\Delta g\} = M\{\Delta g \, \Delta g'\} \quad (PP' = s) \,. \tag{9-6}$$

For $s = 0$, we have

$$C(0) = M\{\Delta g^2\} = \text{var}\{\Delta g\} \tag{9-7}$$

according to (9–2). The covariance for $s = 0$ is the variance.

A typical form of the function $C(s)$ is shown in Fig. 9.1. For small distances s (1 km, say), $\Delta g'$ is almost equal to Δg, so that the covariance is almost equal to the variance; in other words, there is a very strong correlation. The covariance $C(s)$ decreases with increasing s because then the

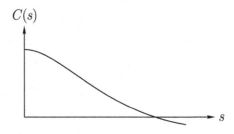

Fig. 9.1. The covariance function

anomalies Δg and $\Delta g'$ become more and more independent. For very large distances, the covariance will be very small but not in general exactly zero because the gravity anomalies are affected not only by local mass disturbances but also by regional factors. Therefore, we may expect an oscillation of the covariance between small positive and negative values.

Note that positive covariances mean that Δg and $\Delta g'$ tend to have the same size and the same sign; negative covariances mean that Δg and $\Delta g'$ tend to have the same size and opposite sign. The stronger this tendency, the larger is $C(s)$; the absolute value of $C(s)$ can, however, never exceed the variance $C(0)$.

The practical determination of the covariance function $C(s)$ is somewhat problematic. If we were to determine it exactly, we should have to know gravity at every point of the earth's surface. This we obviously do not know; and if we knew it, then the covariance function would have lost most of its significance because then we could solve our problems rigorously without needing statistics. As a matter of fact, we can only estimate the covariance function from samples distributed over the whole earth. But even this is not quite possible at present because of the imperfect or completely missing gravity data over the oceans. For a discussion of sampling and related problems see Kaula (1963, 1966 b).

The first comprehensive estimate of the covariance function was made by Kaula (1959). Some of his values are given in Table 9.1 for historical interest. They refer to free-air anomalies. The argument is the spherical distance

$$\psi = \frac{s}{R} \qquad (9\text{--}8)$$

corresponding to a linear distance s measured on the earth's surface; R is a mean radius of the earth. The rms free-air anomaly is

$$\text{rms}\{\Delta g\} = \sqrt{1201} = \pm\, 35 \text{ mgal}\,. \qquad (9\text{--}9)$$

We see that $C(s)$ decreases with increasing s and that, for $s/R > 30°$, very small values oscillate between plus and minus.

For some purposes we need a *local* covariance function rather than a global one; then the average M is extended over a limited area only, instead of over the whole earth as above. Such a local covariance function is useful for more detailed studies in a limited area – for instance, for interpolation problems. As an example we mention that Hirvonen (1962), investigating the local covariance function of the free-air anomalies in Ohio, found numerical values that are well represented by an analytical expression of the form

$$C(s) = \frac{C_0}{1 + (s/d)^2}\,, \qquad (9\text{--}10)$$

Table 9.1. Estimated values of the covariance
function for free-air anomalies [unit mgal^2]

ψ	$C(\psi)$	ψ	$C(\psi)$	ψ	$C(\psi)$
0.0°	+1201	8°	+124	27°	+18
0.5°	751	9°	104	29°	+6
1.0°	468	10°	82	31°	+8
1.5°	356	11°	76	33°	+5
2.0°	332	13°	54	35°	−8
2.5°	306	15°	47	40°	−12
3.0°	296	17°	45	50°	−20
4.0°	272	19°	34	60°	−30
5.0°	246	21°	35	90°	−4
6.0°	214	23°	10	120°	+12
7.0°	174	25°	20	150°	−21

where

$$C_0 = 337 \text{ mgal}^2, \quad d = 40 \text{ km}. \tag{9-11}$$

This function is valid for $s < 100$ km.

In the meantime it has been recognized that a proper determination of global and local covariance functions is a central practical problem in this context.

The Tscherning–Rapp covariance model and the COVAXN subroutine

The fundamental covariance model by Tscherning and Rapp (1974) and the subroutine COVAXN (Tscherning 1976) are still very much up to date, as the following quotation from Kühtreiber (2002 b) shows:

"The global covariance function of the gravity anomalies $C_g(P,Q)$ given by Tscherning and Rapp (1974, p. 29) is written as

$$C_g(P,Q) = A \sum_{n=3}^{\infty} \frac{n-1}{(n-2)(n+B)} s^{n+2} P_n(\cos \psi), \tag{9-12}$$

where $P_n(\cos \psi)$ denotes the Legendre polynomial of degree n; ψ is the spherical distance between P and Q; and A, B and s are the model parameters. A closed expression for (9–12) is available in (ibid., p. 45).

The local covariance function of gravity anomalies $C(P,Q)$ given by Tscherning–Rapp can be defined as

$$C(P,Q) = A \sum_{n=N+1}^{\infty} \frac{n-1}{(n-2)(n+B)}\, s^{n+2} P_n(\cos\psi)\,. \tag{9–13}$$

Modeling the covariance function means in practice fitting the empirically determined covariance function (through its three essential parameters: the variance C_0, the correlation length ξ and the variance of the horizontal gradient G_0) to the covariance function model. Hence the four parameters A, B, N and s are to be determined through this fitting procedure. A simple fitting of the empirical covariance function was done using the COVAXN-subroutine (Tscherning 1976).

The essential parameters of the empirical covariance parameters for 2489 gravity stations in Austria are $740.47\,\mathrm{mgal}^2$ for the variance C_0 and $43.5\,\mathrm{km}$ for the correlation length ψ_1. The value of the variance for the horizontal gradient G_0 was roughly estimated as $100\,\mathrm{E}^2$ (note that E indicates the Eötvös unit, where $1\,\mathrm{E} = 10^{-9}\,\mathrm{s}^{-2}$).

With a fixed value $B = 24$, the following Tscherning–Rapp covariance function model parameters were fitted: $s = 0.997\,065$, $A = 746.002\,\mathrm{mgal}^2$ and $N = 76$. The parameters were used for the astrogeodetic, the gravimetric as well as the combined geoid solution." (End of quotation.)

The Tscherning–Rapp model can be summed to get closed expressions. Its popularity is due to its comprehensiveness: there are expressions for co-variances of various quantities derived by covariance propagation (Sect. 10.1), and to its flexibility since it contains several parameters which can be given various numerical values.

Remark. The spherical-harmonic expression of the covariance function is considered in Sect. 9.3. The theory of global and local covariance functions is described in great detail in Moritz (1980 a: Sects. 22 and 23). The three essential parameters of a local covariance function (variance C_0, correlation length ξ, and curvature parameter G_0) are also defined there. Fundamental numerical studies on local covariance functions have been made by Kraiger (1987, 1988).

9.3 Expansion of the covariance function in spherical harmonics

The more or less complicated integral formulas of physical geodesy frequently take on a much simpler form if they are rewritten in terms of spherical harmonics. A good example is Stokes' formula (see Sect. 2.15).

Unfortunately, this theoretical advantage is in most cases balanced by the practical disadvantage that the relevant series converge very slowly. In certain cases, however, the convergence is good. Then the use of spherical harmonics is very convenient practically; we consider such a case in the next section.

The spherical-harmonic expansion of the gravity anomalies Δg may be written in different ways, such as

$$\Delta g(\vartheta, \lambda) = \sum_{n=2}^{\infty} \Delta g_n(\vartheta, \lambda), \qquad (9\text{--}14)$$

where $\Delta g_n(\vartheta, \lambda)$ is the Laplace surface harmonic of degree n; or, more explicitly,

$$\Delta g(\vartheta, \lambda) = \sum_{n=2}^{\infty} \sum_{m=0}^{n} \left[a_{nm} \mathcal{R}_{nm}(\vartheta, \lambda) + b_{nm} \mathcal{S}_{nm}(\vartheta, \lambda) \right], \qquad (9\text{--}15)$$

where

$$\begin{aligned}
\mathcal{R}_{nm}(\vartheta, \lambda) &= P_{nm}(\cos \vartheta) \cos m\lambda, \\
\mathcal{S}_{nm}(\vartheta, \lambda) &= P_{nm}(\cos \vartheta) \sin m\lambda
\end{aligned} \qquad (9\text{--}16)$$

are the conventional spherical harmonics; or in terms of fully normalized harmonics (see Sect. 1.10):

$$\Delta g(\vartheta, \lambda) = \sum_{n=2}^{\infty} \sum_{m=0}^{n} \left[\bar{a}_{nm} \bar{\mathcal{R}}_{nm}(\vartheta, \lambda) + \bar{b}_{nm} \bar{\mathcal{S}}_{nm}(\vartheta, \lambda) \right]. \qquad (9\text{--}17)$$

Here ϑ is the polar distance (complement of geocentric latitude) and λ is the longitude.

Let us now find the average products of two Laplace harmonics

$$\Delta g_n(\vartheta, \lambda) = \sum_{m=0}^{n} \left[\bar{a}_{nm} \bar{\mathcal{R}}_{nm}(\vartheta, \lambda) + \bar{b}_{nm} \bar{\mathcal{S}}_{nm}(\vartheta, \lambda) \right]. \qquad (9\text{--}18)$$

These average products are

$$M\{\Delta g_n \Delta g_n'\} = \frac{1}{4\pi} \int_{\lambda=0}^{2\pi} \int_{\vartheta=0}^{\pi} \Delta g_n(\vartheta, \lambda) \, \Delta g_n'(\vartheta, \lambda) \sin \vartheta \, d\vartheta \, d\lambda, \qquad (9\text{--}19)$$

since the averaging is extended over the whole earth, that is, over the whole unit sphere. Take first $n' = n$, which gives the average square of the Laplace harmonic of degree n:

$$M\{\Delta g_n^2\} = \frac{1}{4\pi} \int_{\lambda=0}^{2\pi} \int_{\vartheta=0}^{\pi} \left[\Delta g_n(\vartheta, \lambda) \right]^2 \sin \vartheta \, d\vartheta \, d\lambda. \qquad (9\text{--}20)$$

Substituting (9–18) and taking into account the orthogonality relations (1–83) and the normalization (1–91), we easily find

$$M\{\Delta g_n^2\} = \sum_{m=0}^{n} (\bar{a}_{nm}^2 + \bar{b}_{nm}^2) \, . \tag{9–21}$$

Consider now the average product (9–19) of two Laplace harmonics of different degree, $n' \neq n$. Owing to the orthogonality of the spherical harmonics, the integral in (9–19) is zero:

$$M\{\Delta g_n \Delta g_n'\} = 0 \quad \text{if } n' \neq n \, . \tag{9–22}$$

In statistical terms this means that two Laplace harmonics of different degrees are *uncorrelated* or, broadly speaking, *statistically independent*.

In a way similar to that used for the gravity anomalies, we may also expand the covariance function $C(s)$ into a series of spherical harmonics. Let us take an arbitrary, but fixed, point P as the pole of this expansion. Thus spherical polar coordinates ψ (angular distance from P) and α (azimuth) are introduced (Fig. 9.2). The angular distance ψ corresponds to the linear distance s according to (9–8). If we expand the covariance function, with argument ψ, into a series of spherical harmonics with respect to the pole P and coordinates ψ and α, we have

$$C(\psi) = \sum_{n=2}^{\infty} \sum_{m=0}^{n} \left[c_{nm} \mathcal{R}_{nm}(\psi, \alpha) + d_{nm} \mathcal{S}_{nm}(\psi, \alpha) \right] , \tag{9–23}$$

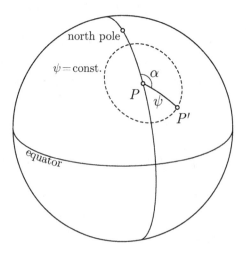

Fig. 9.2. Spherical coordinates ψ, α

which is of the same type as (9–15). But since C depends only on the distance ψ and not on the azimuth α, the spherical harmonics cannot contain any terms that explicitly depend on α. The only harmonics independent of α are the zonal functions

$$\mathcal{R}_{n0}(\psi, \alpha) \equiv P_n(\cos \psi) \,, \tag{9–24}$$

so that we are left with

$$C(\psi) = \sum_{n=2}^{\infty} c_n P_n(\cos \psi) \,. \tag{9–25}$$

The $c_n \equiv c_{n0}$ are the only coefficients that are not equal to zero. We also use the equivalent expression in terms of fully normalized harmonics:

$$C(\psi) = \sum_{n=2}^{\infty} \bar{c}_n \bar{P}_n(\cos \psi) \,. \tag{9–26}$$

The coefficients in these series, according to Sects. 1.9 and 1.10, are given by

$$\begin{aligned} c_n &= \frac{2n+1}{4\pi} \int_{\alpha=0}^{2\pi} \int_{\psi=0}^{\pi} C(\psi) \, P_n(\cos \psi) \sin \psi \, d\psi \, d\alpha \\ &= \frac{2n+1}{2} \int_{\psi=0}^{\pi} C(\psi) \, P_n(\cos \psi) \sin \psi \, d\psi \end{aligned} \tag{9–27}$$

and

$$\bar{c}_n = \frac{c_n}{\sqrt{2n+1}} \,. \tag{9–28}$$

We now determine the relation between the coefficients c_n of $C(\psi)$ in (9–25) and the coefficients $\bar{a}_{nm}$ and $\bar{b}_{nm}$ of Δg in (9–18). For this purpose we need an expression for $C(\psi)$ in terms of Δg, which is easily obtained by writing (9–27) more explicitly. Take the two points $P(\vartheta, \lambda)$ and $P'(\vartheta', \lambda')$ of Fig. 9.2. Their spherical distance ψ is given by

$$\cos \psi = \cos \vartheta \, \cos \vartheta' + \sin \vartheta \, \sin \vartheta' \cos(\lambda' - \lambda) \,. \tag{9–29}$$

Here ψ and the azimuth α are the polar coordinates of $P'(\vartheta', \lambda')$ with respect to the pole $P(\vartheta, \lambda)$.

The symbol M in (9–6) denotes the average over the unit sphere. Two steps are required to find it. First, we average over the spherical circle of radius ψ (denoted in Fig. 9.2 by a broken line), keeping the pole P fixed and letting P' move along the circle so that the distance PP' remains constant. This gives

$$C^* = \frac{1}{2\pi} \int_{\alpha=0}^{2\pi} \Delta g(\vartheta, \lambda) \, \Delta g(\vartheta', \lambda') \, d\alpha \,, \tag{9–30}$$

where C^* still depends on the point P chosen as the pole $\psi = 0$. Second, we average C^* over the unit sphere:

$$
\frac{1}{4\pi} \int_{\lambda=0}^{2\pi} \int_{\vartheta=0}^{\pi} C^* \sin \vartheta \, d\vartheta \, d\lambda
$$
$$
= \frac{1}{8\pi^2} \int_{\lambda=0}^{2\pi} \int_{\vartheta=0}^{\pi} \int_{\alpha=0}^{2\pi} \Delta g(\vartheta, \lambda) \, \Delta g(\vartheta', \lambda') \sin \vartheta \, d\vartheta \, d\lambda \, d\alpha \,. \tag{9-31}
$$

This is equal to the covariance function $C(\psi)$, the symbol M in (9–6) now being written explicitly:

$$
C(\psi) = \frac{1}{8\pi^2} \int_{\lambda=0}^{2\pi} \int_{\vartheta=0}^{\pi} \int_{\alpha=0}^{2\pi} \Delta g(\vartheta, \lambda) \, \Delta g(\vartheta', \lambda') \sin \vartheta \, d\vartheta \, d\lambda \, d\alpha \,. \tag{9-32}
$$

The coordinates ϑ', λ' in this formula are understood to be related to ϑ, λ by (9–29) with $\psi = $ constant, but to be arbitrary otherwise; this expresses the fact that in (9–6) the average is extended over all pairs of points P and P' for which $PP' = \psi = $ constant.

To compute the coefficients c_n, substitute (9–32) into (9–27), obtaining

$$
c_n = \frac{2n+1}{2} \int_{\psi=0}^{\pi} C(\psi) \, P_n(\cos \psi) \sin \psi \, d\psi
$$
$$
= \frac{1}{4\pi} \frac{2n+1}{4\pi} \int_{\lambda=0}^{2\pi} \int_{\vartheta=0}^{\pi} \int_{\alpha=0}^{2\pi} \int_{\psi=0}^{\pi} \Delta g(\vartheta, \lambda) \, \Delta g(\vartheta', \lambda') \cdot \tag{9-33}
$$
$$
\cdot P_n(\cos \psi) \sin \psi \, d\psi \, d\alpha \cdot \sin \vartheta \, d\vartheta \, d\lambda \,.
$$

Consider first the integration with respect to α and ψ. According to (1–89), we have

$$
\frac{2n+1}{4\pi} \int_{\alpha=0}^{2\pi} \int_{\psi=0}^{\pi} \Delta g(\vartheta', \lambda') \, P_n(\cos \psi) \sin \psi \, d\psi \, d\alpha
$$
$$
= \frac{2n+1}{4\pi} \int_{\lambda'=0}^{2\pi} \int_{\vartheta'=0}^{\pi} \Delta g(\vartheta', \lambda') \, P_n(\cos \psi) \sin \vartheta' \, d\vartheta' \, d\lambda' = \Delta g_n(\vartheta, \lambda) \,, \tag{9-34}
$$

the change of integration variables being evident. Hence (9–33) becomes

$$
c_n = \frac{1}{4\pi} \int_{\lambda=0}^{2\pi} \int_{\vartheta=0}^{\pi} \Delta g(\vartheta, \lambda) \, \Delta g_n(\vartheta, \lambda) \sin \vartheta \, d\vartheta \, d\lambda \,. \tag{9-35}
$$

This may also be written

$$
c_n = M\{\Delta g \, \Delta g_n\} \,. \tag{9-36}
$$

Into this we now insert (9–14), which we write

$$\Delta g(\vartheta, \lambda) = \sum_{n'=2}^{\infty} \Delta g_{n'}(\vartheta, \lambda), \qquad (9\text{--}37)$$

denoting the summation index by n' instead of n. We get

$$c_n = M\left\{\sum_{n'=2}^{\infty} \Delta g_{n'} \Delta g_n\right\} = \sum_{n'=2}^{\infty} M\{\Delta g_n \, \Delta g_{n'}\}. \qquad (9\text{--}38)$$

According to (9–22), only the term with $n' = n$ is different from zero so that from (9–21) we finally obtain

$$c_n = M\{\Delta g_n^2\} = \sum_{m=0}^{n} (\bar{a}_{nm}^2 + \bar{b}_{nm}^2). \qquad (9\text{--}39)$$

Hence, c_n is the average square of the Laplace harmonic $\Delta g_n(\vartheta, \lambda)$ of degree n, or its variance. For these reasons the c_n are also called *degree variances*. The "degree covariances" are zero because of (9–22).

Equation (9–39) relates the coefficients $\bar{a}_{nm}$ and $\bar{b}_{nm}$ of Δg and c_n of $C(s)$ in the simplest possible way. Note that $\bar{a}_{nm}$ and $\bar{b}_{nm}$ are coefficients of fully normalized harmonics, whereas c_n are coefficients of conventional harmonics. As a matter of fact, we may also use the a_{nm} and b_{nm} (conventional) or the $\bar{c}_n$ (fully normalized); but then (9–39) will obviously become slightly more complicated. It should be mentioned that the mathematics behind the statistical description of the gravity anomalies is the theory of *stochastic processes*. The gravity anomaly field is treated as a stationary stochastic process on a sphere; the spherical-harmonic expansions of this section are nothing but the spectral analysis of that process. A comprehensive treatment of this topic is found in Moritz (1980 a).

9.4   Interpolation and extrapolation of gravity anomalies

As pointed out in Sect. 9.1, the purpose of prediction (interpolation and extrapolation) is to supplement the gravity observations, which can be made at only relatively few points, by estimating the values of gravity or of gravity anomalies at all the other points P of the earth's surface.

If P is surrounded by gravity stations, we must interpolate; if the gravity stations are far away from P, we extrapolate. Evidently, there is no sharp distinction between these two kinds of prediction and the mathematical formulation is the same in both cases.

In order to predict a gravity anomaly at P, we must have information about the gravity anomaly function. The values observed at certain points are the most important information. In addition, we need some information on the form of the anomaly function. If the gravity measurements are very dense, then the continuity or "smoothness" of the function is sufficient – for instance, for linear interpolation. Otherwise we may try to use statistical information on the general structure of the gravity anomalies. Here we must consider two kinds of statistical correlation: the *autocorrelation* – the correlation between each other – of gravity anomalies and the *correlation* of the gravity anomalies *with height*.

Correlation with height will for the moment be disregarded; Sect. 9.7 will be devoted to this topic. The autocorrelation is characterized by the covariance function considered in Sect. 9.2.

Mathematically, the purpose of prediction is to find a function of the observed gravity anomalies $\Delta g_1, \Delta g_2, \ldots, \Delta g_n$ in such a way that the unknown anomaly Δg_P at P is approximated by the function

$$\Delta g_P \doteq F(\Delta g_1, \Delta g_2, \ldots, \Delta g_n). \qquad (9\text{--}40)$$

Here Δg_i denotes the value of Δg at a point i, not a spherical harmonic! In practice, only linear functions of the Δg_i are used. If we denote the predicted value of Δg_P by $\widetilde{\Delta g}_P$, such a linear prediction has the form

$$\widetilde{\Delta g}_P = \alpha_{P1}\,\Delta g_1 + \alpha_{P2}\,\Delta g_2 + \ldots + \alpha_{Pn}\,\Delta g_n \equiv \sum_{i=1}^{n}\alpha_{Pi}\,\Delta g_i. \qquad (9\text{--}41)$$

The coefficients α_{Pi} depend only on the relative position of P and the gravity stations $1, 2, \ldots, n$; they are independent of the Δg_i. Depending on the way we choose these coefficients, we obtain different interpolation or extrapolation methods. Here are some examples.

Geometric interpolation

The "gravity anomaly surface", as represented by a gravity anomaly map, may be approximated by a polyhedron by dividing the area into triangles whose corners are formed by the gravity stations and passing a plane through the three corners of each triangle (Fig. 9.3). This is approximately what is done in constructing the contour lines of a gravity anomaly map by means of graphical interpolation.

Analytically, this interpolation may be formulated as follows. Let point P be situated inside a triangle with corners 1, 2, 3 (Fig. 9.3). To each point

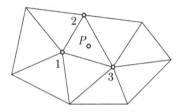

Fig. 9.3. Geometric interpolation

we assign its value Δg as its z-coordinate, so that the points 1, 2, and 3 have "spatial" coordinates (x_1, y_1, z_1), (x_2, y_2, z_2), and (x_3, y_3, z_3); x and y are ordinary plane coordinates. The plane through 1, 2, 3 has the equation

$$
\begin{aligned}
z = {}& \frac{(x_2 - x)(y_3 - y_2) - (y_2 - y)(x_3 - x_2)}{(x_2 - x_1)(y_3 - y_2) - (y_2 - y_1)(x_3 - x_2)} \, z_1 \\
& + \frac{(x_3 - x)(y_1 - y_3) - (y_3 - y)(x_1 - x_3)}{(x_3 - x_2)(y_1 - y_3) - (y_3 - y_2)(x_1 - x_3)} \, z_2 \\
& + \frac{(x_1 - x)(y_2 - y_1) - (y_1 - y)(x_2 - x_1)}{(x_1 - x_3)(y_2 - y_1) - (y_1 - y_3)(x_2 - x_1)} \, z_3 \, .
\end{aligned}
\tag{9–42}
$$

If we replace z_1, z_2, z_3 by Δg_1, Δg_2, Δg_3, then z is the interpolated value $\widetilde{\Delta g}_P$ at point P, which has the plane coordinates x, y. Thus,

$$
\widetilde{\Delta g}_P = \alpha_{P1} \, \Delta g_1 + \alpha_{P2} \, \Delta g_2 + \alpha_{P3} \, \Delta g_3 \, ,
\tag{9–43}
$$

where the α_{Pi} are the coefficients of z_i in the preceding equation.

Representation

Often the measured anomaly of a gravity station 1 is made to represent the whole neighborhood so that

$$
\widetilde{\Delta g}_P \equiv \Delta g_1
\tag{9–44}
$$

as long as P lies within a certain neighborhood of point 1. Then

$$
\alpha_{P1} = 1, \qquad \alpha_{P2} = \alpha_{P3} = \ldots = \alpha_{Pn} = 0 \, .
\tag{9–45}
$$

This method is rather crude but simple and accurate enough for many purposes.

Zero anomaly

If there are no gravity measurements in a large area – for instance, on the
oceans –, then the estimate

$$\widetilde{\Delta g}_P \equiv 0 \qquad (9\text{--}46)$$

is used in this area. In this trivial case all α_{Pi} are zero.

If all known gravity stations are far away, and if we know of nothing
better, then this primitive extrapolation method is applied, although the
accuracy is poor. At best, this method may work with isostatic anomalies.

None of these three methods gives optimum accuracy. In the next section we
investigate the accuracy of the general prediction formula (9–41) and find
those coefficients α_{Pi} that yield the most accurate results.

9.5   Accuracy of prediction methods

In order to compare the various possible methods of prediction, to determine
their range of applicability, and to find the most accurate method, we must
evaluate their accuracy.

Consider the general case of Eq. (9–41). The correct gravity anomaly at
P is Δg_P, the predicted value is

$$\widetilde{\Delta g}_P = \sum_{i=1}^{n} \alpha_{Pi}\, \Delta g_i\,. \qquad (9\text{--}47)$$

The difference is the error ε_P of prediction,

$$\varepsilon_P = \Delta g_P - \widetilde{\Delta g}_P = \Delta g_P - \sum_i \alpha_{Pi}\, \Delta g_i\,. \qquad (9\text{--}48)$$

By squaring we find

$$
\begin{aligned}
\varepsilon_P^2 &= \left(\Delta g_P - \sum_i \alpha_{Pi}\, \Delta g_i\right)\left(\Delta g_P - \sum_k \alpha_{Pk}\, \Delta g_k\right) \\
&= \Delta g_P^2 - 2\sum_i \alpha_{Pi}\, \Delta g_P\, \Delta g_i + \sum_i \sum_k \alpha_{Pi}\, \alpha_{Pk}\, \Delta g_i\, \Delta g_k\,.
\end{aligned}
\qquad (9\text{--}49)
$$

Let us now form the average M of this formula over the area considered
(either a limited region or the whole earth). Then we have from (9–6),

$$
\begin{aligned}
M\{\Delta g_i\, \Delta g_k\} &= C(i\,k) \equiv C_{ik}\,, \\
M\{\Delta g_P\, \Delta g_i\} &= C(P\,i) \equiv C_{Pi}\,, \\
M\{\Delta g_P^2\} \quad &= C(0) \equiv C_0\,.
\end{aligned}
\qquad (9\text{--}50)
$$

These are particular values of the covariance function $C(s)$, for $s = i\,k$, $s = Pi$, and $s = 0$; for instance, $i\,k$ is the distance between the gravity stations i and k. The abbreviated notations C_{ik} and C_{Pi} are self-explanatory.

We further set

$$M\{\varepsilon_P^2\} = m_P^2. \tag{9-51}$$

Thus m_P is the root mean square error of a predicted gravity anomaly at P, or briefly, the standard *error of prediction* (interpolation or extrapolation).

Taking all these relations into account, we find the average M of (9–49) to be

$$m_P^2 = C_0 - 2 \sum_{i=1}^{n} \alpha_{Pi}\, C_{Pi} + \sum_{i=1}^{n} \sum_{k=1}^{n} \alpha_{Pi}\, \alpha_{Pk}\, C_{ik}. \tag{9-52}$$

This is the fundamental formula for the standard error of the general prediction formula (9–41). For the special cases described in the preceding section, the particular values of α_{Pi} are to be inserted.

Einstein's summation convention

At least at this point the reader will be grateful to Albert Einstein for having invented not only the theory of relativity – well, even the general theory of relativity has been used in geodesy (Moritz and Hofmann-Wellenhof 1993), but the reader of the present book will be saved from it – but also the very practical summation convention which has eradicated myriads of unnecessary summation signs from the mathematical literature. This convention simply says that, if an index occurs twice in a product, summation is automatically implied. Using this convention, the preceding equation is simply written

$$m_P^2 = C_0 - 2\,\alpha_{Pi}\, C_{Pi} + \alpha_{Pi}\, \alpha_{Pk}\, C_{ik}. \tag{9-53}$$

In the future we shall take this equation for granted unless stated otherwise. Such formulas are also handsome for programming (a loop).

Now back to reality in the form of examples.

As an example consider the case of representation, Eq. (9–44); all α are zero except one. Here (9–53) yields

$$m_P^2 = C_0 - 2C_{P1} + C_0 = 2C_0 - 2C_{P1}. \tag{9-54}$$

For the case of zero anomaly, there is $m_p^2 = C_0$, as should be expected.

Often we need not only the standard error m_P of prediction but also the correlation of the prediction errors ε_P and ε_Q at two different points P and Q, expressed by the *"error covariance"* σ_{PQ}, which is defined by

$$\sigma_{PQ} = M\{\varepsilon_P\, \varepsilon_Q\}. \tag{9-55}$$

If the errors ε_P and ε_Q are uncorrelated, then the error covariance $\sigma_{PQ} = 0$. From (9–48) we have generally

$$\sigma_{PQ} = M\{(\Delta g_P - \alpha_{Pi}\,\Delta g_i)(\Delta g_Q - \alpha_{Qk}\,\Delta g_k)\}$$

$$= M\{\Delta g_P\,\Delta g_Q - \alpha_{Pi}\,\Delta g_Q\,\Delta g_i - \alpha_{Qk}\,\Delta g_P\,\Delta g_k + \alpha_{Pi}\,\alpha_{Pk}\,\Delta g_i\,\Delta g_k\}$$

$$(9\text{--}56)$$

and finally

$$\sigma_{PQ} = C_{PQ} - \alpha_{Pi}\,C_{Qi} - \alpha_{Qi}\,C_{Pi} + \alpha_{Pi}\,\alpha_{Qk}\,C_{ik}\,. \qquad (9\text{--}57)$$

The notations are self-explanatory; for instance, $C_{PQ} = C(PQ)$.

The error covariance function

The values of the error covariance σ_{PQ}, for different positions of the points P and Q, form a continuous function of the coordinates of P and Q. This function is called the *error covariance function*, or briefly, the *error function*, and is denoted by $\sigma(x_P, y_P, x_Q, y_Q)$. If P and Q are different, then we simply have

$$\sigma(x_P, y_P, x_Q, y_Q) = \sigma_{PQ}\,; \qquad (9\text{--}58)$$

if P and Q coincide, then (9–57) reduces to (9–53) so that

$$\sigma(x_P, y_P, x_P, y_P) = m_P^2 \qquad (9\text{--}59)$$

is the square of the standard prediction error at P.

Thus the error covariances σ_{PQ} may be considered as special values of the error covariance function, just as the covariances C_{PQ} of the gravity anomalies may be considered as special values of the covariance function $C(s)$. To repeat, the error function is the covariance function of the prediction errors, defined as

$$M\{\varepsilon_P\,\varepsilon_Q\}\,, \qquad (9\text{--}60)$$

whereas $C(s)$ is the covariance function of the gravity anomalies, defined as

$$M\{\Delta g_P\,\Delta g_Q\}\,. \qquad (9\text{--}61)$$

The term "covariance function" in the narrower sense will be reserved for $C(s)$ – in contrast to least-squares adjustment, where "covariances" automatically mean error covariances. Covariances are "isotropic", which means independent of directions; the error covariances are nonisotropic.

From (9–53) and (9–57) *the error function can be expressed in terms of the covariance function*; we may write more explicitly

$$\sigma(x_P, y_P, x_Q, y_Q) = C(PQ) - \alpha_{Pi}\,C(Qi) - \alpha_{Qi}\,C(Pi) + \alpha_{Pi}\,\alpha_{Qk}\,C(ik)\,.$$

$$(9\text{--}62)$$

Thus we recognize the basic role of the covariance function in accuracy studies. The error function, on the other hand, is fundamental for problems of error propagation.

9.6   Least-squares prediction

The values of α_{Pi} for the most accurate prediction method are obtained by minimizing the standard prediction error expressed by (9–53) as a function of the α. The familiar necessary conditions for a minimum are

$$\frac{\partial m_P^2}{\partial \alpha_{Pi}} \equiv -2C_{Pi} + 2\alpha_{Pk}\,C_{ik} = 0 \quad (i = 1, 2, \ldots, n) \qquad (9\text{–}63)$$

or

$$C_{ik}\,\alpha_{Pk} = C_{Pi}. \qquad (9\text{–}64)$$

This is a system of n linear equations in the n unknowns α_{Pk}; the solution is

$$\alpha_{Pk} = C_{ik}^{(-1)}\,C_{Pi}, \qquad (9\text{–}65)$$

where $C_{ik}^{(-1)}$ denote the elements of the inverse of the symmetric matrix $[C_{ik}]$.

Substituting (9–65) into (9–41) gives

$$\widetilde{\Delta g}_P = \alpha_{Pk}\,\Delta g_k = C_{ik}^{(-1)}\,C_{Pi}\,\Delta g_k. \qquad (9\text{–}66)$$

In matrix notation this is written

$$\widetilde{\Delta g}_P = \begin{bmatrix} C_{P1}, & C_{P2}, & \ldots, & C_{Pn} \end{bmatrix} \begin{bmatrix} C_{11} & C_{12} & \ldots & C_{1n} \\ C_{21} & C_{22} & \ldots & C_{2n} \\ \vdots & \vdots & & \vdots \\ C_{n1} & C_{n2} & \ldots & C_{nn} \end{bmatrix}^{-1} \begin{bmatrix} \Delta g_1 \\ \Delta g_2 \\ \vdots \\ \Delta g_n \end{bmatrix}. \qquad (9\text{–}67)$$

We see that *for optimal prediction we must know the statistical behavior of the gravity anomalies* through the covariance function $C(s)$.

There is a close connection between this optimal prediction method and the method of least-squares adjustment. Although they refer to somewhat different problems, both are designed to give most accurate results. The linear equations (9–64) correspond to the "normal equations" of adjustment computations. Prediction by means of formula (9–67) is therefore called *"least-squares prediction"*. A generalization to heterogeneous data is "least-squares collocation" to be treated in Chap. 10. In its most general form, least-squares collocation also includes parameter estimation by least-squares adjustment. This is an advanced subject treated in great detail in Moritz (1980 a).

It is easy to determine the accuracy of least-squares prediction. Insert the α of Eq. (9–65) into (9–53), after appropriate changes in the indices of summation. This gives

$$m_P^2 = C_0 - 2\alpha_{Pk}\,C_{Pk} + \alpha_{Pk}\,\alpha_{Pl}\,C_{kl}$$
$$= C_0 - 2C_{ik}^{(-1)}C_{Pi}C_{Pk} + C_{ik}^{(-1)}C_{Pi}C_{jl}^{(-1)}C_{Pj}C_{kl}\,. \tag{9–68}$$

For the reader to appreciate the Einstein summation convention, we give this equation in its original form:

$$m_P^2 = C_0 - 2\sum_k \alpha_{Pk}\,C_{Pk} + \sum_k\sum_l \alpha_{Pk}\,\alpha_{Pl}\,C_{kl}$$
$$= C_0 - 2\sum_i\sum_k C_{ik}^{(-1)}C_{Pi}C_{Pk} + \sum_i\sum_j\sum_k\sum_l C_{ik}^{(-1)}C_{Pi}C_{jl}^{(-1)}C_{Pj}C_{kl}\,. \tag{9–69}$$

But now back to normal! We have

$$C_{jl}^{(-1)}\,C_{kl} = \delta_{jk} = \begin{cases} 1 & \text{if } j = k \\ 0 & \text{if } j \neq k\,. \end{cases} \tag{9–70}$$

The matrix $[\delta_{kl}]$ is the unit matrix. This formula states that the product of a matrix and its inverse is the unit matrix. Thus, we further have

$$C_{ik}^{(-1)}\,C_{jl}^{(-1)}\,C_{kl} = C_{ik}^{(-1)}\,\delta_{jk} = C_{ij}^{(-1)} \tag{9–71}$$

because a matrix remains unchanged on multiplication by the unit matrix. Hence, we get

$$m_P^2 = C_0 - 2C_{ik}^{(-1)}\,C_{Pi}\,C_{Pk} + C_{ij}^{(-1)}\,C_{Pi}\,C_{Pj}$$
$$= C_0 - 2C_{ik}^{(-1)}\,C_{Pi}\,C_{Pk} + C_{ik}^{(-1)}\,C_{Pi}\,C_{Pk} \tag{9–72}$$
$$= C_0 - C_{ik}^{(-1)}\,C_{Pi}\,C_{Pk}\,.$$

Thus, the standard error of least-squares prediction is given by

$$m_P^2 = C_0 - C_{ik}^{(-1)}\,C_{Pi}\,C_{Pk}$$
$$= C_0 - [C_{P1}, C_{P2}, \ldots, C_{Pn}] \begin{bmatrix} C_{11} & C_{12} & \cdots & C_{1n} \\ C_{21} & C_{22} & \cdots & C_{2n} \\ \vdots & \vdots & & \vdots \\ C_{n1} & C_{n2} & \cdots & C_{nn} \end{bmatrix}^{-1} \begin{bmatrix} C_{P1} \\ C_{P2} \\ \vdots \\ C_{Pn} \end{bmatrix}. \tag{9–73}$$

In the same way we find the error covariance in the points P and Q:

$$\sigma_{PQ} = C_{PQ} - C_{ik}^{(-1)} C_{Pi} C_{Qk}$$

$$= C_{PQ} - \begin{bmatrix} C_{P1}, C_{P2}, \ldots, C_{Pn} \end{bmatrix} \begin{bmatrix} C_{11} & C_{12} & \cdots & C_{1n} \\ C_{21} & C_{22} & \cdots & C_{2n} \\ \vdots & \vdots & & \vdots \\ C_{n1} & C_{n2} & \cdots & C_{nn} \end{bmatrix}^{-1} \begin{bmatrix} C_{Q1} \\ C_{Q2} \\ \vdots \\ C_{Qn} \end{bmatrix}.$$

$$(9\text{--}74)$$

These two formulas give the error covariance function for least-squares prediction. Both formulas have a form similar to that of (9–67) and are equally well suited for computations so that $\widetilde{\Delta g}$ and its accuracy can be calculated at the same time.

It is clear that, after appropriate slight changes, this theory applies automatically to *gravity disturbances* δg.

Practical considerations

Geometric interpolation (Sect. 9.4) is suited for the interpolation of point anomalies in a dense gravity net, with station distances of 10 km or less. If mean anomalies for blocks of $5' \times 5'$ or larger are needed rather than point anomalies, then some kind of representation, such as that considered in the previous section, may be simpler and hardly less accurate.

Least-squares prediction is, by its very definition, more accurate than either geometric interpolation or representation, but the improvement in accuracy is not striking. The main advantage of least-squares prediction is that it permits a systematic, purely numerical, digital processing of gravity data; gravity anomalies are stored in data bases, and gravity anomaly maps, if necessary, are generated automatically. The same formula applies to both interpolation and extrapolation so that gaps in the gravity data make no difference in the method of computation, which becomes completely schematic (Moritz 1963). For practical and computational details see Rapp (1964) and many other papers published since.

For larger station distances, of 50 km or more, prediction of individual point values becomes meaningless. In this case we must work with mean anomalies of, say, $1° \times 1°$ blocks.

9.7   Correlation with height

So far we have taken into account only the mutual correlation of the gravity anomalies, their autocorrelation, disregarding the correlation with height,

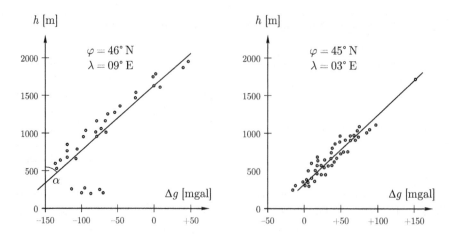

Fig. 9.4. Correlation of the free-air anomalies with height

which is important in many cases. Therefore our formulas were valid only
for gravity anomalies uncorrelated with height, such as isostatic or, to a
certain extent, Bouguer anomalies; or for free-air anomalies in moderately
flat areas. Free-air anomalies in mountains must be treated differently.

Figure 9.4 due to U.A. Uotila shows the correlation of free-air anomalies
with height. The gravity anomalies Δg are plotted against the height h. If
there were an exact functional dependence between Δg and h, then all points
would lie on a straight line (or, more generally, on a curve). In reality, there
is only an approximate functional relation, a general trend or tendency of
the free-air anomalies to increase linearly with height; exceptions, even large
ones, are possible. This shows very well the meaning of correlation.

We have characterized the mutual correlation of the gravity anomalies
by the "autocovariance function" (9–6),

$$C(s) = M\{\Delta g \, \Delta g'\},\tag{9–75}$$

where $s = PP'$. Similarly, we may form the "cross-covariance function"

$$B(s) = M\{\Delta g \, \Delta h'\} = M\{\Delta g' \, \Delta h\},\tag{9–76}$$

expressing the correlation between gravity and height, and

$$A(s) = M\{\Delta h \, \Delta h'\},\tag{9–77}$$

which is the autocovariance function of the height differences

$$\Delta h = h - M\{h\},\tag{9–78}$$

where the symbol $M\{h\}$ denotes the mean height of the whole area considered.

If Δg and Δh are not correlated, then the function $B(s)$ is identically zero. If this is not the case, then we should also take the height into account in our interpolation.

It is easy to extend the prediction formula (9–41) for this purpose, but this has turned out to be of little practical importance.

Application to Bouguer anomalies

Of great practical importance, however, is the question whether it is possible to render the free-air anomalies independent of height by adding a term that is proportional to the height. In other words, when is the quantity

$$z = \Delta g - b\,\Delta h\,, \tag{9–79}$$

with a certain coefficient b, uncorrelated with height? In statistical terminology, correlation with height is a *trend*, which may be capable of being removed.

The trend z has the form of a Bouguer anomaly; for a real Bouguer anomaly we have, according to Sect. 3.4,

$$b = 2\pi\,G\,\varrho\,. \tag{9–80}$$

For the density $\varrho = 2.67$ g/cm^3 we get

$$b = +0.112 \text{ mgal/m}\,. \tag{9–81}$$

Let us form the covariance function $Z(s)$ between the "Bouguer anomaly" z of (9–79) and height difference Δh

$$Z(s) \equiv M\{z\,\Delta h'\} = M\{\Delta g\,\Delta h' - b\,\Delta h\,\Delta h'\} = B(s) - b\,A(s)\,. \tag{9–82}$$

If z is to be uncorrelated with h, then $Z(s)$ must be identically zero. The condition is

$$B(s) - b\,A(s) \equiv 0\,, \tag{9–83}$$

which must be satisfied for all s and a certain constant b at least approximately.

We see that the "Bouguer anomaly" z is uncorrelated with height if the functions $A(s)$ and $B(s)$ are proportional for the area considered; the constant b is then represented by

$$b = \frac{B(s)}{A(s)}\,. \tag{9–84}$$

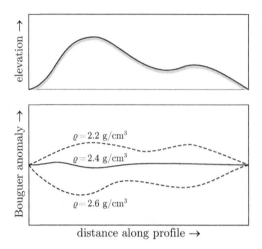

Fig. 9.5. Bouguer anomalies corresponding to different densities ϱ: the best density is $\varrho = 2.4$ g/cm^3 (no correlation); for other densities the Bouguer anomalies are correlated with height (positive correlation for $\varrho = 2.2$ g/cm^3, negative correlation for $\varrho = 2.6$ g/cm^3)

It may be shown that this is equivalent to the condition that the points of Fig. 9.4 lie approximately on a straight line. The coefficient b is then given by

$$b = \tan \alpha \qquad (9\text{--}85)$$

as the inclination of the line towards the h-axis.

In practice these conditions are very often fulfilled to a good approximation. Furthermore, by computing b from Eq. (9–84) or determining it graphically by means of (9–85), we often get a value that is close to the normal Bouguer gradient (9–81).

If we assume that b depends only on the rock density ϱ, then we obtain a means for determining the average density, which is often difficult to measure directly. This is the "*Nettleton method*", used in geophysical prospecting: the coefficient b is found statistically by means of Eqs. (9–84) or (9–85), and the rock density ϱ is then computed from (9–80). Figure 9.5 illustrates the principle of this method; see also Jung (1956: p. 600).

If the condition (9–83) is fulfilled, then we may consider the "Bouguer anomaly" z as a gravity anomaly that is completely uncorrelated with height; we can directly apply to it the whole theory of the preceding sections. But even when this condition is not quite satisfied, Bouguer anomalies will in general be far less correlated with height than free-air anomalies. The fact that in (9–79) gravity is reduced to a mean height and not to sea level, is quite irrelevant in this connection because this is only a question of an additive constant. More recent developments are discussed by Moritz (1990: p. 244).

It is thus possible to consider the Bouguer reduction as a means of obtaining gravity anomalies that are less dependent on height and hence more representative than free-air anomalies. More precisely, the Bouguer anomalies take care of the dependence on the local irregularities of height. The isostatic anomalies are, in addition, also largely independent of the regional features of topography. See also Chaps. 3 and 8.

10 Least-squares collocation

10.1 Principles of least-squares collocation

The principle of collocation is very simple. The anomalous potential T outside the earth is a harmonic function, that is, it satisfies Laplace's differential equation

$$\Delta T = \frac{\partial^2 T}{\partial x^2} + \frac{\partial^2 T}{\partial y^2} + \frac{\partial^2 T}{\partial z^2} = 0 \,. \qquad (10\text{--}1)$$

An approximate analytical representation of the external potential T is obtained by

$$T(P) \doteq f(P) = \sum_{k=1}^{q} b_k \, \varphi_k(P) \,, \qquad (10\text{--}2)$$

a linear combination f of suitable base functions $\varphi_1, \varphi_2, \ldots, \varphi_q$ with appropriate coefficients b_k. All these are functions of the space point P under consideration.

As T is harmonic outside the earth's surface, it is natural to choose base functions φ_k which are likewise harmonic, so that

$$\Delta \varphi_k = 0 \,, \qquad (10\text{--}3)$$

in correspondence to (10–1).

There are many simple systems of functions satisfying the harmonicity condition (10–3), and thus we have many possibilities for a suitable choice of base functions φ_k. We might, for instance, choose spherical harmonics or potentials of suitably distributed point masses, depending on whether we emphasize global or local applications.

The coefficients b_k may be chosen such that the given *observational values are reproduced exactly* – for instance, all deflections of the vertical in a given area. This means that the assumed approximating function f in (10–2) gives the same deflections of the vertical at the observation stations as the actual potential and hence may well be considered a suitable approximation for T. Let us now try to put these ideas into a mathematical form.

Interpolation

Let errorless values of T be given at q spatial points $P_1, P_2, \ldots, P_q$; these points may lie on the earth's surface or in space above the earth's surface. We put

$$T(P_i) = f_i \,, \qquad i = 1, 2, \ldots, q \qquad (10\text{--}4)$$

and postulate that in approximating $T(P)$ by $f(P)$, the observations (10–4) will be reproduced exactly. The condition for this is

$$\sum_{k=1}^{q} b_k \, \varphi_k(P_i) = T(P_i) = f_i \,, \tag{10–5}$$

which may be written as a system of linear equations

$$\sum_{k=1}^{q} A_{ik} \, b_k = f_i \quad \text{with} \quad A_{ik} = \varphi_k(P_i) \tag{10–6}$$

or in matrix notation

$$\mathbf{A} \, \mathbf{b} = \mathbf{f} \,. \tag{10–7}$$

If the square matrix $\mathbf{A}$ is regular, then the coefficients b_k are uniquely determined by

$$\mathbf{b} = \mathbf{A}^{-1} \mathbf{f} \,. \tag{10–8}$$

This model is suitable, for instance, for a determination of the geoid by satellite altimetry, since this method, rather directly, yields geoidal heights N_i and hence, by Bruns' theorem (2–236), $T(P_i) = \gamma_i \, N_i$. For the astrogeodetic geoid determination, we must generalize this model, which leads us to collocation.

Collocation

Here we wish to reproduce, by means of the approximation (10–2), q measured values which again are assumed to be errorless (this assumption is not essential and will be dropped later). These measured values are assumed to be linear functionals $L_1T, L_2T, \ldots, L_qT$ of the anomalous potential T. "Linear functional" means nothing else than a quantity LT that depends linearly on T but need not be an ordinary function but may, say, also contain a differentiation or an integral; essentially, it is the same as a "linear operator".

In fact, deflections of the vertical,

$$\xi = -\frac{1}{\gamma} \frac{\partial T}{\partial x} \,, \quad \eta = -\frac{1}{\gamma} \frac{\partial T}{\partial y} \,, \tag{10–9}$$

but also gravity anomalies,

$$\Delta g = -\frac{\partial T}{\partial z} - \frac{2}{R} \, T \,, \tag{10–10}$$

and gravity disturbances

$$\delta g = -\frac{\partial T}{\partial z} \tag{10–11}$$

are such linear functionals of T; here, x, y, z denotes a local coordinate system in which the z-axis is vertical upwards and the x- and y-axes are directed towards north and east, and $R = 6371$ km is a mean radius of the earth. Equation (10–9) is a consequence of equations such as (2–377), with $\partial s = \partial x$ or ∂y; normal gravity γ may be considered constant with respect to horizontal derivation. Equation (10–10) is the well-known fundamental equation of physical geodesy in spherical approximation (2–263). Equations (10–9) and (10–10) refer to the earth's surface.

To repeat, by saying that deflections of the vertical and gravity disturbances and anomalies are linear functionals of T, we simply indicate the fact that $\xi, \eta, \delta g, \Delta g$ depend on T by the expressions (10–9) and (10–10), which clearly are linear; they are the linear terms of a Taylor expansion, neglecting quadratic and higher terms. In the above notation $L_i T$, the symbol L_i denotes, for instance, the operation

$$L_i = \frac{1}{\gamma} \frac{\partial}{\partial x} \tag{10–12}$$

applied to T at some point.

Putting

$$L_i f = L_i T = \ell_i \tag{10–13}$$

and substituting (10–2), we get

$$\sum_{i=1}^{q} B_{ik}\, b_k = \ell_i \quad \text{with} \quad B_{ik} = L_i \varphi_k, \tag{10–14}$$

where $L_i \varphi_k$ denotes the number obtained by applying the operation L_i to the base function φ_k; the coefficient B_{ik} obtained in this way does not depend on the measured values. Equation (10–14) is a linear system of q equations for q unknowns, which is quite similar to (10–6). This method of fitting an analytical approximating function to a number of given linear functionals is called collocation and is frequently used in numerical mathematics.

It is clear that interpolation is a simple special case of collocation in which

$$L_i f = f(P_i) \tag{10–15}$$

is the "evaluation functional", giving the value of f at a point P_i. Thus we see that in both interpolation and collocation the coefficients b_k require the solution of a linear system of equations (which in general will not be symmetric).

Least-squares interpolation

Let us consider a function

$$K = K(P, Q), \tag{10–16}$$

in which two points P and Q are the independent variables. Let this function K be

- symmetric with respect to P and Q,
- harmonic with respect to both points, everywhere outside a certain sphere, and
- positive-definite (the positive definitiveness of a function is defined similarly as in the case of a matrix).

Then the function $K(P, Q)$ is called a (harmonic) kernel function (Moritz 1980 a: p. 205). A kernel function $K(P, Q)$ may serve as "building material" from which we can construct base functions. Taking for the base functions the form

$$\varphi_k(P) = K(P, P_k), \tag{10–17}$$

where P denotes the variable point and P_k is a fixed point in space, we obtain *least-squares interpolation* already treated by a quite different approach in Chap. 9.

This name originates from the statistical interpretation of the kernel function as a *covariance function* (Sect. 9.2); then least-squares interpolation has some minimum properties (least-error variance, similarly as in least-squares adjustment). This interpretation is not essential, however; one may also work with arbitrary analytical kernel functions, considering the procedure as a purely analytical mathematical approximation technique. Normally one tries to combine both aspects in a reasonable way.

Substituting (10–17) into (10–6), we get

$$A_{ik} = K(P_i, P_k) = C_{ik} ; \tag{10–18}$$

this square matrix now is symmetric (in the general case, A_{ik} is not symmetric!) and positive definite because of the corresponding properties of the function $K(P, Q)$. Then the coefficients b_k follow from (10–8) and may be substituted into (10–2). With the notation

$$\varphi_k(P) = K(P, P_k) = C_{Pk} , \tag{10–19}$$

the result may be written in the form

$$f(P) = \begin{bmatrix} C_{P1} & C_{P2} & \cdots & C_{Pq} \end{bmatrix} \begin{bmatrix} C_{11} & C_{12} & \cdots & C_{1q} \\ C_{21} & C_{22} & \cdots & C_{2q} \\ \vdots & \vdots & & \vdots \\ C_{q1} & C_{q2} & \cdots & C_{qq} \end{bmatrix}^{-1} \begin{bmatrix} f_1 \\ f_2 \\ \vdots \\ f_q \end{bmatrix}, \tag{10–20}$$

formally identical with Eq. (9–67) obtained in a completely different way.

Least-squares collocation

Here we again derive the base functions from a kernel function $K(P, Q)$, but in a way slightly different from (10–17): we put

$$\varphi_k(P) = L_k^Q K(P, Q), \tag{10-21}$$

where L_k^Q means that the functional L_k is applied to the variable Q; the result no longer depends on Q (since the application of a functional results in a definite number). Thus, in (10–14) we must put

$$B_{ik} = L_i^P L_k^Q K(P, Q) = C_{ik}, \tag{10-22}$$

which gives a matrix which again is symmetric. Solving (10–14) for b_k and substituting into (10–2) gives with

$$\varphi_k(P) = L_k^Q K(P, Q) = C_{Pk} \tag{10-23}$$

the formula

$$f(P) = \begin{bmatrix} C_{P1} & C_{P2} & \cdots & C_{Pq} \end{bmatrix} \begin{bmatrix} C_{11} & C_{12} & \cdots & C_{1q} \\ C_{21} & C_{22} & \cdots & C_{2q} \\ \vdots & \vdots & & \vdots \\ C_{q1} & C_{q2} & \cdots & C_{qq} \end{bmatrix}^{-1} \begin{bmatrix} \ell_1 \\ \ell_2 \\ \vdots \\ \ell_q \end{bmatrix}. \tag{10-24}$$

This is formally the same expression as (10–20), but with f_i replaced by ℓ_i and with "covariances" C_{ik} and C_{Pi} defined by "*covariance propagation*" (10–22) and (10–23). The concept of covariance propagation is a straightforward generalization of the formal structure of error propagation known from adjustment computations. However, this structure as such is purely mathematical rather than statistical. We know that a "linear functional" is the continuous analogue (in infinite-dimensional Hilbert space) to the usual concept of a linear function in n-dimensional vector space. We try not to burden the reader with too much mathematical formalism, but this is treated in great detail in Moritz (1980 a) and in Moritz and Hofmann-Wellenhof (1993: Chap. 10). We cannot, however, resist the temptation to compare the structure

$$b_i = L_i^j a_j \tag{10-25}$$

leading to

$$\text{cov}(b_i, b_j) = L_i^k L_j^l \, \text{cov}(a_k, a_l) \tag{10-26}$$

for finite-dimensional vectors a and matrix L using the usual summation over two equal indices, and $N_i = L_i^P \Delta g_P$ leading to

$$\text{cov}(N_P, N_Q) = L_i^P L_j^Q \, \text{cov}(\Delta g_P, \Delta g_Q), \qquad (10\text{--}27)$$

where N_i denotes the geoidal height at point i and Δg is the gravity anomaly at point P, and L denotes the Stokes formula. Explicit expressions are found in Moritz (1980 a: Sect. 15).

In this statistical interpretation, we take the kernel function $K(P, Q)$ as the covariance function $C(P, Q)$. Then $f(P)$ is an optimal estimate (in the sense of least variance) for the anomalous potential T and hence for the height anomaly $\zeta = T/\gamma$, on the basis of arbitrary measurement data. For geoid determination in mountainous areas, relevant terrestrial measurement data primarily are ξ, η, and Δg. The covariances C_{ik} and C_{Pi} are given by known analytical expressions, see Tscherning and Rapp (1974) or Moritz (1980 a: Sect. 15). A general computer program for collocation is described in Sünkel (1980).

Least-squares collocation may easily be generalized to observational data affected by random errors; systematic effects may also be taken into consideration. In addition to the estimated quantities (f in our present case) we may also compute their standard error by a formula similar to (10–24). A comprehensive presentation of a least-squares collocation may be found in Moritz (1980 a). *You cannot learn collocation from this slight chapter only!*

Harmonicity of the covariance functions.

In three-dimensional space, the covariance functions, being kernel functions and their linear functional transformations L, are always harmonic. If we have (9–25),

$$C(\psi) = \sum_{n=2}^{\infty} c_n P_n(\cos \psi) \qquad (10\text{--}28)$$

on the sphere, then in space there will be

$$C(r, r', \psi) = \sum_{n=2}^{\infty} c_n \left(\frac{R^2}{r \, r'} \right)^{n+2} P_n(\cos \psi) \qquad (10\text{--}29)$$

(Moritz 1980 a: Sect. 23, Eq. (32-1)). The point $P(r, \theta, \lambda)$ is the computation point, and $Q(r', \theta', \lambda')$ is a current data point; ψ is the spherical distance between (θ, λ) and (θ', λ'), and R is the mean radius of the earth. The dependence on r is given by the factor

$$r^{-(n+2)} \qquad (10\text{--}30)$$

because $r\Delta g$ is harmonic, and similarly for r'. The factor

$$\left(\frac{R^2}{r\,r'}\right) \tag{10–31}$$

is chosen to become equal to 1 if both points P and Q lie on sea level; in this case, Eq. (10–29) reduces to (10–28).

So, each of the terms of (10–29) is harmonic, that is, it satisfies Laplace's equation. Thus, the whole series (10–29) is harmonic (if it converges), being a linear combination of harmonic terms. This is a well-known consequence of the linearity of Laplace's equation: the linear combination of solutions of any linear equation is itself a solution of this equation.

Thus, also the spherical harmonics series of $T = r\Delta g$ is harmonic down to the reference sphere $r = R$, with respect both to r and r'. Harmonic functions, by their very definition, are *regular analytic functions* down to $r = R$, so T and all its linear combinations are regular and thus *admit downward continuation down to the reference sphere* (cf. Sect. 8.6).

10.2 Application of collocation to geoid determination

It is well known that the direct interpolation of free-air gravity anomalies, which essentially are surface gravity anomalies (8–128) in high mountains, e.g., by least-squares interpolation, leads to relatively poor results because of the correlation of the free-air anomalies with elevation (Sect. 9.7). This correlation with elevation constitutes a considerable trend which must be removed before the interpolation. Bouguer anomalies take care of the dependence on the local irregularities of elevation; isostatic anomalies are, in addition, also largely independent on the regional features of topography; in Sect. 11.1 we shall consider, in addition, also the removal of global trends by spherical-harmonic earth gravity models (e.g., EGM 96, see www.iges .polimi.it/index/geoid_repo/global_models.htm) obtainable from the internet.

In exactly the same way we must remove the main trend of the vertical deflections ξ, η and the gravity anomalies Δg by an isostatic reduction before applying collocation. Thus, isostatic reduction, pragmatically regarded as trend removal, is essential for the practical application of least-squares collocation in mountainous regions (Forsberg and Tscherning 1981).

Physically speaking, we transport the topographic masses to the interior of the geoid in such a way that the isostatic mass deficiencies are filled. The observation point P remains in its position on the earth's surface. In this way, not only the harmonic character of the anomalous potential T outside the earth's surface is preserved, but in addition, the computational removal of the topographic masses above sea level makes the function T harmonic down to sea level. Hence, the collocation formula (10–24) can be applied also at sea level, giving cogeoid heights N^c. By applying the inverse reduction (the indirect effect) to the computed height anomalies ζ^c and cogeoid heights N^c, we get actual ζ and N. It can be expected that errors in the isostatic model used (e.g., an Airy–Heiskanen model) will largely cancel in this combined procedure of reduction and "anti-reduction" (remove-restore technique; see Sect. 11.1).

The procedure is theoretically optimal and practically well suited for computer use. The integrability conditions, which in Helmert integration are represented by the closures of the individual triangles (see Sect. 5.14), are automatically taken into account. The fact that the deflections of the vertical are given only in a certain region has the effect that the geoid can only be computed in that region. Since, even by collocation, differences in geoidal heights between two neighboring stations A and B depend essentially only on the deflections in those two stations, the lack of data outside the region under consideration will hardly cause a noticeable distortion. Note, however, that the addition of a constant to all geoidal heights N will not affect the deflections of the vertical; hence, astrogeodetic data determine the geoidal heights only up to an additive constant. This constant may be chosen such that the average value of the computed N is zero, and the result of collocation comes near to this case.

To get immediately almost geocentric geoidal heights, it is appropriate to take into consideration a global trend which mainly affects ζ and N itself, by subtracting the effect of a suitable global gravity field, e.g., the gravity earth model given as a spherical-harmonic expansion up to degree $180° \times 180°$ of Rapp (1981), say, following Sünkel (1983). This will be described in the next section; in the present section we limit ourselves to the isostatic reduction.

Computational procedure

The computational procedure consists of the following steps:

1. Transformation of the astrogeodetic surface deflections ξ, η from the local datum used for the geocentric Geodetic Reference System 1980 by the well-known differential formulas of Vening Meinesz (see Heiskanen and Moritz 1967: Eq. (5-59)). This is necessary since collocation requires a reference system which is as realistic as possible.

2. Application of the normal plumb line curvature (8–137) to the "geometric" surface deflections ξ, η gives the "dynamic" surface deflections

$\bar{\xi}$, $\bar{\eta}$ by (8–136).

3. Computation of the gravity anomalies Δg, also referred to the earth's surface according to (8–128).

4. The topographic-isostatic reduction of $\bar{\xi}$, $\bar{\eta}$, Δg by (8–154) and (8–101) gives values ξ^c, η^c, Δg^c which continue to refer to the surface point P.

5. The application of collocation to ξ^c, η^c, Δg gives height anomalies ζ^c and cogeoid heights N^c, by simply varying the elevation parameter (h and zero, respectively) in the collocation program (see Sünkel 1983).

6. By applying the indirect effect (10–2) and (8–153), we get actual height anomalies ζ and geoidal heights N.

11 Computational methods

11.1 The remove-restore principle

Let us start with gravity reduction according to the modern view of measuring and calculating the gravity field in principle always *at the earth's surface*, or briefly, *on the ground*, or equivalently, *at point level*. This is used in the sense of Sects. 8.9 and 8.14. More precisely, it is *topographic-isostatic reduction at ground level*.

The most practical way to realize this idea is least-squares collocation, because it automatically works in three-dimensional space, by simply putting the desired topographic height h as parameters for input (measurements: gravity anomalies, deflections of the vertical, etc.) and output (potential T or its functionals to be computed). Symbolically, this means

$$T = \mathcal{L}(\ell) \tag{11–1}$$

or

$$\text{output} = \mathcal{L}(\text{input})\,, \tag{11–2}$$

where $\mathcal{L}$ denotes the linear operation of least-squares collocation (not to be confused with a linear functional L as used, e.g., in Eq. (10–13)).

In Sect. 8.9 we have introduced gravity reduction from the point of view of the modern theory. To repeat, immediately specializing to topographic-isostatic reduction, we have

- measured gravity anomalies Δg at ground level,
- reduced topographic-isostatic anomalies Δg^{c} obtained by *removing* the attraction of the topographic-isostatic masses δg_{TI},
- "co-potential" $T^{\mathrm{c}} = \mathcal{L}(\Delta g^{\mathrm{c}})$ computed by collocation, and
- "real potential" T by *restoring* the "indirect effect" of the topographic-isostatic masses δT_{TI}.

Mathematically this may be written

$$T = \mathcal{L}(\Delta g - \delta g_{\mathrm{TI}}) + \delta T_{\mathrm{TI}}\,. \tag{11–3}$$

This is a reinterpretation of the gravity reduction of Sect. 8.9. It must be correct since if

$$\delta T_{\mathrm{TI}} = \mathcal{L}(\delta g_{\mathrm{TI}}) \tag{11–4}$$

then Eq. (11–3) gives

$$\delta T = \mathcal{L}(\Delta g), \tag{11-5}$$

as it should be.

The same principle works also with deflections of the vertical ξ, η at the earth's surface, both as input data and as output results (Sects. 8.14 and 10.2).

The underlying isostatic model is in principle arbitrary. For practical purposes it should provide a good approximation (small residuals δT) and be computationally convenient.

We see, however, a change of perspective. Collocation is no longer applied to the "real" anomalous gravity field as in (11–5) but to the residual field, removing the field generated by the assumed topographic-isostatic model. The *model* is arbitrary, but the *derived quantities* must be computed in a rigorous consistent fashion. (Consistency for the quantities computed by collocation as guaranteed by a correct covariance propagation; see Sect. 10.2.)

This change of perspective may not seem important because it is just a change of nomenclature: what formerly was importantly called "isostatic anomaly" is now degraded to a miserable "residual". However, the remove-restore principle permits also the use of other approximate fields to remove trends; especially one of the numerous existing "earth (gravity) models" (EM or EGM) consisting of spherical-harmonic expansions of the potential T up to degree 180 or higher.

Therefore, we "remove" from the observations ℓ – gravity anomalies, gravity disturbances, deflections of the vertical, etc. – the effect ℓ_{EM} computed from the earth model used, and after collocation "restore" the effect of the EM on the result. The mathematics is the same as in (11–4) and (11–5):

$$\delta T_{\mathrm{EM}} = \mathcal{L}(\delta \ell_{\mathrm{EM}}) \tag{11-6}$$

and

$$\delta T = \mathcal{L}(\ell). \tag{11-7}$$

We have only slightly generalized from Δg to ℓ.

Now we proceed an important step further. The remove-restore principle has only two requirements:

1. the removed auxiliary potentials must be harmonic, precomputable, and used in a mathematically consistent way: what is removed in the input must be restored in the output;

2. in the usual case of linearity, two or more different auxiliary potentials may be used (removed-restored) simultaneously.

Thus, we use simultaneously the earth model EM for the longer wavelengths and the topographic-isostatic geological model TI for the shorter wavelengths. Since the spherical-harmonic expansions are generalizations, for the sphere, of Fourier series for the circle, we can speak of wavelengths. Denoting the maximum degree of the spherical-harmonic expansion with N, this can be associated with a shortest resolvable wavelength λ according to

$$\lambda = \frac{2\pi}{N} = \frac{360°}{N} . \tag{11–8}$$

For an expansion to degree $N = 180$ (say), we have $\lambda = 360°/180 = 2°$, which roughly corresponds to 200 km on a meridian or on the equator. In many cases, the half wavelength $\lambda/2$ is considered (see Seeber 2003: p. 469).

Since EM (approximately) takes care of the long waves up to a certain maximum degree N, it is resonable to represent the remaining short waves from N to infinity. This sequence $N+1, N+2, \ldots, \infty$ will be denoted by CN, where CN is the abbreviation of the "complement" of the sequence from 2 to N.

Thus, we may write for the residuals

$$\begin{aligned}
\delta T &= T - T_{\text{EM}}^{N} - T_{\text{TI}}^{CN} , \\
\delta \ell &= \ell - \ell_{\text{EM}}^{N} - \ell_{\text{TI}}^{CN} .
\end{aligned} \tag{11–9}$$

The collocation procedure will be applied to these residuals.

Remark

As we have noted at the beginning of Sect. 10.2, the remove-restore process aims at removing all known major trends:

- the local topography produces *Bouguer anomalies*,
- the regional features (i.e., their isostatic compensation), in addition to the Bouguer effect, produce *topographic-isostatic anomalies*,
- the global irregularities are expressed by an earth model and lead to what is modestly called the *"residual anomalies"*.

It is clear that what is "removed" before the computation, must be fully "restored" after the computation.

11.2   Geoid in Austria by collocation

Austria is a nice country, and in spite of being small, it has all types of topography: flat, hilly, and alpine, up to an elevation of 3800 m. Thus, beyond being a pleasant place to live, it is an interesting geodetic test area.

The pioneering work has been done by Sünkel (1983). Later work, especially by Sünkel et al. (1987), Kühtreiber (1998, 2002 a, 2002 b), and Erker et al. (2003) has refined, extended and perfected the gravity field in Austria, but the 1983 work is good for an introduction.

Sünkel (1983) used least-squares collocation to calculate the geoid for the main part of Austria from a very good material of deflections of the vertical. Gravity anomalies of a comparable quality were not yet available in 1983. In addition to an isostatic reduction (Sect. 8.14) according to Airy–Heiskanen ($T = 30$ km), he also removed a global trend by means of an earth gravity model, represented by a spherical-harmonic expansion up to a certain degree N. In particular, he used the model of Rapp (1981) with $N = 180$.

After removing the topographic-isostatic trend T_{TI} and this global trend T_{EM}^N (remember, EM denotes earth model), there remains a residual anomalous potential δT, given by

$$\delta T = T - T_{\text{TI}} - T_{\text{EM}}^N + T_{\text{TI}}^N. \qquad (11\text{--}10)$$

Since the earth model potential T_{EM}^N is represented by a spherical-harmonic expansion up to degree N, it may be appropriate to consider, for the isostatic reduction, only the effect for degrees $N > 180$ (or, say, $N > 360$), replacing T_{TI} by

$$T_{\text{TI}}^{CN} = (T_{\text{TI}})_{N>180} = T_{\text{TI}} - T_{\text{TI}}^N, \qquad (11\text{--}11)$$

where T_{TI}^N represents a spherical-harmonic expansion for T_{TI} truncated at degree $N = 180$. This explains Eq. (11–11).

The observations $\ell_i = [\xi, \eta, \Delta g]$, which represent linear functionals $L_i T$, are reduced in the same way, obtaining

$$\ell_i - L_i T_{\text{TI}} - L_i T_{\text{EM}}^N + L_i T_{\text{TI}}^N = L_i \delta T. \qquad (11\text{--}12)$$

Adding the earth model reduction to the computational procedure outlined at the end of the preceding section, we thus have the flow diagram of Table 11.1.

Data

The topography in Austria is rather varied, with elevations up to 3800 m. The density of astrogeodetic stations was 10 to 20 km; the total number of deflections data used was 521. No gravity anomalies were used in this first computation.

The topographic-isostatic reduction of the deflections of the vertical was made using a rather crude digital terrain model consisting of mean elevations for $20'' \times 20''$ rectangles. It has been obtained by digitizing a map 1 : 500 000.

Table 11.1. From observations to the geoid

$(L_i T)$ observations referred to
Geodetic Reference System 1980

↓

reduction TI, EM

$-L_i(T_{\mathrm{TI}} + T_{\mathrm{EM}}^N - T_{\mathrm{TI}}^N)$

↓

$(L_i \delta T)$

↓

collocation

↓

δT

↓

inverse reduction TI, EM

$+ \quad (T_{\mathrm{TI}} + T_{\mathrm{EM}}^N - T_{\mathrm{TI}}^N)$

↓

T

↓

$$N = (T/\gamma)_0 , \quad \zeta = (T/\gamma)_h$$

The standard error of this model is on the order of 100 m. Investigations have shown that, in spite of its poor accuracy, the model is reasonably adequate for reduction of deflections of the vertical; it is, however, totally inadequate for gravity! In fact, the reduction error for ξ, η is approximately proportional to the terrain inclination; it is thus very small if the deflection station is situated in an area of inclination zero. This is the case not only if the station lies in a horizontal plane but also if it lies on the top of a mountain, as most deflection stations do.

Results

It turned out that almost all of the signal (T, N, ζ) comes from the topo-graphic-isostatic model and the $N = 180$ gravity model used. This part, TI + EM, lies between 41.5 m and 49.5 m. The contribution of collocation $(\gamma^{-1} T)$ lies between -0.5 m and 1.5 m, after removal of a pronounced trend on the order of 3 m.

The efficiency of topographic-isostatic reduction can also be seen from the fact that it has reduced the variance of the deflections of the vertical in

Austria (the square of the average size of ξ and η) from 30 (arc second)^2 to 5 (arc second)^2.

So we may say that we can determine the Austrian geoid to 1–2 m without measurements (deflections of the vertical) and without collocation, knowing only a topographic map! This is even more surprising since Austria is not particularly well isostatically compensated.

Of considerable interest is the effect of analytical continuation on the isostatically (plus earth model) reduced anomalous potential T_{TI}. It is expressed by the difference $\gamma^{-1}T$ at the earth's surface minus $\gamma^{-1}T$ at sea level. This difference reaches a maximum of 13 cm in the Central Alps and is otherwise positive and negative. In the terminology of the present book, this is the *separation between the real geoid and the harmonic geoid* (Sect. 8.15).

Of the same interest is the difference between the height anomalies ζ $(=\gamma^{-1}T$ at the earth's surface) and the geoidal heights N $(=\gamma^{-1}T$ at sea level). The maximum of 35 cm for $\zeta - N$ is reached at the Grossglockner mountain (the highest peak in Austria, $H = 3797\,\mathrm{m}$). The results are in excellent agreement with the approximate formula

$$\zeta - N = -(981\ \mathrm{gal})^{-1}\,\Delta g_B\,H\,,\qquad\qquad(11\text{–}13)$$

where Δg_B is the Bouguer anomaly in gal and H is the elevation in the same units as ζ and N. The agreement may easily be verified, since the Bouguer anomalies in the investigated area range from 10 mgal to -170 mgal, corresponding to topographic heights from 200 m to 3000 m (Sünkel 1983: p. 140). In Sünkel et al. (1987: p. 69), the differences $\zeta - N$ for the whole of Austria range between -2 cm and $+56$ cm.

All this has been computed only from the measured deflections of the vertical. Gravity observations have been included by Kühtreiber (2002 a, 2002 b) and Erker et al. (2003), leading to what might be a "few-centimeter geoid".

Important: the astrogeodetic geoid and the gravimetric geoid are compared and finally combined after systematic trends have been eliminated by Kühtreiber (2002 b) and Erker et al. (2003).

11.3   Molodensky corrections

In Sect. 8.6 we have given a solutions of Molodensky's problem by means of a series obtained on the basis of analytical continuation. It can be written in the form of Eqs. (8–68), (8–69), (8–67),

$$\zeta = \zeta_0 + \zeta_1 + \zeta_2 + \zeta_3 + \cdots\,,\qquad\qquad(11\text{–}14)$$

$$\zeta_i = \frac{R}{4\pi \, \gamma} \iint\limits_{\sigma} g_i \, S(\psi) \, d\sigma \, , \tag{11–15}$$

$$\Delta g^* = \Delta g + g_1 + g_2 + g_3 + \cdots \, . \tag{11–16}$$

The correction terms g_n are evaluated recursively by

$$g_n = -\sum_{r=1}^{n} z^r \, L_r(g_{n-r}) \, , \tag{11–17}$$

starting from

$$g_0 = \Delta g \, . \tag{11–18}$$

Here the operator L_n is also defined recursively:

$$L_n(\Delta g) = n^{-1} L_1 \left[L_{n-1}(\Delta g) \right] \tag{11–19}$$

starting with

$$L_1 = L \tag{11–20}$$

with the gradient operator L defined by the integral (8–60), that is,

$$L(f) = \frac{R^2}{2\pi} \iint\limits_{\sigma} \frac{f - f_Q}{l_0^3} \, d\sigma \, . \tag{11–21}$$

This means: take $g_0 = \Delta g$, where Δg is the free-air anomaly at ground level in the sense of Molodensky, then compute g_1 by (11–17) with $n = 1$, then compute g_2 by (11–17) with $n = 2$ and L_2 by (11–19), then g_3 by (11–17) with $n = 3$ and L_3 by (11–19), etc.

The operator L behaves like differentiation $(L(f) = \frac{\partial \Delta g}{\partial r})$ and, therefore, "roughens" the function f; this means that each successive L becomes rougher and rougher. This is not conducive to the convergence of Molodensky's series unless the original Δg is very smooth, which cannot be assumed in mountainous areas.

In such cases, some smoothing of Δg is inevitable. Numerical analysis is constantly confronted with problems of smoothing, so many techniques of smoothing have been developed such as the sliding average. For evaluating the integral L, fast Fourier methods are available. The problem is to find an appropriate degree of smoothing which makes consecutive corrections $g_1, g_2, g_3, \ldots$ decrease in order to achieve practical convergence without "oversmoothing". At any rate, smoothing must ensure that $g_5, g_6, \ldots$ are practically negligible since they cannot be meaningfully computed because of the inevitable accumulation of round-off errors, which finally tends to producing pure noise.

Table 11.2. Characteristic values in arc seconds for Molodensky corrections ξ_i and η_i for deflections of the vertical until $i = 4$, computed from free-air gravity anomalies

	ξ_1	η_1	ξ_2	η_2	ξ_3	η_3	ξ_4	η_4
min	−2.44	−1.94	−0.92	−0.84	−0.35	−0.24	−0.08	−0.12
max	2.36	3.654	0.88	0.86	0.21	0.20	0.05	0.09
mean	0.19	0.32	−0.02	−0.02	−0.01	−0.01	0.00	0.00
rms	0.90	0.96	0.29	0.27	0.06	0.06	0.02	0.02

Table 11.3. Characteristic values in arc seconds for Molodensky corrections ξ_i and η_i for deflections of the vertical until $i = 4$, computed from isostatic gravity anomalies

	ξ_1	η_1	ξ_2	η_2	ξ_3	η_3	ξ_4	η_4
min	−0.57	−0.36	−0.06	−0.07	−0.01	−0.02	0.00	0.00
max	0.33	0.46	0.09	0.05	0.01	0.01	0.00	0.00
mean	−0.04	0.01	0.00	−0.01	0.00	0.00	0.00	0.00
rms	0.11	0.09	0.02	0.02	0.00	0.00	0.00	0.00

As Kühtreiber (1990) showed in his thorough work, there is no rough-and-ready prescription for finding an optimal smoothing. Trial and error may be the best approach.

Isostatic reduction might be considered a smoothing method on a geophysical basis, cf. Tables 11.2 and 11.3.

Just to give an idea of the order of magnitudes, we take some typical sizes of the Molodensky corrections in high mountains.

We take two tables from Kühtreiber (1990): the following Tables 11.2 and 11.3 are Kühtreiber's Tables (8-3) and (8-6). The gravity data are assumed to be given in a rectangular grid of size $11.25'' \times 18.75''$. A suitable smoothing is presupposed. Much better is, of course, the use of *isostatic reduction*, which should provide a physically meaningful and efficient smoothing. This is shown by Table 11.3.

To provide some contrast and to include also Molodensky corrections for the height anomaly ζ, we quote also Table 11.4 of a somewhat earlier work by Kraiger et al. (1987) (denoted as Table 6.1 there). The values are not directly comparable because test areas and selected methods of integration, smoothing, data density, etc., are different. Still, they lead to interesting

Table 11.4. Comparison of direct numerical integration and fast Fourier transform (FFT): maximum and (arithmetic) mean values of Molodensky corrections ζ_i, ξ_i, η_i for $i = 1, 2$; test area: $46.788° \leq \varphi \leq 46.512°$, $13.438° \leq \lambda \leq 14.646°$, $600\,\mathrm{m} \leq$ topographic height $\leq 2400\,\mathrm{m}$

	ζ_1 [cm]	ξ_1 ["]	η_1 ["]	ζ_2 [cm]	ξ_2 ["]	η_2 ["]	
maximum	40.8	2.0	2.0	0.8	0.2	0.2	direct int.
values	47.6	1.5	1.4	0.7	0.1	0.1	FFT
mean	31.3	0.4	0.4	0.2	0.03	0.03	direct int.
values	36.7	0.3	0.4	0.5	0.02	0.02	FFT

conclusions:

1. The method of Molodensky corrections depends very much on the details of numerical integration (data density, smoothing, etc.).

2. The corrections decrease for increasing $i = 1, 2, 3, \ldots$. This is what they have to do. Higher corrections may be expected finally to consist of "pure noise" because of general roughening and increasing round-off errors, so that the question of convergence becomes practically as well as theoretically meaningless: higher terms must simply be put equal to zero by higher force.

3. The Molodensky correction ζ_1 may reach a few decimeters, ζ_2 and higher-order terms might frequently be negligible.

4. At the end of Sects. 2.21 and 8.8, we have remarked a curious phenomenon. Using the same data, gravimetric methods seem to furnish the vertical position (expressed by ζ or N) roughly by one order of magnitude better than the horizontal position (as expressed by ξ, η). If we take the old astronomer's rule that $1'' \cong 30\,\mathrm{m}$ in position, then $1\,\mathrm{m}$ corresponds to $0.03''$. Assume that we get $1\,\mathrm{m}$ in vertical position and wish to get the same accuracy for horizontal position. This would mean that we have to get the astronomical measurements Φ, Λ and the deflections of the vertical ξ, η with better than $0.03''$. This also seems to apply with the order of magnitude of the Molodensky corrections, where a Molodensky correction $\zeta_1 = 0.41\,\mathrm{m}$ comes along with a $\xi_1 = \eta_1 = 2''$, which corresponds to $60\,\mathrm{m}$.

   In this sense, gravimetry is weaker by one order of magnitude in determining the horizontal than the vertical position. This is an admittedly one-sided perspective, but it was used against scientists who claimed, still around 1960, that the gravimetric method was able to do everything that satellite geodesy could. With GPS now we know better, and without ideological scruples we combine satellite data with terrestrial gravity.

(A second perspective of astronomical observations is the astrogeodetic geoid determination. Here the accuracy of astronomy is sufficient; cf. Sec. 5.14.)

Final remark

The computation of Molodensky reductions is heavy work. So in mountainous areas, least-squares collocation is definitely preferable to integration, except for certain test computations (Sideris 1987, 1990).

Collocation also permits comparison and combination of astrogeodetic and gravimetric data; a key paper is Kühtreiber (2002 b).

All this, however, builds on the fundamental ideas of M.S. Molodensky. In his landmark publication (Krarup 1969) one clearly sees the transition from Molodensky's problem to least-squares collocation.

11.4   The geoid on the internet

The International Association of Geodesy (IAG) has a very active International Geoid Service (IGeS–IAG). Before you try to compute your own geoid, look at www.iges.polimi.it to see what is available there. You can find global and regional geoids, data, software, references, plans for future work, etc. We particularly mention the geoid repositories:

- www.iges.polimi.it/index/geoid_repo/global_models.htm ,
- www.iges.polimi.it/index/geoid_repo/regional_models.htm .

In the latter file you can find:

- USA gravimetric Geoid 1996 (Dru Smith),
- European Geoid/Quasigeoid EGG97 (H. Denker),
- Austrian Geoid 1996 (H. Sünkel).

Other important internet addresses:

- International Gravity Bureau (Toulouse):
  http://bgi.cnes.fr8110/bgi_a.html ,
- International Association of Geodesy:
  www.iag_aig.org .

References

Abd-Elmotaal H (1995): Attraction of the topographic masses. Bulletin Géodésique, 69: 304–307.

Anger G, Moritz H (2003): Inverse problems and uncertainties in science and medicine. Sitzungsberichte der Leibniz-Sozietät, 61(5): 171–212.

Anger G, Gorenflo R, Jochmann H, Moritz H, Webers W (1993): Inverse problems: principles and applications in geophysics, technology, and medicine. Mathematical Research 74: 37–44.

Bomford G (1962): Geodesy, 2nd edition. Oxford University Press.

Brouwer D, Clemence GM (1961): Methods of celestial mechanics. Academic Press, New York.

Bruns H (1878): Die Figur der Erde. Publikation des Preussischen Geodätischen Instituts, Berlin.

Cassinis G (1930): Sur l'adoption d'une formule internationale pour la pesanteur normale. Bulletin Géodésique, 26: 40–49.

Civil GPS Service Interface Committee (2002): Summary report of the 40th meeting of the Civil GPS Service Interface Committee (CGSIC), Oregon, March 22–24. Available at www.navcen.uscg.gov/cgsic/default.htm.

Daxinger W, Stirling R (1995): Kombinierte Ausgleichung von terrestrischen und GPS-Messungen. Österreichische Zeitschrift für Vermessung und Geoinformation, 83(1+2): 48–55.

DeLoach SR, Remondi B (1991): Decimeter positioning for dredging and hydrographic surveying. In: Proceedings of the First International Symposium on Real Time Differential Applications of the Global Positioning System, vol 1. TÜV Rheinland, Köln: 258–263.

Department of Defense (2001): Global Positioning System standard positioning service performance standard. Available from the U.S. Assistant for GPS, Positioning and Navigation, Defense Pentagon, Washington (DC).

Dobrin MB, Savit CH (1988): Introduction to geophysical prospecting, 4th edition. McGraw Hill, New York.

Drinkwater MR, Floberghagen R, Haagmans R, Muzi D, Popescu A (2003): GOCE: ESA's first earth explorer core mission. In: Beutler GB, Drinkwater MR, Rummel R, Steiger R von (eds): Earth gravity from space – from sensors to earth sciences. Space Sciences Series of ISSI, 18: 419–432. Kluwer, Dordrecht.

Elmiger A (1969): Studien über Berechnung von Lotabweichungen aus Massen, Interpolation von Lotabweichungen und Geoidbestimmung in der Schweiz. Doctoral Dissertation, ETH Zürich.

Erker E, Höggerl N, Imrek E, Hofmann-Wellenhof B, Kühtreiber N (2003): The Austrian geoid – recent steps to a new solution. Österreichische Zeitschrift für Vermessung & Geoinformation, 91(1): 4–13.

ESA (1999): Gravity field and steady-state ocean circulation mission. Reports for mission selection. The four candidate earth explorer core missions, SP-1233(1). Available at http://esamultimedia.esa.int/docs/goce_sp1233_1.pdf.

Fontana RD, Cheung W, Stansell T (2001): The modernized L2 civil signal. GPS World, 12(9): 28–34.

Forsberg R, Tscherning CC (1981): The use of height data in gravity field approximation by collocation. Journal of Geophysical Research, 86 (B9): 7843–7854.

Forsberg R, Tscherning CC (1997): Topographic effects in gravity field modelling for BVP. Available at www.gfy.ku.dk/~cct/comored.htm.

Frank P, Mises R von (eds) (1930): Die Differential- und Integralgleichungen der Mechanik und Physik, 2nd edition, part 1: Mathematischer Teil. Vieweg, Braunschweig (reprint 1961 by Dover, New York and Vieweg, Braunschweig).

Galle A (1914): Das Geoid im Harz. Veröffentlichung des Geodätischen Instituts Potsdam, vol 61.

Grafarend E, Offermanns G (1975): Eine Lotabweichungskarte Westdeutschlands nach einem geodätisch konsistenten Kolmogorov-Wiener-Modell. Deutsche Geodätische Kommission bei der Bayerischen Akademie der Wissenschaften, Reihe A: Theoretische Geodäsie, vol 82.

Groten E (2004): Fundamental parameters and current (2004) best estimates of the parameters of common relevance to astronomy, geodesy, and geodynamics. Available at www.gfy.ku.dk/~iag/HB2004/part5/51-groten.pdf.

Gurtner W (1978): Das Geoid in der Schweiz. Institut für Geodäsie und Photogrammetrie der ETH Zürich, vol 20.

Gurtner W, Elmiger H (1983): Computation of geoidal heights and vertical deflections in Switzerland. Presented at the XVIII General Assembly of the IUGG at Hamburg, August 15–27.

Heiskanen W (1924): Untersuchungen über Schwerkraft und Isostasie. Finnish Geodetic Institute, Helsinki, vol 4.

Heiskanen W (1928): Ist die Erde ein dreiachsiges Ellipsoid? Gerlands Beiträge zur Geophysik, 19: 356–377.

Heiskanen WA (1957): The Columbus geoid. EOS, Transactions, American Geophysical Union, 38: 841–848.

Heiskanen WA, Moritz H (1967): Physical geodesy. Freeman, San Francisco London.

Heiskanen WA, Vening Meinesz FA (1958): The earth and its gravity field. McGraw-Hill, New York.

Helmert FR (1880): Die mathematischen und physikalischen Theorien der höheren Geodäsie, part 1. Teubner, Leipzig (reprint 1962).

Helmert FR (1884): Die mathematischen und physikalischen Theorien der Höheren Geodäsie, part 2. Teubner, Leipzig (reprint 1962).

Hirvonen RA (1960): New theory of the gravimetric geodesy. Publications of the Isostatic Institute of the International Association of Geodesy, Helsinki, vol 32.

Hirvonen RA (1961): The reformation of geodesy. Journal of Geophysical Research, 66: 1471–1478.

Hirvonen RA (1962): On the statistical analysis of gravity anomalies. Publications of the Isostatic Institute of the International Association of Geodesy, Helsinki, vol 37.

Hofmann-Wellenhof B, Kienast G, Lichtenegger H (1994): GPS in der Praxis. Springer, Wien New York.

Hofmann-Wellenhof B, Lichtenegger H, Collins J (2001): GPS – theory and practice, 5th edition. Springer, Wien New York.

Hofmann-Wellenhof B, Legat K, Wieser M (2003): Navigation – principles of positioning and guidance. Springer, Wien New York.

Hotine M (1969): Mathematical geodesy. ESSA Monograph 2, U.S. Department of Commerce, Washington (reprint 1992 by Springer).

Ilk KH (1999): Energiebetrachtungen für die Bewegung zweier Satelliten im Gravitationsfeld der Erde. In: Krumm F, Schwarze VS (eds): Quo vadis geodesia? Part 1. Schriftenreihe des Instituts des Studienganges Geodäsie und Geoinformatik der Universität Stuttgart, Report 1999/6: 191–205.

Jeffreys H (1931): An application of the free-air reduction of gravity. Gerlands Beiträge zur Geophysik, 31: 378–386.

Jekeli C (1999): The determination of gravitational potential differences from satellite-to-satellite tracking. Celestial Mechanics and Dynamical Astronomy, 75: 85–101.

Jung K (1956): Figur der Erde. In: Flügge S (ed): Handbuch der Physik, vol 47, geophysics 1. Springer, Berlin, 534–639.

Jung K (1961): Schwerkraftverfahren in der Angewandten Geophysik. Akademische Verlagsgesellschaft Geest & Portig, Leipzig.

Kaula WM (1959): Statistical and harmonic analysis of gravity. Journal of Geophysical Research, 64: 2401–2421.

Kaula WM (1963): Determination of the earth's gravitational field. Reviews of Geophysics, 1: 507–551.

Kaula WM (1966 a): Theory of satellite geodesy. Blaisdell, Waltham (Massachusetts) Toronto London.

Kaula WM (1966 b): Global harmonic and statistical analysis of gravimetry. In: Orlin H (ed): Proceedings of the Symposium on Extension of Gravity Anomalies to Unsurveyed Areas. American Geophysical Union, Geophysical Monograph, 9: 58–67.

Kellogg OD (1929): Foundations of potential theory. Springer, Berlin (reprint 1954 by Dover, New York, and 1967 by Springer, Berlin Heidelberg New York).

Koch KR (1971): Die geodätische Randwertaufgabe bei bekannter Erdoberfläche. Zeitschrift für Vermessungswesen, 96: 218–224.

Kraiger G (1987): Untersuchungen zur Prädiktion nach kleinsten Quadraten mittels empirischer Kovarianzfunktion unter besonderer Beachtung des Krümmungsparameters. Mitteilungen der geodätischen Institute der Technischen Universität Graz, vol 53.

Kraiger G (1988): Influence of the curvature parameter on least-squares prediction. Manuscripta geodaetica, 13: 164–171.

Kraiger G, Kühtreiber N, Wang YM (1987): The correction terms of the solution of Molodensky's problem by analytical continuation in the Central Alps of Austria. In: Austrian Geodetic Commission (ed): The gravity field in Austria. Geodätische Arbeiten Österreichs für die Internationale Erdmessung, Neue Folge, vol IV: 95–109.

Krarup T (1969): A contribution to the mathematical foundation of physical geodesy. Danish Geodetic Institute, Copenhagen, vol 44.

Kühtreiber (1990): Untersuchungen zur gravimetrischen Bestimmung von Lotabweichungen im Hochgebirge nach Molodensky mittels Fast-Fourier-Transfor-
mation. PhD thesis, Department of Physical Geodesy, Graz University of
Technology, Graz, Austria.

Kühtreiber (1998): Improved gravimetric geoid AGG97 of Austria. In: Forsberg R,
Feissel M, Dietrich R (eds): Geodesy on the move: gravity, geoid, geodynamics
and Antarctica. IAG Scientific Assembly, Rio de Janeiro, Brazil, September
3–9, 1997. Springer, Berlin Heidelberg, 306–311 [Schwarz K-P (ed): International Association of Geodesy Symposia, vol 119].

Kühtreiber (2002 a): High precision geoid determination for Austria. Habilitation,
Department of Positioning and Navigation, Graz University of Technology,
Graz, Austria.

Kühtreiber N (2002 b): High precision geoid determination of Austria using heterogeneous data. In: Tziavos IN (ed): Gravity and geoid 2002. Proceedings
of the Third Meeting of the International Gravity and Geoid Commission,
Thessaloniki, Greece, August 26–30, 144–149.

Lachapelle G, Cannon ME, Erickson C, Falkenberg W (1992): High precision C/A
code technology for rapid static DGPS surveys. In: Proceedings of the Sixth
International Geodetic Symposium on Satellite Positioning, Columbus, Ohio,
March 17–20, vol 1: 165–173.

Ledersteger K (1955): Der Schwereverlauf in den Lotlinien und die Berechnung der
wahren Geoidschwere. Publication dedicated to Weikko A. Heiskanen. Finnish
Geodetic Institute, Helsinki, vol 46: 109–124.

Lerch FJ, Klosko SM, Lauhscher RE, Wagner CA (1979): Gravity model improvement using Geos 3 (GEM 9 and 10). Journal of Geophysical Research, 84(B8):
3897–3916.

Levallois JJ (1963): La réhabilitation de la géodésique classique et la géodésie tridi-
mensionelle. Bulletin Géodésique, 68: 193–199.

Mader K (1954): Die orthometrische Schwerekorrektion des Präzisions-Nivellements
in den Hohen Tauern. Österreichische Zeitschrift für Vermessungswesen, Son-
derheft, vol 15.

Malys S, Slater J (1994): Maintenance and enhancement of the World Geodetic
System 1984. In: Proceedings of ION GPS-94, 7th International Technical
Meeting of the Satellite Division of the Institute of Navigation, Salt Lake
City, Utah, September 20–23, part 1: 17–24.

March JG, Chang ES (1979): Global detailed gravimetric geoid. Marine Geodesy,
2(2): 145–159.

Marquis W (2001): M is for modernization. GPS World, 12(9): 36–42.

Marussi A (1985): Intrinsic geodesy. Springer, Berlin Heidelberg New York Tokyo.

McCarthy DD (ed) (1996): IERS Conventions. Observatoire de Paris, IERS Technical Note 21.

Misra P, Enge P (2001): Global Positioning System – signals, measurements, and
performance. Ganga-Jamuna, Lincoln (Mass.).

Molodenski MS (1958): Grundbegriffe der geodätischen Gravimetrie. VEB Verlag
Technik, Berlin (Russian originals 1945 and 1948).

Molodenskii MS, Eremeev VF, Yurkina MI (1962): Methods for study of the external gravity field and figure of the earth. Israel Program of Scientific Translations, Jerusalem (Russian original 1960).

Montenbruck O, Gill E (2001): Satellite orbits – models, methods, and applications, corrected 2nd printing. Springer, Berlin.

Moritz H (1962): Studies on the accuracy of the computation of gravity in high elevations. Publications of the Isostatic Institute of the International Association of Geodesy, Helsinki, vol 38.

Moritz H (1963): Interpolation and prediction of point gravity anomalies. Publications of the Isostatic Institute of the International Association of Geodesy, Helsinki, vol 40.

Moritz H (1964): Zur Bestimmung des Geoides und seiner Verwendung als Reduktionsfläche. Zeitschrift für Vermessungswesen, 89: 200–202.

Moritz H (1965): Schwerevorhersage und Ausgleichungsrechnung. Zeitschrift für Vermessungswesen, 90: 181–184.

Moritz H (1980 a): Advanced physical geodesy. Wichmann, Karlsruhe (reprint 2001 by Civil and Environmental Engineering and Geodetic Science, Ohio State University, Columbus, Ohio).

Moritz H (1980 b): Geodetic Reference System 1980. Bulletin Géodésique, 54: 395–405.

Moritz H (1990): The figure of the earth – theoretical geodesy and the earth's interior. Wichmann, Karlsruhe.

Moritz H (1995): Science, mind and the universe – an introduction to natural philosophy. Wichmann, Heidlberg.

Moritz H, Hofmann-Wellenhof B (1993): Geometry, relativity, geodesy. Wichmann, Karlsruhe.

Moritz H, Mueller II (1987): Earth rotation – theory and observation. Ungar, New York.

Moritz H, Yurkina MI (eds) (2000): M.S. Molodensky – in memoriam. Mitteilungen der geodätischen Institute der Technischen Universität Graz, vol 88.

Mueller II (1985): Reference coordinate systems and frames: concepts and realization. Bulletin Géodésique, 59: 181–188.

Müller J (2001): Die Satellitengradiometriemission GOCE – Theorie, technische Realisierung und wissenschaftliche Nutzung. Deutsche Geodätische Kommission bei der Bayerischen Akademie der Wissenschaften, Reihe C, vol 541.

National Imagery and Mapping Agency (2000): Department of Defense World Geodetic System 1984 – its definition and relationships with local geodetic systems, 3rd edition, amendment 1. NIMA Technical Report TR 8350.2, Bethesda, Maryland. Available as PDF file at www.nima.mil.

Neumann F (1887): Vorlesungen über die Theorie des Potentials und der Kugelfunktionen (edited by C. Neumann). Teubner, Leipzig.

Österreichische Kommission für die Internationale Erdmessung (ed) (1983): Das Geoid in Österreich. Geodätische Arbeiten Österreichs für die Internationale Erdmessung. Neue Folge, vol III.

Pail R (2003): Satellitengeodäsie. Lecture manuscript available at the Institute of Navigation and Satellite Geodesy of the Graz University of Technology.

Pellinen LP (1962): Accounting for topography in the calculation of quasigeoidal heights and plumb-line deflections from gravity anomalies. Bulletin Géodésique, 63: 57–65.

Pizzetti P (1894): Sulla espressione della gravita alla superficie del geoide, supposto ellissoidico. Atti della Reale Accademia dei Lincei, Rome, V(3): 166.

Pizzetti P (1911): Sopra il calcolo teorico delle deviazioni del geoide dall' ellissoide. Atti della Reale Accademia della Scienze di Torino, 46: 331.

Plummer HC (1918): An introductory treatise on dynamical astronomy. Cambridge University Press (reprint 1960 by Dover, New York).

Rapp RH (1963): A consideration of Hayford's best fitting ellipsoid data using the differential change equations of Vening Meinesz. Geofisica pura e applicata, 54: 1–5.

Rapp R (1964): The prediction of point and mean gravity anomalies through the use of a digital computer. Institute of Geodesy, Photogrammetry and Cartography, Ohio State University, vol 43.

Rapp RH (1981): The earth's gravity field to degree and order 180 using SEASAT altimeter data, terrestrial gravity data, and other data. Ohio State University, Department of Geodetic Sciences, Columbus, Ohio, vol 322.

Rebhan H, Aguirre M, Johannessen J (2000): The gravity field and steady-state ocean circulation explorer mission – GOCE. ESA Earth Observation Quarterly, 66: 6–11.

Reigber C, Muller H, Rizos C, Bosch W (1983): An improved earth gravity model (GRIM 3B). Presented at the XVIII General Assembly of the IUGG at Hamburg, August 15–27.

Reigber C, Lühr H, Schwintzer P (eds) (2003): First CHAMP mission results for gravity, magnetic and atmospheric studies. Springer, Berlin Heidelberg New York.

Rinner K (1956): Über die Reduktion grosser elektronisch gemessener Entfernungen. Zeitschrift für Vermessungswesen, 81: 47–55.

Rummel R (1986): Satellite gradiometry. In: Sünkel H (ed): Mathematical and numerical techniques in physical geodesy. Springer, Berlin Heidelberg New York London Paris Tokyo, 317–363 [Bhattacharji S, Friedman GM, Neugebauer HJ, Seilacher A (eds): Lecture Notes in Earth Sciences, vol 7].

Rummel R, Balmino G, Johannessen J, Visser P, Woodworth P (2002): Dedicated gravity field missions – principles and aims. Journal of Geodynamics, 33: 3–20.

Sagrebin DW (1956): Die Theorie des regularisierten Geoids. Veröffentlichung des Geodätischen Instituts Potsdam, vol 9.

Schmitt G, Illner M, Jäger R (1991): Transformationsprobleme. Deutscher Verein für Vermessungswesen, special issue: GPS und Integration von GPS in bestehende geodätische Netze, vol 38: 125–142.

Schwarz KP (1976): Least-squares collocation for large systems. Bollettino di Geodesia e Scienze Affini, 35(3): 309–324.

Seeber G (2003): Satellite geodesy, 2nd edition. Walter de Gruyter, Berlin New York.

Sideris M (1987): Spectral methods for the numerical solution of Molodensky's problem. University of Calgary, Department of Surveying Engineering, vol 20024.

Sideris M (1990): Rigorous gravimetric terrain modelling using Molodensky's operator. Manuscripta geodaetica, 15: 97–106.

Sneeuw N, Gerlach C, Svehla D, Gruber C (2002): A first attempt at time-variable gravity recovery from CHAMP using the energy balance approach. In: Tziavos IN (ed): Gravity and geoid 2002. Proceedings of the Third Meeting of the International Gravity and Geoid Commission, Thessaloniki, Greece, August 26–30, 237–242.

Spilker JJ (1996): GPS signal structure and theoretical performance. In: Parkinson BW, Spilker JJ (eds): Global Positioning System: theory and applications, vol 1. American Institute of Aeronautics and Astronautics, Washington DC, 57–119 (Progress in Astronautics and Aeronautics, vol 163).

Sünkel H (1977): Ein nichtiteratives Verfahren zur Transformation geodätischer Koordinaten. Österreichische Zeitschrift für Vermessungswesen und Photogrammetrie, 64(1): 29–33.

Sünkel H (1980): A general surface representation module designed for geodesy. Ohio State University, Department of Geodetic Sciences, Columbus, Ohio, vol 292.

Sünkel H (1983): Geoidbestimmung: Berechnungen an der TU Graz, 2. Teil. In: Österreichische Kommission für die Internationale Erdmessung (ed): Das Geoid in Österreich. Geodätische Arbeiten Österreichs für die Internationale Erdmessung, Neue Folge, vol III: 125–143.

Sünkel H, Bartelme N, Fuchs H, Hanafy M, Schuh W-D, Wieser M (1987): The gravity field in Austria. In: Austrian Geodetic Commission (ed): The gravity field in Austria. Geodätische Arbeiten Österreichs für die Internationale Erdmessung, Neue Folge, vol IV: 47–75.

Telford WM, Geldart LP, Sheriff RE (1990): Applied geophysics, 2nd edition. Cambridge University Press, Cambridge New York Port Chester Melbourne Sydney.

Todhunter I (1873): A history of the mathematical theories of attraction and of the figure of the earth from the time of Newton to that of Laplace. Macmillan, London (reprint 1962 by Dover Publications, New York).

Tscherning CC (1976): Implementation of Algol-procedures for covariance computation on the RC 4000-computer. Danish Geodetic Institute, Copenhagen, vol 12.

Tscherning CC, Rapp RH (1974): Closed covariance expressions for gravity anomalies, geoid undulations, and deflections of the vertical implied by anomaly degree variance models. Ohio State University, Department of Geodetic Sciences, Columbus, Ohio, vol 208.

Tziavos IN (ed): Gravity and geoid 2002. Proceedings of the Third Meeting of the International Gravity and Geoid Commission, Thessaloniki, Greece, August 26–30.

Uotila UA (1960): Investigations on the gravity field and shape of the earth. Publications of the Isostatic Institute of the International Association of Geodesy, Helsinki, vol 33.

Vening Meinesz FA (1928): A formula expressing the deflection of the plumb-line in the gravity anomalies and some formulae for the gravity field and the gravity potential outside the geoid. Proceedings of the Koninklijke Nederlandse Akademie van Wetenschappen, 31(3): 315–331.

Vollath U, Birnbach S, Landau H, Fraile-Ordoñez JM, Martín-Neira M (1999): Analysis of three-carrier ambiguity resolution technique for precise relative positioning in GNSS-2. Navigation, 46(1): 13–23.

Vorhies J (2000): WRC 2000 results – GPS. Available at www.igeb.gov.

Zhu J (1993): Exact conversion of earth-centered, earth-fixed coordinates to geodetic coordinates. Journal of Guidance, Control, and Dynamics, 16: 389–391.

Subject index

SpringerGeosciences

Bernhard Hofmann-Wellenhof,
Klaus Legat, Manfred Wieser

Navigation

Principles of Positioning and Guidance

With a contribution by H. Lichtenegger.
2003. XXIX, 427 pages. 99 figures.
Softcover **EUR 54,–**
(Recommended retail price)
Net-price subject to local VAT.
ISBN 3-211-00828-4

Global positioning systems like GPS or the future European Galileo are influencing the world of navigation tremendously. Today, everybody is concerned with navigation even if unaware of this fact. Therefore, the interest in navigation is steadily increasing.

This book provides an encyclopedic view of navigation. Fundamental elements are presented for a better understanding of the techniques, methods, and systems used in positioning and guidance.

The book consists of three parts. Beside a historical review and maps, the first part covers mathematical and physical fundamentals. The second part treats the methods of positioning including terrestrial, celestial, radio- and satellite-based, inertial, image-based, and integrated navigation. Routing and guidance are the main topics of the third part. Applications on land, at sea, in the air, and in space are considered, followed by a critical outlook on the future of navigation.

This book is designed for students, teachers, and people interested in entering the complex world of navigation.

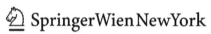 Springer Wien NewYork

P.O. Box 89, Sachsenplatz 4–6, 1201 Vienna, Austria, Fax +43.1.330 24 26, books@springer.at, **springer.at**
Haberstraße 7, 69126 Heidelberg, Germany, Fax +49.6221.345-4229, SDC-bookorder@springer-sbm.com, springeronline.com
P.O. Box 2485, Secaucus, NJ 07096-2485, USA, Fax +1.201.348-4505, orders@springer-ny.com, springeronline.com
Eastern Book Service, 3–13, Hongo 3-chome, Bunkyo-ku, Tokyo 113, Japan, Fax +81.3.38 18 08 64, orders@svt-ebs.co.jp
Prices are subject to change without notice. All errors and omissions excepted.

SpringerGeosciences

Bernhard Hofmann-Wellenhof,
Herbert Lichtenegger, James Collins

Global Positioning System

Theory and Practice

Fifth, revised edition.

2001. XXIII, 382 pages. 45 figures.

Softcover **EUR 62,95**

(Recommended retail price)

Net-price subject to local VAT.

ISBN 3-211-83534-2

This new edition accommodates the most recent advances in GPS technology. Updated or new information has been included although the
overall structure essentially conforms to the former editions. The textbook explains in comprehensive manner the concepts of GPS as well as
the latest applications in surveying and navigation. Description of project planning, observation, and data processing is provided for novice
GPS users. Special emphasis is put on the modernization of GPS covering the new signal structure and improvements in the space and the
control segment. Furthermore, the augmentation of GPS by satellite-
based and ground-based systems leading to future Global Navigation
Satellite Systems (GNSS) is discussed.

SpringerWien NewYork

P.O. Box 89, Sachsenplatz 4–6, 1201 Vienna, Austria, Fax +43.1.330 24 26, books@springer.at, **springer.at**
Haberstraße 7, 69126 Heidelberg, Germany, Fax +49.6221.345-4229, SDC-bookorder@springer-sbm.com, springeronline.com
P.O. Box 2485, Secaucus, NJ 07096-2485, USA, Fax +1.201.348-4505, orders@springer-ny.com, springeronline.com
Eastern Book Service, 3–13, Hongo 3-chome, Bunkyo-ku, Tokyo 113, Japan, Fax +81.3.38 18 08 64, orders@svt-ebs.co.jp
Prices are subject to change without notice. All errors and omissions excepted.

Springer and the Environment

WE AT SPRINGER FIRMLY BELIEVE THAT AN INTER-
national science publisher has a special obligation to
the environment, and our corporate policies consistently reflect this conviction.

WE ALSO EXPECT OUR BUSINESS PARTNERS – PRINTERS,
paper mills, packaging manufacturers, etc. – to commit
themselves to using environmentally friendly mate-
rials and production processes.

THE PAPER IN THIS BOOK IS MADE FROM NO-CHLORINE
pulp and is acid free, in conformance with international standards for paper permanency.